T0332905

Cloud Data Center Network Architectures and Technologies

Data Communication Series

For more information on this series please visit: https://www.routledge.com/
Data-Communication-Series/book-series/DCSHW

Cloud Data Center Network Architectures and Technologies

Lei Zhang and Le Chen

CRC Press
Taylor & Francis Group
Boca Raton London New York

CRC Press is an imprint of the
Taylor & Francis Group, an **informa** business

人民邮电出版社
POSTS & TELECOM PRESS

First edition published 2021
by CRC Press
6000 Broken Sound Parkway NW, Suite 300, Boca Raton, FL 33487-2742

and by CRC Press
2 Park Square, Milton Park, Abingdon, Oxon, OX14 4RN

Library of Congress Cataloging-in-Publication Data
Names: Zhang, Lei (Engineering teacher), author. | Chen, Le
(Writer on computer networking), author.
Title: Cloud data center network architectures and technologies / Lei Zhang, Le Chen.
Description: First edition. | Boca Raton : CRC Press, 2021. | Summary: "This book has been written with the support of Huawei's large accumulation of technical knowledge and experience in the data center network (DCN) field, as well as its understanding of customer service requirements. This book describes in detail the architecture design, technical implementation, planning and design, and deployment suggestions for cloud DCNs based on the service challenges faced by cloud DCNs. This book starts by describing the overall architecture and technical evolution of DCNs, with the aim of helping readers understand the development of DCNs. It then proceeds to explain the design and implementation of cloud DCNs, including the service model of a single data center (DC), construction of physical and logical networks of DCs, construction of multiple DCNs, and security solutions of DCs. Next, this book dives deep into practices of cloud DCN deployment based on real-world cases to help readers better understand how to build cloud DCNs. Finally, this book introduces DCN openness and some of the hottest forward-looking technologies. In summary, you can use this book as a reference to help you to build secure, reliable, efficient, and open cloud DCNs. It is intended for technical professionals of enterprises, research institutes, information departments, and DCs, as well as teachers and students of computer network-related majors in colleges and universities"— Provided by publisher.
Identifiers: LCCN 2020048556 (print) | LCCN 2020048557 (ebook) |
ISBN 9780367695705 (hardcover) | ISBN 9781003143185 (ebook)
Subjects: LCSH: Cloud computing. | Computer network architectures.
Classification: LCC QA76.585 .Z429 2021 (print) | LCC QA76.585 (ebook) |
DDC 004.67/82—dc23
LC record available at https://lccn.loc.gov/2020048556
LC ebook record available at https://lccn.loc.gov/2020048557

ISBN: 978-0-367-69570-5 (hbk)
ISBN: 978-0-367-69775-4 (pbk)
ISBN: 978-1-003-14318-5 (ebk)

Typeset in Minion
by codeMantra

Contents

Summary

This book has been written with the support of Huawei's large accumulation of technical knowledge and experience in the data center network (DCN) field as well as its understanding of customer service requirements. It describes in detail the architecture design, technical implementation, planning and design, and deployment suggestions for cloud DCNs based on the service challenges faced by cloud DCNs. It starts by describing the overall architecture and technical evolution of DCNs, with the aim of helping readers understand the development of DCNs. It then proceeds to explain the design and implementation of cloud DCNs, including the service model of a single data center (DC), construction of physical and logical networks of DCs, construction of multiple DCNs, and security solutions of DCs. Next, it dives deep into practices of cloud DCN deployment based on real-world cases to help readers better understand how to build cloud DCNs. Finally, it introduces DCN openness and some of the hottest forward-looking technologies.

In summary, you can use this book as a reference to help you to build secure, reliable, efficient, and open cloud DCNs. It is intended for technical professionals of enterprises, research institutes, information departments, and DCs, as well as teachers and students of computer network-related majors in colleges and universities.

DATA COMMUNICATION SERIES

Technical Committee

Director

Kevin Hu, President of Huawei Data Communication Product Line

Deputy Director

Kaisheng Zhong, Vice President of Marketing & Enterprise Network Technical Sales Department, Huawei Enterprise Business Group

Members

Shaowei Liu, President of Huawei Data Communication Product Line R&D Department

Zhipeng Zhao, Director of Huawei Data Communication Marketing Department

Xing Li, President of Campus Network Domain, Huawei Data Communication Product Line

Xiongfei Gu, President of WAN Domain, Huawei Data Communication Product Line

Leon Wang, President of Data Center Network Domain, Huawei Data Communication Product Line

Mingsong Shao, Director of Switch & Enterprise Gateway Product Department, Huawei Data Communication Product Line

Mingzhen Xie, Director of Information Digitalization and Experience Assurance Department, Huawei Data Communication Product Line

Jianbing Wang, Director of Architecture & Design Department, Huawei Data Communication Product Line

INTRODUCTION

This book first looks at the service characteristics of cloud computing and describes the impact of cloud computing on DCNs, evolution of the overall architecture and technical solution of DCs, and physical network, logical network, multi-DC, and security design solutions of DCs. Then, based on practical experiences of cloud DC deployment, it provides the recommended planning before deployment and key steps in implementation. Finally, it explains the hottest technologies of DCNs and the construction solution of Huawei cloud DCNs.

This book is a useful guide during SDN DCN planning and design, as well as engineering deployment, for ICT practitioners such as network engineers. For network technology enthusiasts and students, it can also be used as a reference for learning and understanding the cloud DCN architecture, common technologies, and cutting-edge technologies.

How Is the Book Organized

This book consists of 12 chapters. Chapter synopses follow below.

Chapter 1: Introduction to Cloud DCNs

This chapter covers basic features of cloud computing, development and evolution of virtualization technologies, and basics of cloud DCNs. It also describes characteristics of SDN network development and the relationship between orchestration and control, before explaining the concerns of enterprises in selecting different solutions from the service perspective.

Chapter 2: DCN Challenges

This chapter describes the five challenges faced by DCNs in the cloud computing era: large pipes for big data; pooling and automation for networks; security as a service (SECaaS) deployment; reliable foundation for networks; and intelligent O&M of DCs.

Chapter 3: Architecture and Technology Evolution of DCNs

This chapter describes the general architecture of physical networks in DCs and the evolution of major network technologies. The architecture and evolution process of typical DCNs are then described, with financial services companies and carriers used as examples.

Chapter 4: Functional Components and Service Models of Cloud DCNs

This chapter describes the orchestration model and service process of network services provided by the cloud platform and by the SDN controller (for when no cloud platform is available).

Chapter 5: Constructing a Physical Network (Underlay Network) in a DC

This chapter describes the typical architecture and design principles of the physical network, as well as comparison and selection of common network technologies.

Chapter 6: Constructing a Logical Network (Overlay Network) in a DC

This chapter describes basic concepts of logical networks, basic principles of mainstream VXLAN technologies, and how to use VXLAN to build logical networks.

Chapter 7: Constructing a Multi-DC Network

Multiple DCs need to be deployed to meet service requirements due to service scale expansion and service reliability and continuity requirements. This chapter describes the service requirement analysis and recommended network architecture design for multiple DCNs.

Chapter 8: Building E2E Security for Cloud DCNs

This chapter describes security challenges faced by cloud DCs and the overall technical solution at the security layer. It also describes specific security technologies and implementation solutions in terms of virtualization security, network security, advanced threat detection and defense, border security, and security management.

Chapter 9: Best Practices of Cloud DCN Deployment

This chapter describes the recommended planning methods for service and management networks of typical cloud DCs based on cloud DC deployment practices, and also explains key configuration processes during deployment, operation examples, and common service provisioning process based on the SDN controller.

Chapter 10: Openness of DCN

This chapter describes the necessity of DCN openness and the capabilities and benefits brought by openness of SDN controllers and forwarders.

Chapter 11: Cutting-Edge Technologies

This chapter describes some new DCN technologies and development trends that attract industry attention, including basic concepts and mainstream solutions of container networks, hybrid clouds, and AI Fabric.

Chapter 12: Components of the Cloud DCN Solution

This chapter describes the positioning, features, and functional architecture of some components that can be used during cloud DCN construction, including CloudEngine DC switches, CloudEngine virtual switches, HiSecEngine series firewalls, iMaster NCE-Fabric, and the SecoManager.

Icons Used in This Book

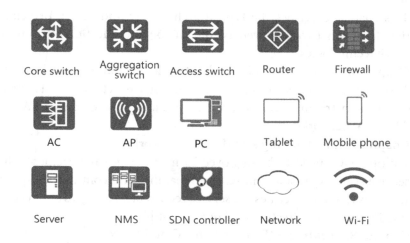

Core switch	Aggregation switch	Access switch	Router	Firewall
AC	AP	PC	Tablet	Mobile phone
Server	NMS	SDN controller	Network	Wi-Fi

Acknowledgments

This book has been jointly written by the Data Communication Digital Information and Content Experience Department and Data Communication Architecture & Design Department of Huawei Technologies Co., Ltd. During the process of writing the book, high-level management from Huawei's Data Communication Product Line provided extensive guidance, support, and encouragement. We are sincerely grateful for all their support.

The following is a list of participants involved in the preparation and technical review of this book.

Editorial board: Lei Zhang, Le Chen, Fan Zhang, Shan Chen, Xiaolei Zhu, Zhongping Jiang, and Xuefeng Wu

Technical reviewers: Jun Guo, Jianbing Wang, Lei Zhang, Le Chen, and Fan Zhang

Translators: Yongdan Li, Zhenghong Zhu, Kaiyan Zhang, Mengyan Wang, Michael Chapman, and Fionnuala Magee

While the writers and reviewers of this book have many years of experience in ICT and have made every effort to ensure accuracy, it may be possible that minor errors have been included due to time limitations. We would like to express our heartfelt gratitude to the readers for their unremitting efforts in reviewing this book.

Authors

Mr. Lei Zhang is the Chief Architect of Huawei's DCN solution. He has more than 20 years' experience in network product and solution design, as well as a wealth of expertise in product design and development, network planning and design, and network engineering project implementation. He has led the design and deployment of more than ten large-scale DCNs for Fortune Global 500 companies worldwide.

Mr. Le Chen is a Huawei DCN solution documentation engineer with eight years' experience in developing documents related to DCN products and solutions. He has participated in the design and delivery of multiple large-scale enterprise DCNs. He has written many popular technical document series such as *DCN Handbook* and *BGP Topic*.

CHAPTER 1

Introduction to Cloud DCNs

A CLOUD DATA CENTER (DC) is a new type of DC based on cloud computing architecture, where the computing, storage, and network resources are loosely coupled. Within a cloud DC, various IT devices are fully virtualized, while also being highly modularized, automated, and energy efficient. In addition, a cloud DC features virtualized servers, storage devices, and applications, enabling users to leverage various resources on demand. Automatic management of physical and virtual servers, service processes, and customer service charging is also provided. Starting with cloud computing and virtualization, this chapter describes the software-defined networking (SDN) technology used by cloud data center networks (DCNs) to tackle the challenges introduced by this new architecture.

1.1 CLOUD COMPUTING

Before examining cloud DCNs in more detail, we should first take a closer look at cloud computing. The pursuit of advanced productivity is never ending. Each industrial revolution has represented a leap in human productivity, as our society evolved from the mechanical and electric eras through to the current automatic and intelligent era.

Since the 1980s, and owing to the advances of global science and technology, culture, and the economy, we have gradually transitioned from an industrial society to an information society. By the mid-1990s, economic globalization had driven the rapid development of information technologies, with the Internet becoming widely applied by all kinds of businesses. As the global economy continues to grow, cracks have begun to appear in the current processes of enterprise informatization. Constrained by complex management modes, spiraling operational expenses, and weak scale-out support, enterprises require effective new information technology solutions. Such requirements have driven the emergence of cloud computing.

The US-based National Institute of Standards and Technology (NIST) defines the following five characteristics of cloud computing:

- On-demand self-service: Users can leverage self-services without any intervention from service providers.

- Broad network access: Users can access a network through various terminals.

- Resource pooling: Physical resources are shared by users, and resources in a pool are region-independent.

- Rapid elasticity: Resources can be quickly claimed or released.

- Measured service: Resource measurement, monitoring, and optimization are automatic.

"On-demand self-service" and "broad network access" express enterprises' desire for higher productivity and particularly the need for service automation. "Resource pooling" and "rapid elasticity" can be summarized as flexible resource pools, while "measured services" emphasize that operational support tools are required to tackle the considerable challenges of automation and virtualization. More intelligent and refined tools are also required to reduce the operating expense (OPEX) of enterprises.

Cloud computing is no longer just a term specific to the IT field. Instead, it now represents an entirely new form of productivity, as it creates a business model for various industries, drives industry transformation, and reshapes the industry chain. Cloud computing introduces revolutionary changes to traditional operations and customer experience, and seizing the opportunities of cloud computing will boost growth throughout the industry.

1.2 VIRTUALIZATION TECHNOLOGIES INTRODUCED BY CLOUD COMPUTING

Virtualization is a broad term. According to the Oxford Dictionary, "virtual" refers to something that is "physically non-existent, but implemented and presented through software." Put another way, a virtual element is a specific abstraction of an element. Virtualization simplifies the expression, access, and management of computer resources, including infrastructures, systems, and software, and provides standard interfaces for these resources. Virtualization also reduces the dependency of service software on the physical environment, enabling enterprises to achieve higher stability and availability based on simplified operation processes, improve resource utilization, and reduce costs.

Throughout the years, virtualization technologies have flourished in the computing, network, and storage domains, and have become interdependent on one another. The development of computing virtualization technologies is undoubtedly critical, while the development of network and storage virtualization technologies is intended to adapt to the changes and challenges introduced by the former. In computing virtualization, one physical machine (PM) is virtualized into one or more virtual machines (VMs) using a Virtual Machine Manager (VMM), which increases utilization of computer hardware resources and improves IT support efficiency.

A VMM is a software layer between physical servers and user operating systems (OSs). By means of abstraction and conversion, the VMM enables multiple user OSs and applications to share a set of basic physical hardware. Consequently, the VMM can be regarded as a meta OS in a virtual environment. It can allocate the correct amount of logical resources (such as memory, CPU, network, and disk) based on VM configurations,

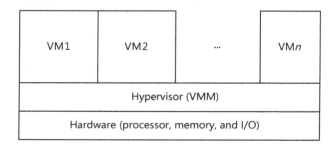

FIGURE 1.1   Virtualization.

load the VM's guest OS, and coordinate access to all physical devices on the VM and server, as shown in Figure 1.1.

The following types of VMMs are available:

- Hypervisor VM: runs on physical hardware and focuses on virtual I/O performance optimization. It is typically used for server applications.

- Hosted VM: runs on the OS of a PM and provides more upper-layer functions such as 3D acceleration. It is easy to both install and use, and is typically utilized for desktop applications.

While multiple computing virtualization technologies exist, they often use different methods and levels of abstraction to achieve the same effect. Common virtualization technologies include the following:

1. **Full virtualization**

   Also known as original virtualization. As shown in Figure 1.2, this model uses a VM as the hypervisor to coordinate the guest OS and original hardware. The hypervisor obtains and processes virtualization-sensitive privileged instructions so that the guest OS can run without modification. As all privileged instructions are processed by the hypervisor, VMs offer lower performance than PMs. While such performance varies depending on implementation, it is usually sufficient to meet user requirements. With the help of hardware-assisted virtualization, full virtualization gradually

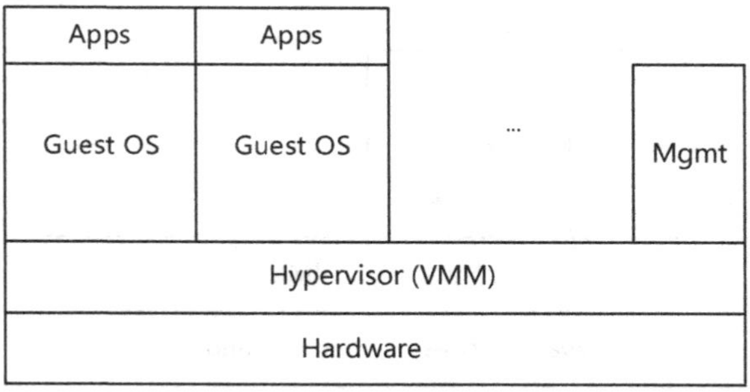

FIGURE 1.2 Full virtualization.

overcomes its bottleneck. Typical hardware products include IBM CP/C MS, Oracle VirtualBox, KVM, VMware Workstation, and ESX.

2. **Paravirtualization**

Also known as hyper-virtualization. As shown in Figure 1.3, paravirtualization, similar to full virtualization, uses a hypervisor to implement shared access to underlying hardware. Unlike full virtualization, however, paravirtualization integrates virtualization-related code into the guest OS so that it can work with the hypervisor to implement virtualization. In this way, the hypervisor does not need to recompile or obtain privileged instructions, and can achieve performance close to that of a PM. The most well-known product of this type is Xen. As Microsoft Hyper-V uses technologies similar to Xen, it can also be classified as paravirtualization. A weakness of paravirtualization is its requirement that a guest OS be modified, and only a limited number of guest OSs are supported, resulting in a poor user experience.

3. **Hardware emulation**

The most complex virtualization technology is undoubtedly hardware emulation. As shown in Figure 1.4, hardware emulation creates a hardware VM program on the OS of a PM in order to emulate the required hardware (VM) and runs this on the VM program. If hardware-assisted virtualization is not available, each instruction must be emulated on the underlying hardware, reducing operational performance to less than one percent of that of a PM in some cases. However, hardware emulation can enable an OS designed for PowerPC to run on an ARM processor host without any

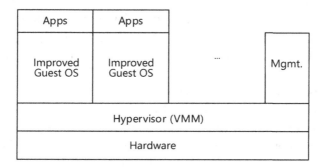

FIGURE 1.3   Paravirtualization.

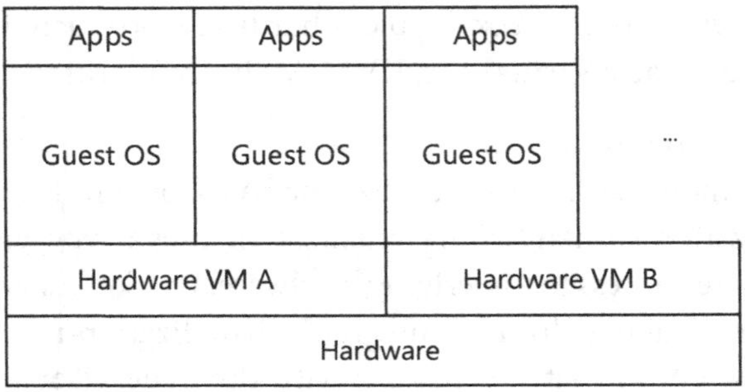

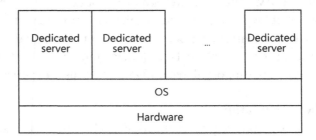

FIGURE 1.4   Hardware emulation.

FIGURE 1.5   OS-level virtualization.

modifications. Typical hardware emulation products include Bochs and quick emulator (QEMU).

4. **OS-level virtualization**

As shown in Figure 1.5, this technique implements virtualization by simply isolating server OSs. As a result, OS-level virtualization can achieve smaller system overheads, preemptive compute resource scheduling, and faster elastic scaling. However, its weaknesses include resource isolation and security. Container technology, as a typical OS-level virtualization technology, is becoming increasingly popular.

5. **Hardware-assisted virtualization**

Hardware vendors such as Intel and AMD improve virtualization performance by implementing software technologies used in full virtualization and paravirtualization based on hardware. Hardware-assisted virtualization is often used to optimize full virtualization and paravirtualization, rather than operating as a parallel. The best-known example of this is VMware Workstation which, as a full-virtualization platform, integrates hardware-assisted virtualization in VMware 6.0 (including Intel VT-x and AMD-V). Mainstream full virtualization and paravirtualization products support hardware-assisted virtualization and include VirtualBox, KVM, VMware ESX, and Xen.

While the above computing virtualization technologies are not perfect, driven by the changing upper-layer application requirements and hardware-assisted virtualization, they have seen widespread application for a number of years. In 2001, VMware launched ESX, which reshaped the virtualization market. Two years later, Xen1.0 was released and open-sourced. In 2007, KVM was integrated into Linux 2.6.20. And in 2008, Microsoft and Citrix joined forces to launch Hyper-V, while Kubernetes was developing into a mature container technology. Today, these virtualization technologies are still developing rapidly, having not yet reached maturity. In terms of business models, the competition between open source and closed source continues unabated.

Table 1.1 describes the strengths and weaknesses of each virtualization technology.

In addition to computing virtualization, DCN virtualization technologies are also evolving rapidly due to the changes in computing and storage. Virtualization technologies evolved from N-to-1 (horizontal/vertical virtualization) and 1-to-N (virtual switch) technologies to overlay technologies, resulting in large-scale virtual Layer 2 networks (multiple Layer 2 networks combined) capable of delivering extensive compute resource pools for VM migration. SDN was then developed, which associates computing and network resources to abstract, automate, and measure network functions. This technological advancement is driving DCNs toward autonomous driving networks or intent-driven networks. The following chapters will elaborate on these technologies.

1.3 SDN FOR CLOUD COMPUTING

In a cloud DC, virtualized resources are further abstracted as services for flexible use. Cloud computing services can be classified into Infrastructure as a Service (IaaS), Platform as a Service (PaaS), and Software as a Service (SaaS), which correspond to hardware, software platform, and application resources, respectively.

TABLE 1.1 Strength and Weakness of Virtualization Technologies

| | | Virtualization Technology | | | |
Item	Full Virtualization	Paravirtualization	Hardware Emulation	OS-level Virtualization	Hardware-Assisted Virtualization
(e) Speed (compared with physical servers)	30%–80%	More than 80%	Less than 30%	80%	More than 80%
Strengths	The Guest OS does not need to be modified. It is fast and easy to use, and provides useful functions	Compared with full virtualization, it offers a more simplified architecture, which enables a faster speed	The Guest OS does not need to be modified. Typically applicable to hardware, firmware, and OS development	Highly cost-effective	Centralized virtualization provides the fastest speed
Weaknesses	Performance, especially I/O, is poor in hosted mode	The Guest OS must be modified, which affects user experience	Very slow speed. (In some cases, speeds are lower than one percent of that of the physical server)	Limited OS support	The hardware implementation requires more optimization
Trend	Becoming mainstream	Significant use	Phasing out, but still in use	Used for specific applications, such as VPS	Widely used

The requirements for quality attributes vary according to the cloud service layer. IaaS is dedicated to providing high-quality hardware services, while SaaS and PaaS emphasize software flexibility and overall availability. Based on the hierarchical decoupling and mutual distrust principles, they decrease the reliability requirement for a single service to 99.9%, meaning a service can only be interrupted for less than 8.8 hours over the course of a year. For example, users may encounter one or two malfunctions in the email system, instant messaging (IM) software, or even the OS, but no real faults in hardware systems or driver software.

In terms of software technologies, the software architecture and technology selection of cloud services offer varying quality attributes. Legacy software can be classified into IT and embedded software. IT software is applicable to the SaaS and PaaS layers, and focuses on elastic expansion and fast rollout. In fault recovery scenarios, or those that require high reliability, IT software uses methods such as overall rollbacks and restarts. Embedded software focuses more on the control of software and hardware statuses to achieve higher reliability, and is more widely applicable to the IaaS layer.

As IaaS systems become more automated and elastic, and as new software technologies such as distribution, service-orientation, Cloud Native, and Service Less continue to emerge, SaaS/PaaS systems are being subverted, and IT software is undergoing an accelerated transformation to Internet software. Based on the DevOps agile development mode, as well as technical methods such as stateless services and distributed computing, SaaS/PaaS systems provide self-service, real-time online, and quick rollout capabilities for services. This transformation drives the development of new Internet business models.

At the same time, users are beginning to re-examine whether IaaS systems have high requirements on real-time performance across all scenarios and whether refined control over systems is required. This kind of thinking also influences the development of SDN, splitting system development along two different paths: control-oriented and orchestration-oriented.

1. In the current phase, upper-layer services cannot be completely stateless, and the network still needs to detect the computing migration status. In addition, a unified management platform is required to implement fine-grained status management for routing protocols and other information, in order to meet the requirements for fast network switchover in fault scenarios and ensure no impact on

upper-layer services. During the enterprise cloudification process, the following three solutions are provided for the control-oriented path, each of which can be chosen depending on software capabilities and the organizational structure of individual enterprises.

- Cloud-network integration solution: This solution utilizes the open-source OpenStack cloud platform and commercial network controllers to centrally manage network, computing, and storage resources, and implement resource pooling. The cloud platform delivers network control instructions to the network controller through the RESTful interface, and the network controller deploys the network as instructed.

- End-to-end cloud-network integration solution: This solution is also based on a combination of the cloud platform and network controller, and is typically used by commercial cloud platforms such as VMware and Azure. Compared with the open-source cloud platform, the commercial cloud platform offers improved availability and maintainability. As a result, most enterprises use this solution, with only those possessing strong technical capabilities preferring the open-source cloud platform.

- Virtualization solution based on a combination of computing virtualization and network controller: This solution is based on a combination of the virtualization platform (such as vCenter and System Center) and network controller, with no cloud platform used to centrally manage computing and network resources. After delivering a computing service, the virtualization platform notifies the network controller, which then delivers the corresponding network service.

2. The orchestration-oriented path depends on future upper-layer service software architecture, and the expectation that IaaS software will be further simplified after SaaS/PaaS software becomes stateless. The status of the IaaS software does not need to be controlled. Instead, IaaS software only needs to be orchestrated using software tools, similar to the upper-layer system.

Selecting the appropriate path or solution depends on many factors such as the enterprise organization structure, existing software architecture, technology and resource investment, and cloudification progress. Although all four solutions have their strengths and weaknesses, they can support enterprise service cloudification over a long term. The following describes what features each solution offers and why an enterprise might select them.

1. Cloud-network integration solution: (OpenStack+network controller): This solution has the following strengths:

   - As OpenStack is the mainstream cloud platform, the open-source community is active and provides frequent updates, enabling rapid construction of enterprise cloud capabilities.

   - Customized development can be applied to meet enterprise business requirements. These are subject to independent intellectual property rights.

   - OpenStack boasts a healthy ecosystem. Models and interfaces are highly standardized, and layered decoupling facilitates multi-vendor interoperability.

   This solution has the following weaknesses:

   - Commercialization of open-source software requires software hardening and customization. As such, enterprises must already possess the required technical reserves and continuously invest in software development.

   - During the current evolution from single to multiple DCs, and to hybrid clouds, a mature multi-cloud orchestrator does not yet exist within the open-source community. This will need to be built by enterprises themselves.

   - Enterprises that choose this solution generally possess advanced software development and integration capabilities, are concerned about differentiated capabilities and independent intellectual property rights, and require standardization and multi-vendor interconnection. Currently, this solution is mainly used for carriers' telco clouds, and for the DCs of large financial institutions and Internet enterprises.

2. End-to-end cloud-network integration solution (Huawei FusionCloud, VMware vCloud, and Microsoft Azure Stack):

   The advantages of this solution are its full support for single/multiple DCs, hybrid clouds, and SaaS/PaaS/IaaS end-to-end delivery, which enable the rapid construction of cloudification capabilities for enterprises.

   However, areas where this solution requires improvement include openness and vendor lock-in.

   Enterprises that choose this solution are often in urgent need of cloud services. They need to quickly build these services in a short period of time to support new business models and quickly occupy the market, but lack the required technical reserves. As such, this solution is typically employed for DCs of small- and medium-sized enterprises.

3. Virtualization solution (computing virtualization + network controller):

   This solution has the following strengths:

   - In the virtualization solution, computing and network resources are independent of one another, and the enterprise organization (IT and network teams) does not require immediate restructuring.

   - A large number of cloud and PaaS/SaaS platforms are currently available, and the industry is considered to be mature. Consequently, the association between computing virtualization and network controllers enables the rapid construction of IaaS platforms capable of implementing automation and satisfying service requirements. This approach is low risk and is easy to initiate.

   - The open architecture of the IaaS layer and layered decoupling allow flexible selection of the cloud and PaaS/SaaS platforms.

   - This solution is based on mature commercial software, which requires no customized development and offers high reliability.

     A weakness of this solution is that IaaS/PaaS/SaaS software models need to be selected in rounds, which slows down the cloudification of enterprises.

Enterprises that choose this solution have complex organizational structures and fixed service applications. They do not want to lock vendors, but have major concerns relating to solution stability and reliability. As such, this solution is typically implemented in the DCs of medium- and large-sized enterprises in the transportation and energy industries.

Following the development of container technologies, compute and storage resources are becoming less dependent on networks. In this solution, the network controller can evolve into an orchestrator, and the overall solution can evolve to tool-based orchestration.

4. Tool-based orchestration solution:
   This solution has the following strengths:

   – Provides customized service orchestration capabilities to adapt to enterprise service applications.

   – Easy to develop and quick to launch, as it is based on script or graphical orchestration tools.

   This solution has the following dependencies or weaknesses:

   – The IaaS, PaaS, and SaaS software must comply with the stateless principle, and service reliability is independent of regions and does not rely on the IaaS layer (VM migration).

   – Service application scenarios are relatively simple and services are independent of each other, preventing conflicts and mutual coverage impacts.

   Enterprises that choose this solution do not rely heavily on legacy service software, or can begin restructuring at low costs. They operate clear service scenarios, and require rapid responses to service changes. In addition, they can apply strict DC construction specifications to ensure service independence. This solution is typically implemented in the DCs of Internet enterprises.

Enterprises have varying concerns about their network capabilities when choosing from the available solutions, as shown in Table 1.2.

TABLE 1.2 Enterprise Concerns When Choosing Cloudification Solutions

Solution Chosen	Single DC		Multiple DCs and Hybrid Cloud	
	Enterprise DC Scenario	Telco Cloud Scenario	Enterprise DC Scenario	Telco Cloud Scenario
Cloud-network integration solution (OpenStack + network controller)	• OpenStack interconnection automation capability • Networking and forwarding performance (network overlay) • VAS multi-vendor automation capability • PaaS software integration capability • O&M capabilities: Zero Touch Provisioning (ZTP), dialing test methods, and underlay network automation • IPv6 capability	• Networking diversity and layered decoupling (hybrid overlay) • OpenStack interconnection automation • Capability (Layer 2, Layer 3, and IPv6) • SFC capability • Routing service (BGP and BFD) • Standardization (BGP-EVPN, and MPLS) • Quality of Service (QoS)	• Multiple OpenStack platforms • Unified resource management • Security automation • Multi-DC O&M capabilities	• Multi-DC inter-operation standardization • Automation of interconnection between MANs and DCs
End-to-end cloud-network integration solution	• Physical server access capability (physical switches connected to the VMware NSX Controller to implement physical server automation) • Rapid VMware vRealize integration capability • Automated the underlay network O&M capability (hardware switches connected to VMware vRNI and Azure Stack) • Hybrid cloud			
Virtualization solution (computing virtualization platform + network controller)	• Fine-grained security isolation (microsegmentation) • Multi-vendor VAS device automation capability (SFC) • Underlay network O&M (ZTP and configuration automation) • Multicast function • IPv6 capability		• Association with VMware and network DR and switchover • Unified management of multiple DCs • Forward compatibility and evolution	
Tool-based orchestration solution	• Interconnection between network devices and orchestration tools such as Ansible and Puppet • Openness of O&M interfaces on network devices • Response speed of customized interfaces on network devices			

To summarize, enterprises should choose a cloud-based transformation solution capable of matching their specific service requirements and technical conditions.

1.4 DCN PROSPECTS

1. Intent-driven network

According to Gartner's technology maturity model, shown in Figure 1.6, SDN/NFV technologies are now ready for large-scale commercial use following many years of development.

In the future, as the development of automation, big data, and cloud technologies continues, autonomous networks (ANs) will gradually be put into practice, once again driving the rapid development and evolution of the entire industry. However, SDN still has a long way to go before reaching the AN goal. While SDN technology activates physical networks through automation, there are still broad gaps separating business intent and user experience. For example, an enterprise's business intent is to quickly expand 100 servers due to the expected service surge of a big event. To address this intent, the enterprise needs to perform a series of operations, such as undoing interface shutdown, enabling LLDP, checking the topology, and enabling the server to manage the network. There are a lot of

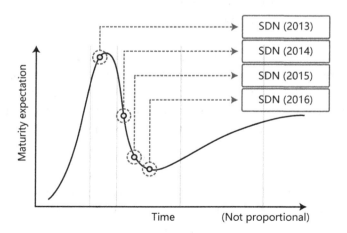

FIGURE 1.6   Gartner's technology maturity model.

expenses involved in implementing this business intent based on the network.

In this case, a digital world must be constructed over the physical network. This not only digitizes a physical network element (NE), but also quickly maps business intents to network requirements and digitizes user experience and applications on the network. Based on automation, big data, and cloud technologies, the digital world transforms from device-centric to user-centric, and bridges customers' business intents and physical networks. Intent-driven networks can quickly provide services based on the digital world, improve user experience, and enable preventive maintenance.

2. AI Fabric

As Internet technologies are developing, upper-layer DC services are shifting their focus from service provisioning efficiency to data intelligence and business values. The three core elements of AI applications are algorithms, computing power, and data. Among the core elements, data is the most critical. All AI applications use advanced AI algorithms to mine intelligence from data and extract useful business value. This poses higher requirements on a DC's IaaS platform.

To start with, AI applications require the IaaS layer to provide a high-performance distributed storage service capable of carrying massive amounts of data, and the AI algorithms require a high-performance distributed computing service capable of massive data computing. It is estimated that by 2025, the amount of data generated and stored worldwide will reach 180 ZB. Such an incredible volume of data will be beyond the processing capacity of humans, leaving 95% to be processed by AI. Service requirements drive the rapid development of Solid-State Disks (SSDs) and AI chips, and the sharp increase in communication services between distributed nodes leads to more prominent network bottlenecks.

- Current storage mediums (SSD) deliver access speeds 100 times faster than conventional distributed storage devices (such as hard disk drives). In addition, network delay rates have increased from less than 5% to about 65%. There are two types of network delay: delay caused by packet loss (about 500 μs) and queuing delay caused by network congestion (about 50 µs). Avoiding packet loss and congestion is a core objective for improving input/output operations per second (IOPS).

- AI chips can be anywhere between 100 and 1000 times faster than legacy CPUs. In addition, the computing volume of AI applications increases exponentially. For example, distributed training for a large-scale speech recognition application results in the training quantity for a computing task reaching approximately 20 exaFLOPS, requiring 40 CPU-installed servers to calculate more than 300 million parameters (4 bytes for a single parameter). In each iterative calculation, the CPU queuing delay (approximately 400 ms) exceeds the CPU calculation delay (approximately 370 ms). If millions of iterative calculations are used during one training session, it will last for an entire month. Reducing the communication waiting time and shortening AI training have become core requirements of AI distributed training.

To meet the requirements of AI applications, network protocols and hardware have been greatly improved. In terms of protocols, Remote Direct Memory Access (RDMA) and RDMA over Converged Ethernet (RoCE) alleviate TCP problems such as slow start, low throughput, multiple copies, high latency, and excessive CPU consumption. In terms of hardware, Ethernet devices have made great breakthroughs in lossless Ethernet.

- Virtual multi-queue technology is used to precisely locate back pressure in congestion flows, which prevents impacts on normal traffic.

- The congestion and back pressure thresholds are dynamically calculated and adjusted in real time, ensuring maximum network throughput without packet loss.

- The devices proactively collaborate with the NIC to schedule traffic to the maximum quota and to prevent congestion. As such, next-generation lossless Ethernet adaptive to AI equals, or even exceeds, the InfiniBand (IB) network in terms of forwarding performance, throughput, and latency. From the perspective of overall DC operation and maintenance, a unified converged network (convergence of the storage network, AI computing network, and service network), which is considerably more cost-effective, can be built based on Ethernet.

AI Fabric is a high-speed Ethernet solution based on lossless network technologies. It provides network support for AI computing, high-performance computing (HPC), and large-scale distributed computing. AI Fabric uses two-level AI chips and a unique intelligent algorithm for congestion scheduling to achieve zero packet loss, high throughput, and ultra-low latency for RDMA service flows, improving both computing and storage efficiency in the AI era. Private network performance is now available at the cost of Ethernet, delivering a 45-fold increase in overall Return on Investment (ROI). For more on AI Fabric, see Chapter 11.

CHAPTER 2

DCN Challenges

I N CLOUD COMPUTING, RESOURCES are allocated on demand and charged by usage. A configurable pool of shared resources is provided, which users can access through the network in addition to services such as storage space, network bandwidth, servers, and application software. Cloud computing pools computing, network, and storage resources, which results in larger service volumes over higher bandwidth but lower latency. In the cloud computing era, as cloud DCs are deployed for large-scale commercial use, and new technologies are shooting up, legacy DCNs face the following challenges:

1. Big data requires wide pipes.

    With the continuous development of Internet technologies, the number of applications and the associated service volume in DCs dramatically increases. Every minute there are more than 1.6 million Google searches, 260 million emails sent, 47,000 apps downloaded, 220,000 photos uploaded to Facebook, and 660 million data packets transmitted. And this explosive growth is set to continue. According to statistics, global DC IP traffic increases five-fold every year. It is also estimated that the total number of servers in global DCs will see a ten times greater increase in 2020 compared with 2015. This results in huge challenges to DCNs as big data requires wide pipes.

    In addition, as a large number of applications are migrated to DCs, the traffic model is also changing. As shown in Figure 2.1, the data from 2015 confirms that the east-west traffic (traffic between

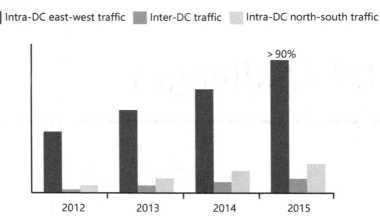

FIGURE 2.1   DCN traffic.

internal servers) in a DC accounted for more than 90% of the total traffic. As such, the tree-shaped network architecture of legacy DCs can no longer meet service requirements, and a new distributed network architecture needs to be built in order to deploy Layer 3 gateways at edge nodes to optimize traffic paths and effectively satisfy the steep increase in new requirements on bandwidth and latency.

2. Quick service rollout requires network pooling and automation.

As shown in Figure 2.2, legacy DCNs are separated and unable to provide large-scale resource pools, unlike cloud DCNs. Their compute resources are limited to modules and cannot be scheduled in a unified manner, which often leads to some resources running out, while others are left idle. In addition, distributed routing policies are used between networks, which results in difficult route optimization and reduces network utilization.

Legacy DCNs feature a low level of automation, which cannot meet the demands for rapid and elastic service rollout. Applications require months for deployment, which is not suitable for new service development. In addition, applications are as complex to scale out as they are to install and deploy, with long interruptions required for installation, storage migration, and service switchover.

DC resources are separated and hard to manage. As services are deployed across DCs, service access relationships and policy management between DCs become more complex. Link bandwidth is

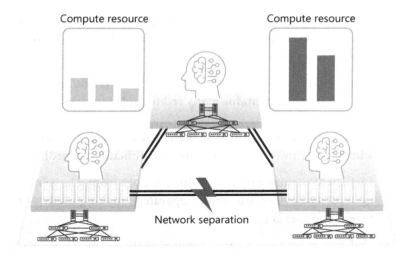

FIGURE 2.2 Separated legacy DCNs.

unevenly allocated between DCs, leading to low overall utilization and high costs.

3. Comprehensive threat escalation requires service-oriented security deployment.

DC security is provided as a service at a low degree and features high OpEx. According to a recent banking industry survey, more than half of the O&M and update operations are related to security services. The following major issues exist:

- Limited types of security services: Open-source OpenStack offers weak security capabilities and only provides basic firewall services.

- Network association required for security services: Network link connectivity cannot be achieved through firewall configurations alone and requires collaboration between network devices.

- Complex manual configurations: Security and network devices must be deployed and configured manually, resulting in heavy workloads.

- Scattered device deployment: Security devices must be deployed at each risk point and are difficult to manage.

- Complex configuration entries: Multiple security devices require separate logins in order to configure specific policies for one-time security protection.

- Waste of security resources: To cope with traffic bursts, each security device maintains many redundant resources.

In addition, constantly escalating threats mean manual threat analysis is inefficient, and countermeasures are difficult to coordinate.

- Inefficient manual analysis of threats: Logs from a large number of security devices are stored separately, making analysis inefficient and ineffective.

- Difficult investigation and countermeasure deployment: Once new attacks occur, security devices cannot detect, trace, or analyze the threats.

- Security silos: Different types of security devices function separately, resulting in security silos, which offer a weak defense.

Visualized analysis of security threats is poor and cannot guide security O&M due to the following issues:

- Excessive threat logs: A large number of security devices use different log formats, making it impractical to view all logs.

- Poor correlative analysis: The correlative analysis software analyzes only a small number of threats, offering unreliable results.

- Non-intuitive display: Security issues cannot be solved based on the unintuitive threat display.

4. Service continuity takes top priority, requiring networks to provide reliable foundations.

   Many countries and regions have developed clear security policies and regulations for key industries such as banking, energy, and transportation. For example, in China, the China Banking Regulatory Commission (CBRC) has issued the *Guidelines on Business Continuity Supervision of Commercial Banks* (CBRC Doc.

No. [2011] 104), which specifies the following criteria for grading business continuity events.

Grade I (extremely critical operation interruption event): Such an event causes interruption to the regular business operations of multiple financial institutions in one or more provinces (autonomous regions or municipalities directly under the Central Government) for 3 hours or more. During this interruption, a single financial institution cannot provide services to two or more provinces (autonomous regions and municipalities under the Central Government) for more than 3 hours, or cannot provide normal services to a province (autonomous regions or municipalities) for more than 6 hours. CBRC will send a special task force to analyze, evaluate, and handle such an event, and related progress shall be reported to the State Council.

Grade II (critical operation interruption event): Such an event causes interruption to the regular business operations of multiple financial institutions in one or more provinces (autonomous regions or municipalities directly under the Central Government) for more than half an hour. During this interruption, a financial institution cannot provide services in two or more provinces (autonomous regions and municipalities directly under the Central Government) for more than 3 hours. CBRC will send a special task force to analyze, evaluate, and handle such an event.

Grade III (major operation interruption event): Such an event causes interruption to the regular business operations of one financial institution in one province (autonomous region or municipality under the Central Government) for more than half an hour.

As shown in Figure 2.3, financial services can be classified by their level of importance as follows: Class A+, Class A, Class B, and Class C. Class A+ and A services must be deployed in active-active mode in the same city to implement Recovery Point Objective (RPO) of 0 and ensure that no data will be lost during DC switchover. As shown in Figure 2.4, service interruptions will result in a great amount of loss to both enterprises and society. The actual requirements for financial service continuity are far beyond the supervision requirements. DC active/standby and active-active disaster recovery (DR) services are indispensable.

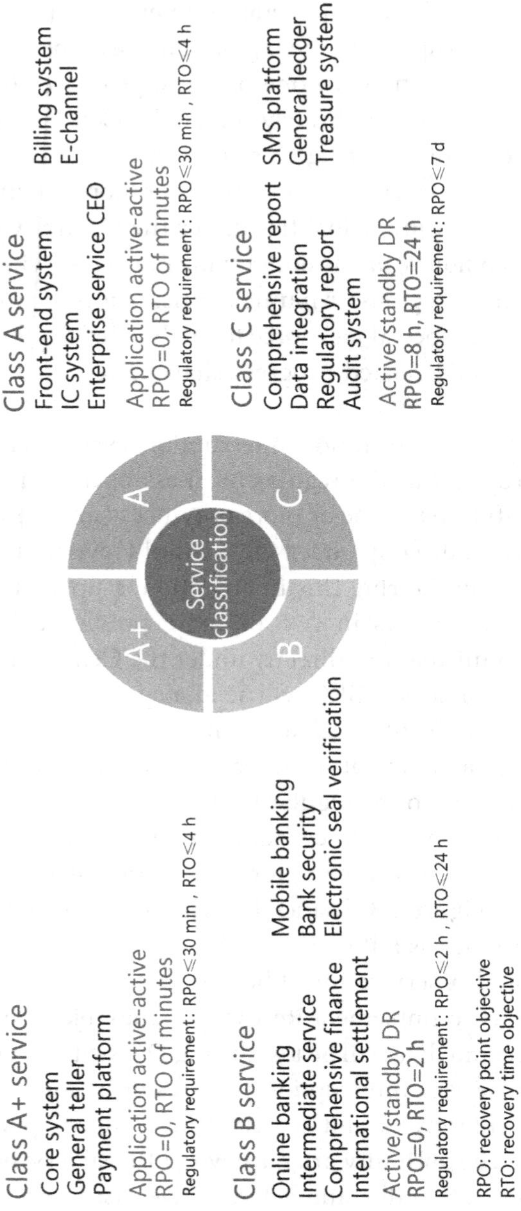

Class A+ service

Core system
General teller
Payment platform

Application active-active
RPO=0, RTO of minutes
Regulatory requirement : RPO≤30 min , RTO≤4 h

Class B service

Online banking   Mobile banking
Intermediate service Bank security
Comprehensive finance Electronic seal verification
International settlement

Active/standby DR
RPO=0, RTO=2 h
Regulatory requirement: RPO≤2 h , RTO≤24 h

RPO: recovery point objective
RTO: recovery time objective

Class A service

Front-end system   Billing system
IC system   E-channel
Enterprise service CEO

Application active-active
RPO=0, RTO of minutes
Regulatory requirement : RPO≤30 min , RTO≤4 h

Class C service

Comprehensive report SMS platform
Data integration   General ledger
Regulatory report   Treasure system
Audit system

Active/standby DR
RPO=8 h, RTO=24 h
Regulatory requirement: RPO≤7 d

Service classification

A+   A

B   C

FIGURE 2.3   Financial service classification.

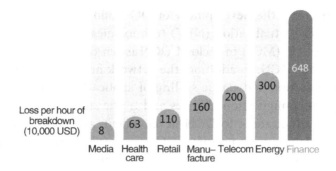

Source: Network Computing, the Meta Group and Contingency Planning Research

FIGURE 2.4 Losses caused by service interruption.

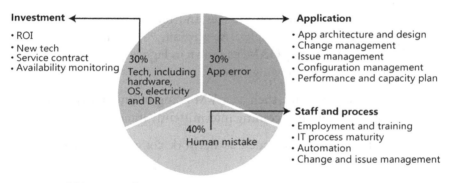

FIGURE 2.5 Gartner's analysis report on unplanned system breakdowns.

As shown in Figure 2.5, according to Gartner's analysis report on unplanned system breakdowns, 40% of a bank's recent service-affected events were caused by human error. In most events, application or system faults spread to the network, which are not well protected. As indicated by the root cause analysis, most faults are related to Layer 2 network technologies. DCNs should provide suppression or self-healing capabilities for Layer 2 loops, unknown unicast flooding, policy switching, and redundancy switchover.

5. Applications are migrated dynamically on demand, traffic increases sharply, and DCNs require intelligent O&M.

Driven by the development of DC cloudification and Network Function Virtualization (NFV) technologies, the number of managed objects (MOs) in a cloud DCN is ten times greater than that of a legacy DCN. In addition, the network needs to detect dynamic VM migration and elastic scaling of applications, which results in frequent configuration changes and traffic surges. Legacy network O&M cannot adapt to DCN development, and O&M pain points become increasingly prominent. Consequently, LinkedIn data shows that the number of network faults saw an 18-fold increase from 2010 to 2015. As the man-machine interface becomes the machine-machine interface, the network becomes invisible. Network, computing, and storage boundaries are blurred, and as larger amounts of data become involved, network faults become more difficult to locate and isolate.

In addition, application policies and mutual access within DCs become increasingly complex. As a result, 70% of faults cannot be identified using legacy O&M, as shown in Figure 2.6. These faults are classified into the following types:

- Connection faults, such as a VM going offline unexpectedly or communication becoming intermittently interrupted.

- Performance faults, such as network congestion during heavy loads.

- Policy faults, such as unauthorized access and port scanning.

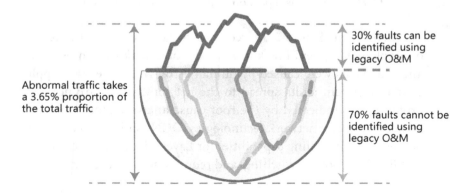

FIGURE 2.6 Abnormal flows.

To address these issues, DCNs require a new intelligent analysis engine. Leveraging big data algorithms, an intelligent analysis engine associates application traffic with the network status in data analysis to accurately and rapidly predict, detect, and isolate network faults, and to form a closed system for network data collection, analysis, and control. Network devices must utilize increased intelligence, as well as the ability to self-heal upon the occurrence of single-NE faults. Based on telemetry and edge intelligence technologies, they collect and pre-process data at a high speed and report to the analysis engine.

Huawei has launched the cloud DCN solution to address the five major challenges faced by DCNs. This powerful solution is capable of building intent-driven cloud DCNs, which are simplified, ultra-broadband, secure, reliable and intelligent.

- Simplified: An SDN controller enables a user to create a logical network in drag-and-drop mode, and cloud-network integration enables service rollout in minutes.

- Ultra-broadband: DC switches featuring a switching capacity three times that of the industry average are used, and which have been running stably at the core layer for 10 years. Access leaf nodes support smooth evolution from 10GE to 25GE Ethernet interface standards. The SDN controller is used to build an elastic large-scale network with zero packet loss and delivers a management scale ten times that of the industry average. It also utilizes a Huawei unique congestion scheduling algorithm to improve AI training efficiency by 40%.

- Security: Seventeen types of value-added services (VASs) are available in PAYG mode. They are flexibly orchestrated using SFC and isolated for security purposes in fine-grained manner using microsegmentation.

- Reliable: The SDN controller enables self-healing of devices, and the solution provides proactive loop prevention. The underlay and overlay protocols isolate resources to prevent crosstalk, and services can be switched over between the active and standby DCs.

- Intelligent: The SDN controller monitors the service running status, logical resource consumption, and physical resource consumption. Real paths between VMs are detected to enable the rapid location of service faults, with intelligent O&M system locating faults in minutes.

In addition, Huawei has launched the HA Fabric, Multi DC Fabric, and AI Fabric solutions for enterprise EDC, carrier telco cloud, and AI distributed computing or storage, addressing the demands for quick cloudification. The following chapters will elaborate on the DCN solutions and demonstrate some solutions in which Huawei has built cloud DCNs.

CHAPTER 3

Architecture and Technology Evolution of DCNs

T HE LEGACY DCN ARCHITECTURE consists of the access layer, aggregation layer, and core layer. To meet large Layer 2 and non-blocking forwarding requirements in DCs, the physical architecture evolves to a two-layer spine-leaf architecture based on the Clos architecture. Forwarding protocols on DCNs also evolve from xSTP and Layer 2 Multipathing (L2MP) to Network Virtualization over Layer 3 (NVO3) protocols, among which Virtual eXtensible Local Area Network (VXLAN) is a typical NVO3 protocol. This chapter uses the DCs of financial services companies and carriers as examples to explain the evolution process and network architecture of DCNs, and summarize DCN design principles for financial services companies and carriers.

3.1 DCN TECHNOLOGY OVERVIEW

DCs have been developing rapidly over the past few years, and because of this, virtualization, cloud (private cloud, public cloud, and hybrid cloud), big data, and SDN have become popular. Whether it is virtualization, cloud computing, or SDN technologies, network data packets must still be transmitted over physical networks. As a result, bandwidth, latency, scalability, and other physical network characteristics greatly affect the performance and function of virtual networks. A performance and scalability test report on OpenStack Neutron by Mirantis, a Silicon Valley startup, showed that transmission efficiency can be significantly improved on virtual networks through network devices with upgrade and adjustment. But before we can understand virtual networks, we must first understand physical networks. To this end, this section describes the physical architecture and technology used in DCNs.

3.1.1 Physical Architecture of DCNs

Network architecture design is a fundamental part of network design. Upgrading or modifying the network architecture brings huge risks and costs. Therefore, exercise caution at the beginning of DC construction when you are selecting and designing the network architecture. The physical network architecture of DCs has evolved from a traditional three-layer network architecture to a two-layer spine-leaf architecture based on the Clos architecture. This section describes how the architecture evolves.

3.1.1.1 Traditional Three-Layer Network Architecture

The three-layer network architecture originates from campus networks. Traditional large DCNs use this architecture, which consists of the following layers:

Aggregation layer: Aggregation switches connect to access switches and provide services such as security, QoS, and network analysis. They function as gateways to collect routing information in a point of delivery (PoD).

Access layer: Access switches connect to PMs and VMs, add or remove virtual local area network (VLAN) tags to and from packets, and forward traffic at Layer 2.

Core layer: Core switches forward traffic entering and leaving a DC at a high speed and provide connectivity for multiple aggregation switches. Figure 3.1 shows a typical three-layer network architecture.

In most cases, aggregation switches form the boundary between Layer 2 and Layer 3 networks. The downstream devices connected to the aggregation switches are on the Layer 2 network, and the upstream devices connected to the aggregation switches are on the Layer 3 network. Each group of aggregation switches manages a PoD, and each PoD has an independent VLAN. When a server is migrated within a PoD, the IP address and default gateway of the server do not need to be changed because one PoD corresponds to one Layer 2 broadcast domain.

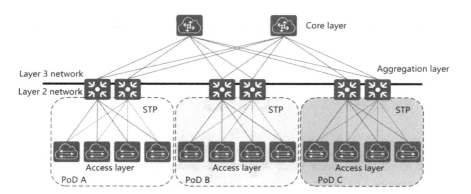

FIGURE 3.1   Three-layer network architecture.

The three-layer network architecture is widely used on legacy DCNs because of its simple implementation, small configuration workload, and strong broadcast control. However, it cannot meet the network requirements of cloud DCs in the cloud computing era for the following reasons:

1. No support for large Layer 2 network construction

The three-layer network architecture has the advantage of effective broadcast control. A broadcast domain can be controlled in a PoD using VLAN technology on aggregation switches. However, compute resources are pooled in the cloud computing era. Compute resource virtualization requires that VMs be created and migrated anywhere without changing their IP addresses or default gateway. This fundamentally changes the network architecture of DCs. A large Layer 2 network, as shown in Figure 3.2, must be constructed to support VM migration and compute resource virtualization. Core switches on the traditional three-layer network architecture must be configured as the boundary between Layer 2 and Layer 3 networks, so that the downstream devices connected to the core switches are on the Layer 2 network. In this way, aggregation switches do not function as gateways, and the network architecture gradually evolves to a two-layer architecture without the aggregation layer.

Another reason for a weakened aggregation layer is that services such as security, QoS, and network analysis are migrated out of a DC.

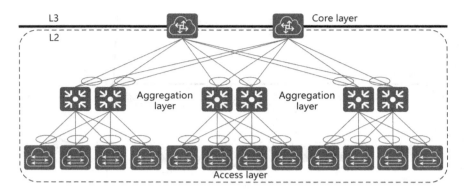

FIGURE 3.2 Large Layer 2 network.

The average server utilization in traditional DCs is only 10%–15%, and network bandwidth is not the main bottleneck. However, the scale and utilization of servers in cloud DCs increase significantly, IP traffic grows exponentially every year, and network forwarding requires corresponding increases in performance. Services such as QoS and network analysis are not deployed inside a DC, and firewalls are deployed near gateways. Therefore, the aggregation layer is dispensable.

A DCN using the three-layer network architecture can be scaled out only by adding PoDs. However, deploying more aggregation switches creates pressure on core switches. As a result, core switches need to provide more and more ports, which depend on device updates by vendors. This network architecture that relies on device capabilities is also criticized by network designers.

2. No support for non-blocking forwarding of traffic (especially east-west traffic)

Data center traffic can be classified into the following types:

- North-south traffic: traffic exchanged between clients outside a DC and servers inside the DC, or traffic from servers in the DC to external networks.

- East-west traffic: traffic exchanged between servers in a DC.

- Cross-DC traffic: traffic exchanged between different DCs.

In traditional DCs, services are usually deployed in dedicated mode. Typically, a service is deployed on one or more physical servers and is physically isolated from other systems. Therefore, the volume of east-west traffic in a traditional DC is low, and the volume of north-south traffic accounts for about 80% of the total traffic volume in the DC.

In cloud DCs, the service architecture gradually changes from a monolithic architecture to a web-application-database architecture, and distribution technologies become popular for enterprise applications. Components of a service are usually distributed in multiple VMs or containers. The service is no longer run by one or more physical servers, but by multiple servers working collaboratively, resulting in rapid growth of east-west traffic.

In addition, the emergence of big data services makes distributed computing a standard configuration in cloud DCs. Big data services can be distributed across hundreds of servers in a DC for parallel computing, which also greatly increases east-west traffic.

East-west traffic replaces north-south traffic because of its sharp increase in cloud DCs. East-west traffic accounts for 90% of all traffic in a cloud DC. Consequently, we must ensure non-blocking forwarding of east-west traffic in cloud DCNs.

The traditional three-layer network architecture is designed for traditional DCs where north-south traffic dominates. It is unsuitable for cloud DCs with a large amount of east-west traffic for the following reasons:

- In the three-layer network architecture designed for north-south traffic, some types of east-west traffic (such as cross-PoD Layer 2 and Layer 3 traffic) must be forwarded by devices at the aggregation and core layers, and unnecessarily passes through many nodes. A bandwidth oversubscription ratio of 1:10 to 1:3 is usually set on traditional networks to improve device utilization. With the oversubscription ratio, performance deteriorates significantly each time traffic passes through a node. In addition, xSTP technologies on the Layer 3 network exacerbate this deterioration.

- If east-west traffic passes through multiple layers of devices, the forward path and return path of the traffic may be different, and the paths have different delays. As a result, the overall traffic delay is difficult to predict, which is unacceptable to delay-sensitive services such as big data.

Therefore, enterprises require high-performance aggregation and core switches to guarantee bandwidth for east-west traffic when deploying DCNs. In addition, to ensure a predictable delay and reduce performance deterioration, enterprises must carefully plan their networks and properly plan services with east-west traffic. These factors undoubtedly reduce the network availability, make DCNs hard to scale, and increase the cost of DCN deployment for enterprises.

3.1.1.2 Spine-Leaf Architecture

Because the traditional three-layer network architecture is not suitable for cloud DCNs, a two-layer spine-leaf architecture based on the Clos architecture becomes popular on cloud DCNs. Figure 3.3 shows a spine-leaf network based on the three-stage Clos architecture. The number of uplinks on each leaf switch is equal to the number of spine switches, and the number of downlinks on each spine switch is equal to the number of leaf switches. That is, spine and leaf switches are fully meshed.

The Clos architecture was proposed by Charles Clos of Bell Labs in "A Study of Non-Blocking Switching Networks." This architecture is widely used on Time-Division Multiplexing (TDM) networks.

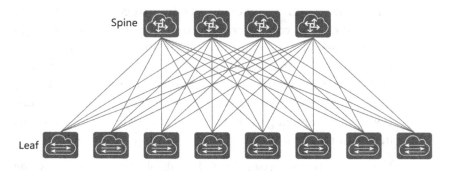

FIGURE 3.3   Spine-leaf network architecture (three-stage Clos architecture).

To commemorate this important achievement, this architecture was named after Charles Clos. The core idea of the Clos architecture is to use a large number of small-scale, low-cost, and replicable network units to construct a large-scale network. Figure 3.4 shows a typical symmetric three-stage Clos architecture. The Clos architecture has three stages: the ingress stage, middle stage, and egress stage. Each stage consists of several identical network elements. In the three-stage Clos switching network shown in Figure 3.4, if the number (m) of middle-stage devices is greater than n after proper rearrangement, the requirement for non-blocking switching can be met from any ingress device to any egress device.

The Clos architecture provides an easy to expand method for constructing a large non-blocking network without depending on hardware specifications of devices. This makes it suitable for DCN architecture design. If you look closely, the spine-leaf architecture is actually a folded three-stage Clos architecture, with the same easy implementation of non-blocking switching. If the rates of all ports are the same and half of the ports can be used as uplink ports, the theoretical bandwidth oversubscription ratio can be 1:1. However, the server utilization cannot reach 100% even in a cloud DC. That is, it is impossible that all servers send traffic at the full rate at any one time. In practice, the ratio of the uplink bandwidth to the downlink bandwidth is designed to be about 1:3, which supports non-blocking forwarding in most cases.

In the spine-leaf architecture, leaf switches are equivalent to access switches in the traditional three-layer architecture, directly connect to physical servers, and usually function as gateways. Spine switches are equivalent to core switches and are the traffic forwarding core of the entire network. Spine and leaf switches use Equal-Cost Multi-Path (ECMP)

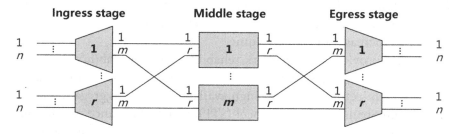

FIGURE 3.4   Symmetric three-stage Clos switching network.

routing to implement multi-path forwarding. Different from core switches on a traditional three-layer network, spine switches are the traffic forwarding core of the entire network and are equivalent to middle-stage devices in the Clos architecture. According to the Clos architecture, north-south traffic can be sent to an external network through leaf switches instead of spine switches. In this way, spine switches can focus on traffic forwarding and do not need to provide other auxiliary functions.

In summary, the spine-leaf architecture has the following advantages over the traditional three-layer network architecture:

First, it supports non-blocking forwarding. The spine-leaf architecture provides the same processing mode for east-west traffic and north-south traffic. Therefore, non-blocking forwarding can be implemented on a properly designed DCN using this architecture. Traffic of any type can be forwarded to the destination through only three nodes: leaf node -> spine node -> leaf node.

Second, it is elastic and scalable. The spine-leaf architecture has a good scale-out capability. To scale out a DCN using this architecture, you only need to ensure that the ratio of spine nodes to leaf nodes is within a certain range and copy the original architecture without the need to redesign the network. The spine-leaf architecture based on the three-stage Clos architecture can meet bandwidth requirements of most DCNs. For ultra-large DCs, a five-stage spine-leaf architecture can be used. That is, a spine-leaf network based on the three-stage Clos architecture is deployed in each PoD, and a layer of core switches is added between PoDs for inter-PoD communication. Cross-PoD traffic can then be forwarded to the destination through five nodes: leaf node -> spine node -> core node -> spine node -> leaf node. Spine and core nodes are fully meshed. In addition, the network design is flexible. When the traffic is light in a DC in the early stage, the number of spine switches can be reduced appropriately. More spine switches can be deployed flexibly to cope with increases in traffic volume.

Third, it provides high network reliability. High availability design is performed for aggregation and core layers in the traditional three-layer network architecture. However, high availability at the aggregation layer is based on Spanning Tree Protocol (STP), and performance of multiple switches cannot be fully used. In addition, if all aggregation switches (typically two) fail, the entire aggregation layer of a PoD breaks down.

In contrast, in the spine-leaf architecture, there are multiple channels for cross-PoD traffic exchanged between two servers. Unless in extreme cases, the reliability of this architecture is higher than that of the traditional three-layer network architecture.

3.1.2 Technology Evolution of DCNs

Although the spine-leaf architecture provides the topology basis for a non-blocking network, a suitable forwarding protocol is required to fully utilize capabilities of the topology. Technologies for fabric networks in DCs evolve from xSTP to virtual chassis technologies including Cluster Switch System (CSS), Intelligent Stacking (iStack), and Virtual Switching System (VSS), and then to L2MP technologies including Transparent Interconnection of Lots of Links (TRILL), Shortest Path Bridging (SPB), and FabricPath. Finally, VXLAN, a typical NVO3 technology, is selected as the de facto technology standard for DCNs. This section describes the reasons for this evolution.

3.1.2.1 xSTP Technologies

Local area networks (LANs) were very popular at the initial stage of network development. A LAN is constructed as a loop-free Layer 2 network, and addressing on the LAN is implemented using broadcast packets. This method is simple and effective. In a broadcast domain, packets are forwarded based on MAC addresses.

Under such circumstances, xSTP technologies were introduced to eliminate loops in a broadcast domain. The original standard for xSTP technologies is IEEE 802.1D. xSTP creates a tree topology to eliminate loops, thereby preventing an infinite loop of packets on a loop network. More in-depth explanations of xSTP technologies can be found in other documents. This section only describes why xSTP technologies are no longer applicable to current DCNs. xSTP technologies have the following problems:

- Slow convergence: When a link or switch on an xSTP network fails, the spanning tree needs to be recalculated and MAC address entries need to be re-generated. If the root node fails, a new root node needs to be elected. As a result, xSTP convergence can take subseconds or even seconds. This is acceptable on a network with 100 Mbit/s or

1000 Mbit/s (1 Gbit/s) access. However, on networks with 10 Gbit/s, 40 Gbit/s, or even 100 Gbit/s access, second-level convergence causes a large number of services to go offline, which is unacceptable. This slow convergence is one of the major disadvantages of xSTP networks.

- Low link utilization: As mentioned above, xSTP constructs a tree network to ensure that no loop exists in the network topology. xSTP blocks some network links to construct a tree topology. While still a part of the network, the blocked links are set to unavailable status and are used only when links that forward traffic fail. As a result, xSTP cannot fully utilize network resources.

- Sub-optimal forwarding paths: There is only one forwarding path between any two switches on an xSTP network because the network uses a tree topology. If a shorter path exists between two non-root switches but the shorter path is blocked due to xSTP calculation, traffic exchanged between the switches has to be forwarded along a longer path.

  - No support for ECMP: On the Layer 3 network, a routing protocol can use the ECMP mechanism to implement traffic forwarding along multiple equal-cost paths between two nodes, achieving path redundancy and improving forwarding efficiency. xSTP technologies have no equivalent mechanism.

  - Broadcast storm: Although xSTP prevents loops by constructing a tree topology, loops may still occur in some fault scenarios. However, xSTP cannot eliminate loops in such scenarios. On the Layer 3 network, the IP header of a packet contains the TTL field. Each time the packet passes through a router, the TTL value is deducted by a preset value. When the TTL value becomes 0, the packet is discarded to prevent it from looping endlessly throughout the network. However, an Ethernet header does not have a similar field. Once a broadcast storm occurs, the load of all involved devices increases sharply, greatly limiting the scale of an xSTP network. In some classic network design documents, the xSTP network diameter is limited to seven hops or fewer, which is useless for a DCN with tens of thousands or even hundreds of thousands of terminals.

- Single-homing access mechanism: Because an xSTP network uses a tree topology, a loop occurs when a server is dual-homed to two xSTP-enabled switches on the network. (A loop does not occur only when the two switches belong to two LANs, and the server may require IP addresses on two network segments.) Therefore, even if the server is dual-homed to the switches through uplinks, xSTP blocks one of the two uplink interfaces, changing the dual-homing networking to single-homing networking.

- Network scale: In addition to the broadcast storms limiting the scale of the xSTP network, so does the number of tenants supported by the xSTP network. xSTP uses VLAN IDs to identify tenants, and the VLAN ID field has only 12 bits. When IEEE 802.1Q was designed, the designers probably thought that 4000 tenants were sufficient for a network. However, in the cloud computing era, there are far more than 4000 tenants on a network.

xSTP technologies are gradually being phased out on DCNs due to the preceding problems. New technologies are emerging to solve these problems.

3.1.2.2 Virtual Chassis Technologies

Virtual chassis technologies are the first type of technologies that can address the preceding problems. This type of technologies can integrate the control planes of multiple devices to form a unified logical device. This logical device has a unified management IP address and also works as one node in various Layer 2 and Layer 3 protocols. Therefore, the network topology after the integration is loop-free for xSTP, which indirectly avoids problems of xSTP.

Virtual chassis technologies implement N:1 virtualization. Different physical devices share the same control plane, which is equivalent to creating a cluster for physical network devices. In addition, master election and a master/standby switchover can also be performed in the cluster. Several virtual chassis technologies exist in the industry, such as Cisco's VSS and

H3C's IRF2. The following describes Huawei's CSS and iStack technologies, which are also called stacking technologies.

1. Virtual chassis technologies: CSS and iStack

Take Huawei switches as an example. A stack of modular switches is called a CSS, and a stack of fixed switches is called an iStack. When multiple switches are stacked, they function as one logical switch with a unified control plane and are managed in a unified manner. A stack consists of switches with one of three roles: master switch, standby switch, and slave switch. As shown in Figure 3.5, the master switch is also called the stack master or system master. It controls and manages the entire system. The standby switch is also called the stack standby or system standby. It takes over services from the master switch when the master switch fails. There is one master switch and one standby switch, whereas all the other member switches in the stack are slave switches.

A modular switch has two MPUs: one active MPU and one standby MPU. As shown in Figure 3.6, two modular switches establish a stack. On the control plane, the active MPU of the master switch becomes the stack master MPU and manages the stack. The active MPU of the standby switch becomes the stack standby MPU and functions as a backup of the stack master MPU. The standby MPUs of the master and standby switches function as cold standby MPUs in the stack and do not have management roles.

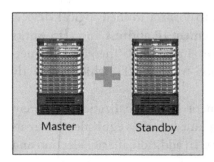
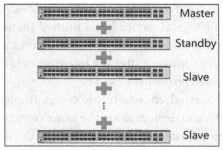

FIGURE 3.5   Roles of member switches in a stack.

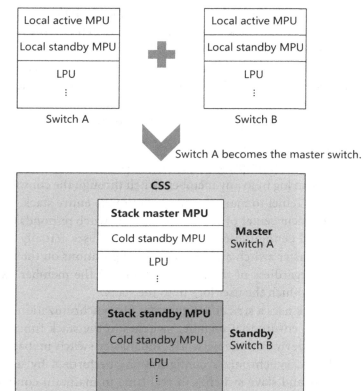

FIGURE 3.6   Roles of MPUs in a stack of modular switches.

2. Stack system management

a. Member management

A stack member ID is used to identify and manage a stack member. Each stack member has a stack member ID, unique within its stack.

Stack member IDs are added to interface numbers to facilitate configuration and identification of interfaces on member switches. When the stacking function is not enabled on a modular switch, interface numbers are in the format of slot ID/subcard ID/port number, for example, 10GE1/0/1. When the stacking function is enabled on the switch, interface numbers are in the format of stack member ID/slot ID/subcard ID/port number. Assume that the stack member ID of the switch is 2, the number of 10GE1/0/1 changes to 10GE2/1/0/1.

Interface numbers on a fixed switch are in the format of stack member ID/subcard ID/port number. (The stacking function is enabled on a fixed switch by default. Therefore, interface numbers on a fixed switch contain the stack member ID.)

b. Configuration management

A stack of multiple switches can be considered a single entity. A user can log in to any member switch through the console port or using Telnet to manage and configure the entire stack. As the management center of a stack, the master switch responds to user login and configuration requests. That is, a user actually logs in to the master switch and performs configurations on the master switch regardless of the login method and the member switch through which the user logs in to the stack.

A stack uses a strict configuration file synchronization mechanism to ensure that member switches in the stack function as a single device on the network. The master switch manages the stack and synchronizes configurations performed by users to standby and slave switches in real time to maintain configuration consistency on all member switches. Real-time configuration synchronization ensures that all member switches in the stack have the same configurations. Even if the master switch fails, the standby switch can provide the same functions based on the same configurations.

c. Version management

All member switches in a stack must use the same version of system software. A stack supports software version synchronization among member switches. Member switches to be stacked do not have to run the same software version and can establish a stack when their software versions are compatible. After the master switch is elected, if the software version running on a member switch is different from that on the master switch, the member switch automatically downloads the system software from the master switch, restarts with the new system software, and joins the stack again.

3. Common stack scenarios

a. Stack establishment

Figure 3.7 shows how a stack is established after all member switches are power-cycled. In this scenario, software stacking configurations are performed on all member switches, the switches are powered off, stack cables are connected, and the switches are powered on. The switches then enter the stack establishment process, which includes the following steps.

   i. Member switches send link detection packets through stack links and check the validity of stack links.

  ii. Member switches send stack competition packets to each other and elect the master switch according to the following election rules. Member switches compare the following items in the listed order to elect the master switch (the election ends when a winning switch is found):

- Running status: The switch that completes startup and enters the stack running state first becomes the master switch.

- Stack priority: The switch with a higher stack priority becomes the master switch.

- Software version: The switch running a later software version becomes the master switch.

- Number of MPUs: A switch with two MPUs takes precedence over a switch with only one MPU and becomes the

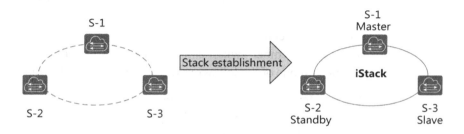

FIGURE 3.7   Stack establishment.

master switch. (This item is compared only in a stack of modular switches.)

- Bridge MAC address: The switch with a smaller bridge MAC address becomes the master switch.

iii. After the master switch is elected, other member switches send member information packets to the master switch. If the IDs of member switches conflict or the software versions of member switches are different from that of the master switch, member switches change their IDs or synchronize their software versions with the master switch.

iv. The master switch collects information about all member switches, calculates the topology, and synchronizes stack member and topology information to other member switches.

v. The master switch elects a standby switch based on election rules and synchronizes information about the standby switch to other member switches.

The stack is established successfully.

b. Joining and leaving a stack

Figure 3.8 illustrates how a new switch joins a running stack. In most cases, a switch joins a stack after being powered off. That is, after software stack configurations are complete on the switch, the switch is powered off, stack cables are connected, and then, the switch is powered on. (If a switch joins a stack after being powered on, the scenario is considered a stack merging scenario, which is described later in this section.)

To not affect the operation of the stack, a new member switch joins the stack as a slave switch, and the original master, standby, and slave switches retain their roles.

In contrast, a member switch can also leave a stack, but will affect the stack in one of the following ways, depending on its role:

i. If the master switch leaves the stack, the standby switch becomes the new master switch. The new master switch then recalculates the stack topology, synchronizes updated topology information to other member switches, and specifies a new standby switch. The stack then enters the running state.

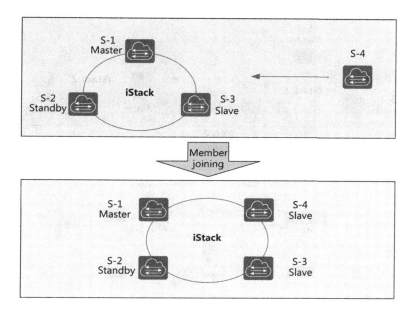

FIGURE 3.8 A new member switch joining a stack.

ii. If the standby switch leaves the stack, the master switch specifies a new standby switch, recalculates the stack topology, and
synchronizes updated topology information to other member switches. The stack then enters the running state.

iii. If a slave switch leaves the stack, the master switch recalculates the stack topology and synchronizes updated topology
information to other member switches. The stack then enters
the running state.

c. Stack merging

Two properly running stacks can merge into a new stack, as
shown in Figure 3.9. Stack merging usually occurs in the following situations:

i. A switch is configured with the stacking function and is connected to a running stack using stack cables when the switch
is powered on.

ii. A stack splits because a stack link or member switch fails.
After the faulty stack link or member switch recovers, the
split stacks merge into one again.

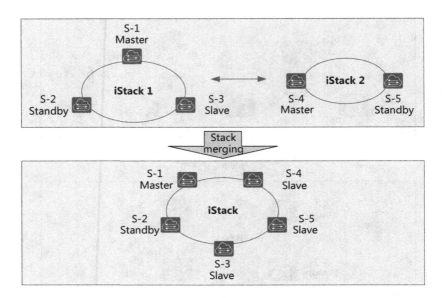

FIGURE 3.9   Stack merging.

When two stacks merge, the master switches in the two stacks compete to be the master switch of the new stack. The master switch election rules used are the same as those used in stack establishment. The member switches that belonged to the same stack as this new master switch retain their roles and configurations after the election, and their services are unaffected. Member switches in the other stack restart and join the new stack through the stack member joining process.

  d. Stack split

A running stack may split into multiple stacks, as shown in Figure 3.10. In most cases, a stack splits because stack connections between member switches are disconnected due to cable or card faults or incorrect configurations, but the split stacks are still running with power on.

Depending on whether the original master and standby switches are in the same stack after a stack splits, either of the following situations occurs:

    i. If the original master and standby switches are in the same stack after the split, the original master switch recalculates

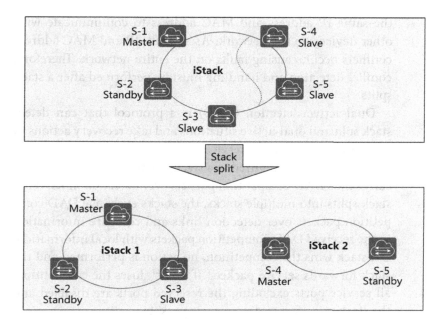

FIGURE 3.10   Stack split.

the stack topology, deletes topology information of the removed member switches, and synchronizes updated topology information to other member switches in the new stack. The removed slave switches automatically restart because there are no master and backup switches, elect a new master switch, and establish a new stack.

ii. If the original master and standby switches are in different stacks after the split, the original master switch specifies a new standby switch in its stack, recalculates the stack topology, and synchronizes updated topology information to other member switches in the stack. The original standby switch becomes the master switch in its stack. It then recalculates the stack topology, synchronizes updated topology information to other member switches in the stack, and specifies a new standby switch.

e. Address conflict detection after a stack splits

After a stack on a network splits, there are multiple stacks with the same global configurations on the network. These stacks use the same IP address and MAC address to communicate with other devices on the network. As a result, IP and MAC address conflicts occur, causing faults on the entire network. Therefore, conflict detection and handling must be performed after a stack splits.

Dual-active detection (DAD) is a protocol that can detect stack split and dual-active situations and take recovery actions to minimize the impact on services.

After DAD is configured in a stack, the master switch periodically sends DAD competition packets over detection links. If the stack splits into multiple stacks, the stacks exchange DAD competition packets over detection links and compare information in the received DAD competition packets with local information. If a stack wins the competition, no action is performed and the stack forwards service packets. If a stack loses the competition, all service ports excluding the reserved ports are disabled and the stack stops forwarding service packets.

The stacks compare the following items in the listed order (the competition ends when a winning stack is found):

iii. Stack priority: The stack with a switch of a higher stack priority wins.

iv. MAC address: The stack with a switch with a smaller MAC address wins.

After the stack split fault is rectified, the stacks merge into one. The stack that loses the competition restarts, the disabled service ports are restored, and the entire stack recovers.

f. Master/Standby switchover in a stack

A master/standby switchover in a stack is a process in which the roles of master and standby switches in the stack change. In most cases, a master/standby switchover occurs in a stack when the master switch fails or restarts or the administrator performs a master/standby switchover.

During a master/standby switchover, the standby switch becomes the master switch and specifies a new standby switch, as shown in Figure 3.11. After the original master switch restarts,

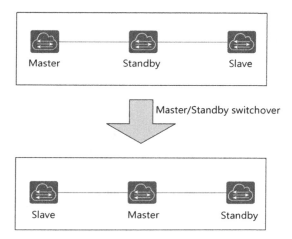

FIGURE 3.11   Role changes of member switches before and after a master/standby switchover.

it joins the stack through the stack member joining process. (If the master/standby switchover occurs in a stack of modular switches because the stack master MPU fails, the original master switch may not restart.)

4. Local preferential forwarding of traffic in a stack

If traffic is not preferentially forwarded by the local switch, some traffic entering a stack through a member switch is sent to an interface on another member switch for forwarding based on a hash algorithm. The traffic passes through stack cables. Because the bandwidth provided by stack cables is limited, this forwarding mode increases the load on stack cables and reduces traffic forwarding efficiency.

To improve traffic forwarding efficiency and reduce inter-device traffic forwarding, local preferential forwarding on Eth-Trunks is required in a stack. This function ensures that traffic entering a member switch is preferentially forwarded through the outbound interface on the local switch, as shown in Figure 3.12. If the local outbound interface fails, the traffic is forwarded through an interface on another member switch.

5. Advantages and disadvantages of virtual chassis technologies

Virtual chassis technologies virtualize multiple devices into one logical device on the control plane. Link aggregation allows this

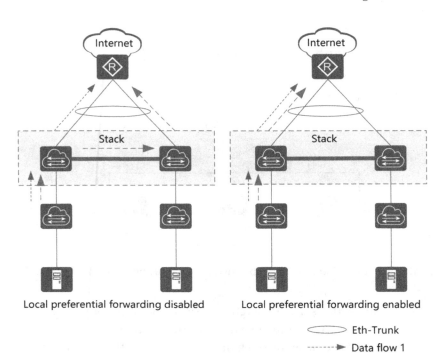

FIGURE 3.12   Local preferential forwarding.

logical device to connect to each physical or logical node at the access layer through only one logical link. In this way, the logical topology of the entire network becomes a loop-free tree topology, meeting loop-free and multi-path forwarding requirements. As shown in Figure 3.13, the entire network uses a tree topology after devices are stacked. The entire network does not even need to use xSTP to eliminate loops.

However, virtual chassis technologies have the following problems.

a. Limited scalability: Because control planes are integrated, the master switch provides the control plane for the entire virtual chassis system. The processing capability of the control plane cannot exceed the processing capability of a single device in the system. Therefore, the scalability of any virtual chassis technology is limited by the performance of the master switch in the system. This is unacceptable for DCNs with explosive traffic growth.

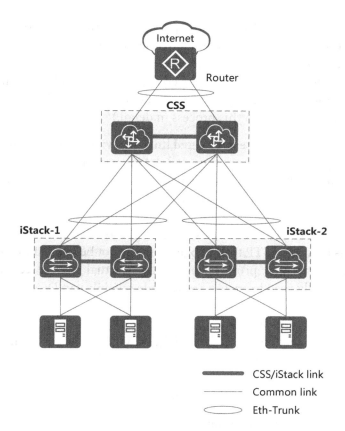

FIGURE 3.13   Constructing a loop-free network through stacking.

b. Reliability: Because the control plane is on the master switch, packet loss may last for a long time or the entire system may stop running if the master switch fails.

c. Packet loss during an upgrade: Control plane integration also makes it difficult to upgrade a virtual chassis system. Common restart and upgrade operations cause interruption of the control plane, resulting in packet loss for a long time. For some services that are sensitive to packet loss, seconds or even minutes of packet loss caused by the upgrade of the virtual chassis system is unacceptable. To address this problem, vendors provide some lossless upgrade solutions such as in-service software upgrade (ISSU). However, these solutions are not effective on live networks because they are very complex, difficult to operate, and have strict network requirements. (These solutions usually implement lossless upgrade through multiple processes on the control plane. When the active process is being restarted, the standby process works properly. An active/standby switchover is performed after the active process is restarted. This document does not explain this process in detail.)

d. Bandwidth waste: Dedicated links in a virtual chassis system are used for status exchange and data forwarding between devices. Generally, the dedicated links may occupy 10% to 15% of the total bandwidth of devices.

In summary, virtual chassis technologies resolve some of the problems found in xSTP technologies, but cannot be applied in DCNs as the sole network protocol. In practice, virtual chassis technologies are usually used to ensure single-node reliability in DCs.

3.1.2.3 L2MP Technologies

The preceding sections describe xSTP and virtual chassis technologies. These technologies have the following major problems:

- They cannot support large DCNs with massive amounts of data.

- Their link utilization is low.

To address the abovementioned problems, it is recommended that link-state routing protocols widely used on Layer 3 networks be employed. These protocols not only support a large number of devices, but they are also loop-free and have high link utilization. Such link-state routing protocols include the intra-AS routing protocols Open Shortest Path First (OSPF) and Intermediate System to Intermediate System (IS-IS). They support ECMP load balancing and use the Shortest Path First (SPF) algorithm to ensure the shortest path for traffic forwarding, and eliminate loops. Both protocols support large-scale networks with hundreds of devices. The basic principle of L2MP technologies is to introduce mechanisms of routing technologies used on Layer 3 networks to Layer 2 networks.

Traditional Ethernet switches transparently transmit packets, do not maintain the link state on the network, and do not require an explicit addressing mechanism. A link-state routing protocol usually requires that each node on a network be addressable. Each node uses the link-state routing protocol to calculate the network topology and then calculates the forwarding database based on the network topology. Therefore, L2MP technologies need to add an addressable identifier, which is similar to an IP address on an IP network, to each device on a network.

On an Ethernet network, MAC addresses of switches and terminals are carried in Ethernet frames. To apply link-state routing protocols to Ethernet networks, a frame header needs to be added to an Ethernet header for the addressing of the link-state routing protocols. Regarding the format of the frame header, TRILL, a standard L2MP protocol defined by the Internet Engineering Task Force (IETF) uses MAC in TRILL in MAC encapsulation. In addition to the original Ethernet header, a TRILL header that provides an addressing identifier and an outer Ethernet header used to forward a TRILL packet on an Ethernet network are added. SPB, another L2MP protocol, uses MAC in MAC encapsulation and directly uses IP addresses as addressing identifiers. Other proprietary L2MP technologies, such as Cisco's FabricPath, are similar. They basically use MAC in MAC encapsulation and differ slightly only in addressing identifier processing.

When selecting link-state routing protocols, almost all vendors and standards organizations in the industry select IS-IS as the control plane protocol for L2MP technologies. This is because it runs at the link layer and works properly on Ethernet networks. This protocol was defined by the Open Systems Interconnection (OSI) protocol suite of the International Organization for Standardization (ISO) and adopted by IETF RFC 1142. IS-IS also has excellent scalability. By defining new Type-Length-Value (TLV) attributes, you can easily extend IS-IS to serve new L2MP protocols.

Huawei CloudEngine switches support TRILL, a standard L2MP protocol defined by the IETF. The following uses TRILL as an example to describe the working principles of L2MP technologies.

1. Basic concepts of TRILL

   The following concepts are used later in this section.

   Router bridge (RB): An RB is a Layer 2 switch that runs TRILL. RBs are classified into ingress RBs, transit RBs, and egress RBs based on their locations on a TRILL network. An ingress RB is an ingress node from which packets enter the TRILL network. A transit RB is an intermediate node through which packets pass on the TRILL network. An egress RB is an egress node from which packets leave the TRILL network.

Designated RB (DRB): A DRB is a transit RB that is designated to perform some special tasks on a TRILL network. A DRB corresponds to a designated intermediate system (DIS) on an IS-IS network. On a TRILL broadcast network, when two RBs in the same VLAN are establishing a neighbor relationship, a DRB is elected from either one of the RBs, based on their interface DRB priorities or MAC addresses. The DRB communicates with each device on the network to synchronize link state databases (LSDBs) of all devices in the VLAN, sparing every pair of devices from communicating with each other to ensure LSDB synchronization. The DRB performs the following tasks:

• Generates pseudonode link state PDUs (LSPs) when more than two RBs exist on the network.

• Sends complete sequence number PDUs (CSNPs) to synchronize LSDBs.

• Specifies a carrier VLAN as the designated VLAN (DVLAN) to forward user packets and TRILL control packets.

• Specifies an appointed forwarder (AF). Only one RB can function as the AF for a customer edge (CE) VLAN.

An AF is an RB elected by the DRB to forward user traffic. Non-AF RBs cannot forward user traffic. On a TRILL network shown in Figure 3.14, a loop may occur if a server is dual-homed to the TRILL network, but its NICs do not work in load balancing mode. In this case, an RB needs to be specified to forward user traffic.

• CE VLAN: A CE VLAN connects to a TRILL network and is usually configured on an edge device of the TRILL network to generate multicast routes.

• Admin VLAN: An admin VLAN is a special CE VLAN that transmits network management traffic on a TRILL network.

• Carrier VLAN: A carrier VLAN forwards TRILL data packets and protocol packets on a TRILL network. A maximum of three carrier VLANs can be configured on an RB. In the

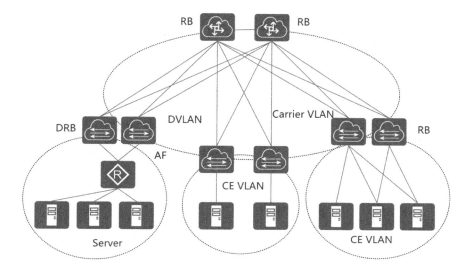

FIGURE 3.14   TRILL network

inbound direction, common Ethernet packets are encapsulated into TRILL packets in carrier VLANs. In the outbound direction, TRILL packets are decapsulated into common Ethernet packets.

- DVLAN: To combine or separate TRILL networks, multiple carrier VLANs may be configured on a TRILL network. However, only one carrier VLAN can be selected to forward TRILL data packets and protocol packets. The selected carrier VLAN is called a DVLAN.

- Nickname: A nickname is similar to an IP address and uniquely identifies a switch. Each RB on a TRILL network can be configured with only one unique nickname.

2. TRILL working mechanism

The working mechanism of the TRILL control plane almost completely reuses the mechanism of IS-IS. RBs establish neighbor relationships and exchange Hello PDUs with each other. After neighbor relationships are established, RBs exchange link state data through LSPs. Finally, all RBs have an LSDB containing information about the entire network and calculate a nickname-based forwarding table based on the LSDB. These functions are implemented by extending IS-IS, and the mechanism is almost the same as that of IS-IS.

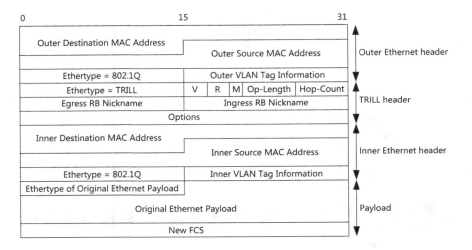

FIGURE 3.15   TRILL data packet.

Figure 3.15 shows a TRILL data packet.

A TRILL data packet consists of four parts: the payload, inner Ethernet header, TRILL header, and outer Ethernet header. On a TRILL network, both the payload and inner Ethernet header are considered payloads. The TRILL header is used for addressing and contains information such as the nickname for table lookup and forwarding. The outer Ethernet header is used to forward the packet at the link layer on the TRILL network. A CE VLAN that connects to the user network is encapsulated in the inner Ethernet header, while a carrier VLAN that is used to carry TRILL protocol packets and data packets is encapsulated in the outer Ethernet header. The TRILL header is similar to an IP header in route-based forwarding.

The TRILL unicast data forwarding process can be briefly summarized as follows: The original payload is encapsulated in the inner Ethernet header and sent to an ingress RB. The ingress RB encapsulates the received packet into a TRILL packet in which the destination nickname is the nickname of the egress RB, and the outer destination MAC address is the MAC address of the next-hop transit RB. The ingress RB then sends the packet to the next-hop transit RB. The packet is forwarded hop by hop based on table lookup, which is similar to the IP forwarding mechanism. The outer destination MAC address is changed to the MAC address of the next-hop transit

RB on each hop, and the nickname remains unchanged until the packet is sent to the egress RB. The egress RB removes the outer Ethernet header and TRILL header, searches the MAC address table based on the destination MAC address in the inner Ethernet header, and sends the original packet to the destination device.

The forwarding mechanism of TRILL broadcast, unknown-unicast, and multicast (BUM) packets is similar to that of IP multicast packets. A multicast distribution tree (MDT) is generated on the entire network. The end (branch) of the MDT extends to the RBs where all CE VLANs on the network reside (similar to the IP multicast mechanism). Multiple MDTs can also be generated, and different CE VLANs select different MDTs to ensure device bandwidth is fully utilized. For example, two MDTs are generated on Huawei CloudEngine switches, and CE VLANs select MDTs based on odd and even VLAN IDs. All the MDT branches corresponding to RBs that are not configured with CE VLANs are pruned. After the network becomes stable, BUM traffic is forwarded along the generated MDT.

The process of establishing an MDT is as follows:

- A root is selected when TRILL neighbor relationships are established. A TRILL Hello PDU carries the root priority and system ID in the TLV field. The device with the highest root priority (32768 by default and configurable) is selected as the root. If the root priorities of devices are the same, the device with the largest system ID is selected as the root.

- After unicast routes are calculated on the TRILL network, each TRILL RB calculates the shortest paths from the root to all other devices based on its LSDB to form a shortest path tree (SPT).

- The MDT is pruned. When a device exchanges LSPs with other devices, the LSPs carry its CE VLAN information. Each device prunes branches in each CE VLAN based on CE VLAN information on the network. When MDT information is delivered as entries, the entries are delivered to corresponding VLANs.

The following uses an example to describe the traffic forwarding process on a TRILL network.

Figure 3.16 shows a typical TRILL large Layer 2 network. On this network, all TRILL RBs have established neighbor relationships

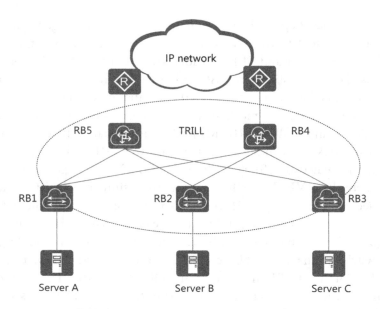

FIGURE 3.16   Traffic forwarding on a TRILL network.

and calculated unicast and multicast routes. The following example describes how Server A accesses Server B. In general, the following three steps are performed if Server A and Server B have not communicated with each other.

- Server A sends an Address Resolution Protocol (ARP) request packet to request the MAC address of Server B.

- Server B sends an ARP reply packet to Server A.

- Server A sends a unicast data packet to Server B.

    1. Server A sends an ARP request packet.
        Figure 3.17 shows how the ARP request packet is transmitted on the network, and Figure 3.18 shows how the packet is changed along the transmission path.

        - Server A broadcasts an ARP request packet to request the MAC address of Server B.

        - RB1 receives the ARP request packet from CE VLAN 1 and learns the MAC address entry based on the source

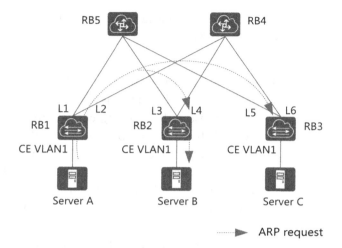

FIGURE 3.17   Server A sending an ARP request packet.

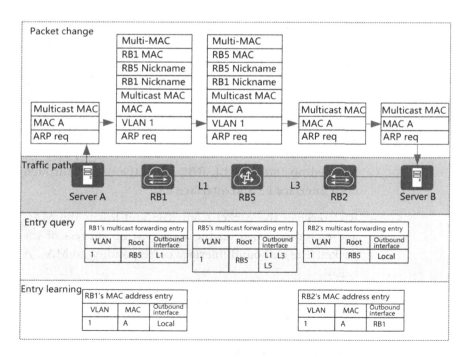

FIGURE 3.18   Change of the ARP request packet sent by Server A.

MAC address of the packet. The outbound interface corresponding to MAC A is a local interface connected to Server A.

- RB1 checks the destination MAC address of the packet and determines that the packet is a broadcast packet. RB1 checks the multicast forwarding entry matching VLAN 1. The root of the MDT is RB5, and the outbound interface is the outbound interface on link L1. Therefore, RB1 encapsulates the packet into a TRILL packet in which the destination nickname is the nickname of RB5 and the destination MAC address is a multicast MAC address. RB1 then sends the packet through interface L1. Note that the multicast forwarding entry matching VLAN 1 on RB1 also contains a local interface in VLAN 1. However, RB1 does not forward the packet through this interface because the packet is received on the interface. This is the same as normal broadcast flooding.

- RB5 receives the packet from RB1, finds that the destination MAC address is a multicast MAC address, and determines that the packet is a multicast packet. RB5 checks the multicast forwarding entry matching VLAN 1 and finds that the outbound interfaces are the interfaces on links L1, L3, and L5. RB5 then changes the source MAC address of the packet to its own MAC address, but retains the destination MAC address. Because the packet is received on interface L1, RB5 sends the packet through both interface L3 and interface L5.

- RB2 receives the packet from RB5 and learns the MAC address entry based on the source MAC address of the packet. The outbound interface corresponding to MAC A is RB1.

- RB2 finds that the destination MAC address of the packet is a multicast MAC address. It checks the multicast forwarding entry matching VLAN 1 and finds that the outbound interface is a local interface in VLAN 1.

RB2 removes the TRILL encapsulation and sends the packet through the interface in VLAN 1. Server B then receives the packet.

2. Server B sends an ARP reply packet.

Figure 3.19 shows how the ARP reply packet is transmitted on the network, and Figure 3.20 shows how the packet is changed along the transmission path.

- After receiving the ARP request packet, Server B unicasts an ARP reply packet to Server A. The destination MAC address of the packet is MAC A, and the source MAC address is MAC B.

- After receiving the ARP reply packet, RB2 learns the MAC address entry based on the source MAC address. The destination MAC address in the MAC address entry is MAC B, and the outbound interface is a local interface in VLAN 1.

- RB2 searches the MAC address table and finds that the outbound interface corresponding to destination MAC A is RB1. RB2 searches the TRILL unicast routing and forwarding table and finds two equal-cost next hops.

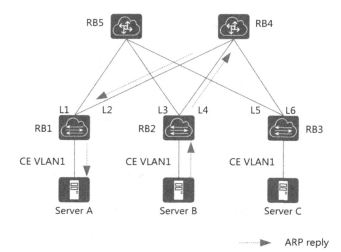

FIGURE 3.19   Server B sending an ARP reply packet.

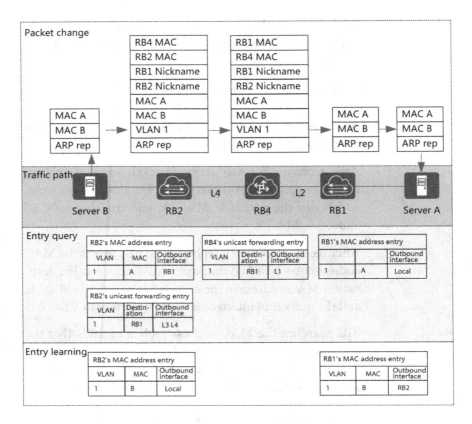

FIGURE 3.20   Change of the ARP reply packet sent by Server B.

RB2 performs hash calculation to obtain the next-hop outbound interface (for example, interface L4 is selected after hash calculation). RB2 then encapsulates the ARP reply packet into a TRILL packet in which the destination nickname is the nickname of RB1 and the destination MAC address is the MAC address of the next hop RB4. Note that per-flow ECMP load balancing is used as an example. Therefore, only one outbound interface is selected for an ARP reply packet.

– After receiving the packet from RB2, RB4 finds that the destination MAC address of the packet is its own MAC address. RB4 removes the outer Ethernet header and finds that the destination nickname is the nickname of RB1.

RB4 then searches the TRILL unicast routing and forwarding table, and finds that the next hop is RB1. RB4 replaces the source MAC address with its own MAC address, sets the destination MAC address to the MAC address of RB1, encapsulates the packet, and sends the encapsulated packet through interface L2.

- After receiving the packet from RB4, RB1 learns the MAC address entry based on the source MAC address of the packet. The outbound interface corresponding to destination MAC B is RB2.

- RB1 checks the destination MAC address of the packet and finds that it is its own MAC address. RB1 removes the outer Ethernet header and checks the destination nickname. RB1 finds that the destination nickname is also its own nickname. RB1 then removes the TRILL header and finds that the destination MAC address in the inner Ethernet header is MAC A. RB1 checks the MAC address table and finds that the outbound interface is an interface in VLAN 1. RB1 then directly sends the inner Ethernet packet to Server A.

3. Server A sends a unicast data packet.

   Figure 3.21 shows how the unicast data packet is transmitted on the network, and Figure 3.22 shows how the packet is changed along the transmission path.

   - After receiving the ARP reply packet, Server A encapsulates a unicast data packet in which the destination MAC address is MAC B and the source MAC address is MAC A.

   - After receiving the packet, RB1 searches the MAC address table based on destination MAC B and finds that the next hop is RB2. RB1 then searches the TRILL unicast routing and forwarding table, and finds two equal-cost unicast next hops. After hash calculation, RB1 selects interface L1 as the next-hop outbound interface, encapsulates the packet into a TRILL packet, and sends the encapsulated packet to RB5.

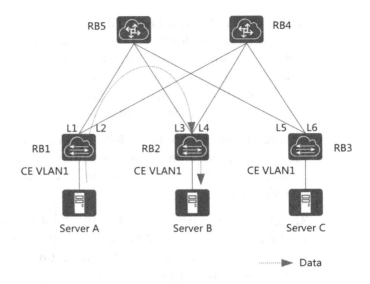

FIGURE 3.21   Server A sending a unicast data packet.

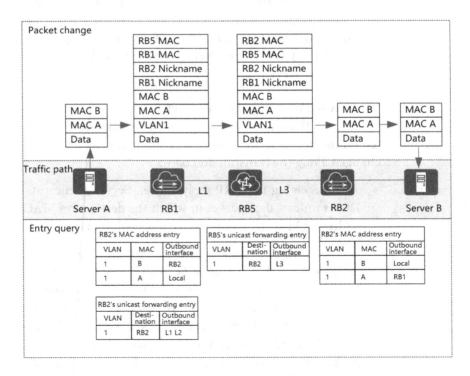

FIGURE 3.22   Change of the unicast data packet sent by Server A.

- After receiving the packet from RB1, RB5 checks the destination MAC address and finds that the MAC address is its own MAC address. RB5 removes the outer Ethernet header, searches the TRILL unicast routing and forwarding table based on the destination nickname, encapsulates the packet, and sends the encapsulated packet to RB2.

- After receiving the packet from RB5, RB2 checks the destination MAC address and destination nickname and finds that they are its own MAC address and nickname. RB2 then removes the TRILL encapsulation and sends the packet to Server B.

3. Advantages and disadvantages of L2MP technologies

The advantages of L2MP technologies, as mentioned above, are that they introduce routing mechanisms to Layer 2 networks to prevent loops, improve network scalability, and increase link utilization.

However, L2MP technologies also have the following disadvantages:

- Limited number of tenants: Similar to xSTP, TRILL uses VLAN IDs to identify tenants. Because the VLAN ID field has only 12 bits, a TRILL network supports only about 4000 tenants.

- Increased deployment costs: L2MP technologies introduce new forwarding identifiers or add new forwarding processes, which inevitably requires the upgrade of forwarding chips. Therefore, devices using old chips on existing networks cannot use L2MP technologies. As such, L2MP technologies increase customers' deployment costs.

- Mechanism-related challenges: The Operations, Administration, and Maintenance (OAM) mechanism and multicast mechanism of TRILL have not been defined into formal standards, restricting further protocol evolution.

Some vendors in the industry have raised objections to L2MP technologies due to the preceding disadvantages. However, in the author's opinion, L2MP technologies are more like a victim of a technology dispute. From 2010 to 2014, L2MP technologies were widely used in DCs of large enterprises in financial services and petroleum industries. However, after the emergence of NVO3 technologies like VXLAN, almost all IT and CT vendors began to use NVO3 technologies (this is because the forwarding mechanisms of NVO3 technologies completely reuse the IP forwarding mechanism, allowing NVO3 technologies to be directly used on existing networks). As a result, the market share of L2MP technologies plummeted.

When reflecting on the disadvantages of L2MP technologies mentioned above, a solution to the tenant problem was considered at the beginning of TRILL design. A field was reserved in the TRILL header for tenant identification, but the problem has not yet been resolved because the protocol has not been continuously evolved. The high deployment cost is not necessarily due to L2MP technologies. In fact, VXLAN also requires device upgrade, but its upgrade process is much smoother than L2MP. Moreover, each new technology suffers from mechanism problems. However, the only way to address this problem is to continuously evolve the technology.

It is worth noting that the abovementioned advantages and disadvantages are only the opinions of the author. What's more, it is important to remember that every technology has both its advantages and disadvantages when compared to other technologies.

3.1.2.4 Multi-Chassis Link Aggregation Technologies

Multi-chassis link aggregation technologies are proposed to address the dual-homing access problem of terminals. Dual-homing of terminals inevitably causes loops on the network. xSTP technologies cannot solve the dual-homing access problem of terminals, and single-homing access of terminals has poor reliability. Among the technologies described in the preceding sections, L2MP technologies focus on traffic forwarding on the network, and there are no explanations provided on how terminals connect to the network. Virtual chassis technologies support dual-homing access because a virtual chassis system is a single device logically. Terminals or switches only need to connect to the virtual chassis system through link aggregation. However, the complete coupling (integration) of control planes in virtual chassis technologies causes some problems.

For example, when two access switches in the same state perform link aggregation negotiation with a connected device, the connected device establishes a link aggregation relationship with one device. The right-hand side of Figure 3.23 shows the logical topology.

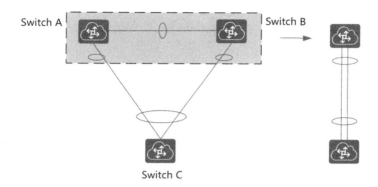

FIGURE 3.23   Multi-chassis link aggregation.

Multi-chassis link aggregation technologies inherit the core principles of virtual chassis technologies and couple the control planes of two devices. The difference is that multi-chassis link aggregation technologies do not require that all status information be synchronized between member devices. Instead, only some link aggregation information is synchronized. Figure 3.23 shows the core principles of multi-chassis link aggregation technologies.

Even though multi-chassis link aggregation technologies are essentially control plane virtualization technologies, they are theoretically more reliable than virtual chassis technologies due to the low coupling degree of control planes. In addition, as two member devices can be upgraded independently, service interruption time is reduced.

Similar to that of virtual chassis technologies, the inter-device information synchronization mechanism of multi-chassis link aggregation technologies is implemented inside a device. Therefore, vendors in the industry have proprietary protocols for multi-chassis link aggregation, such as Cisco's Virtual Port Channel (vPC) and Juniper's Multichassis Link Aggregation Group (MC-LAG). This section uses Multichassis Link Aggregation Group (M-LAG) implemented on Huawei CloudEngine switches as an example to describe the working principles of multi-chassis link aggregation technologies.

1. Basic concepts of M-LAG

   Figure 3.24 shows M-LAG networking. The following concepts are used later in this section:

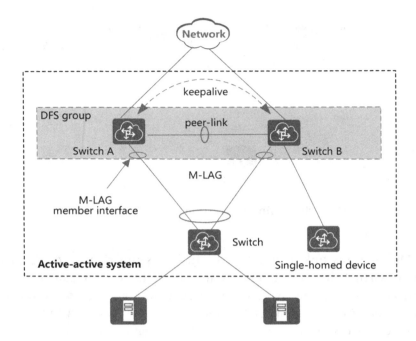

FIGURE 3.24 M-LAG.

- M-LAG: implements link aggregation among multiple devices to achieve device-level link reliability. Two devices configured with M-LAG can establish an active-active system.

- M-LAG master and backup devices: Similar to CSS and iStack, M-LAG also has a master and backup election mechanism. Under normal circumstances, the forwarding behaviors of the master and backup devices are the same. However, when the active-active system splits, the backup device suppresses its forwarding behavior while the master device continues to forward traffic normally.

- Dynamic Fabric Service (DFS) group: M-LAG member devices use the DFS group protocol to synchronize information such as the interface status and entries.

- Peer-link: A direct link configured with link aggregation must exist between the two member devices in an M-LAG, and this link must be configured as the peer-link. The peer-link is a Layer

2 link used to exchange negotiation packets and transmit part of traffic. After an interface is configured as a peer-link interface, other services cannot be configured on the interface.

- Keepalive link: It carries heartbeat data packets and is used for DAD. Note that the keepalive link and peer-link are different links with different functions. Under normal circumstances, the keepalive link does not have any traffic forwarding behaviors in the M-LAG. Instead, it is only used to detect whether two master devices exist when a fault occurs. The keepalive link can be an external link. For example, if the M-LAG is connected to an IP network and the two member devices can communicate through the IP network, the link that enables communication between the member devices can function as the keepalive link. An independent link that provides Layer 3 reachability can also be configured as the keepalive link. For example, a link between management interfaces of the member devices can function as the keepalive link.

- M-LAG member interfaces: They are interfaces through which a device is dual-homed to an M-LAG. The status of the two M-LAG member interfaces needs to be synchronized.

- Single-homed device: It is connected to only one of the two member devices in an M-LAG. If M-LAG is deployed, a single-homed device is not recommended.

2. M-LAG working mechanism

1. M-LAG establishment process
   Figure 3.25 shows the M-LAG establishment process.

   - After the two member devices in an M-LAG are configured, they can be paired. The two devices periodically send Hello packets to each other through the peer-link. The Hello packets carry information such as the DFS group ID, protocol version, and system MAC address.

   - After receiving Hello packets from the remote device, the local device checks whether the DFS group ID in the packets

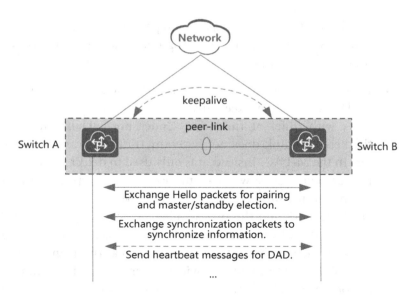

FIGURE 3.25   M-LAG establishment process.

is the same as its DFS group ID. If the DFS group IDs are the same, the two devices are paired successfully.

– After the pairing is successful, the master and backup devices are elected based on their priorities. The device with a higher priority becomes the master device. If the two devices have the same priority, their system MAC addresses are compared, and the device with a smaller system MAC address becomes the master device. By default, the priority is 100 and can be changed manually.

– After the two devices are paired successfully, they send synchronization packets to each other to synchronize information, including the device name, system MAC address, software version, M-LAG status, STP Bridge Protocol Data Unit (BPDU) information, MAC address entries, ARP entries, and Internet Group Management Protocol (IGMP) entries.

The paired devices also send heartbeat messages through the keepalive link. Heartbeat messages are mainly used for DAD when the peer-link is faulty.

Ether Header	Msg Header	Data

FIGURE 3.26 M-LAG protocol packet.

2. M-LAG protocol packets

Figure 3.26 shows an M-LAG protocol packet. Protocol packets (Hello or synchronization packets) are transmitted over the peer-link, which is a Layer 2 link. Therefore, protocol packets are encapsulated in common Ethernet packets. In the outer Ethernet header, the source MAC address is the MAC address of the local device, the destination MAC address is a multicast MAC address, and the VLAN ID is a reserved VLAN ID.

A customized message header is encapsulated after the outer Ethernet header. The customized message header contains the following information:

– Version: indicates the protocol version, which is used to identify the M-LAG version of M-LAG member devices.

– Message type: indicates the type of a packet, which can be Hello or synchronization.

– Node: indicates the device node ID.

– Slot: indicates the slot ID of the card that needs to receive messages. For a fixed switch, the value is the stack ID.

– Serial number: indicates the protocol serial number, which is used to improve reliability.

The normal packet data is encapsulated in the customized message header and includes information that needs to be exchanged or synchronized. For example, the data field of a Hello packet contains information such as the DFS group ID, priority, and device MAC address, and the data field of a synchronization packet contains some entry and status information.

3. M-LAG entry synchronization

An M-LAG is a logical link aggregation group, and entries on the two M-LAG member devices must be the same.

Therefore, the two M-LAG member devices need to synchronize entries. Otherwise, traffic forwarding may be abnormal. The following entries and information need to be synchronized:

- MAC address entries

- ARP entries

- Layer 2 and Layer 3 multicast forwarding entries

- Dynamic Host Configuration Protocol (DHCP) snooping entries

- Link Aggregation Control Protocol (LACP) system ID

- STP configurations in the system view and on interfaces

- Other entries such as Access Control List (ACL) entries.

Entries are encapsulated in the data field of synchronization packets. The data field is encapsulated in the TLV format and can be easily extended. The following uses ARP entry synchronization as an example.

As shown in Figure 3.27, the TLV contains the source M-LAG ID of the received ARP packet and the original ARP packet. M-LAG devices process the original ARP packet in the same way regardless of their protocol versions. Therefore, the compatibility between versions is easy to implement. In addition to ARP entries, M-LAG devices also synchronize entries of protocols such as IGMP based on original packets. After receiving original packets, the remote device synchronizes its own entries based on packet information.

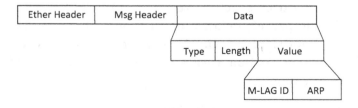

FIGURE 3.27   ARP entry synchronization packet of M-LAG.

4. M-LAG protocol compatibility

Member devices in an M-LAG can be upgraded one by one, facilitating maintenance. For protocols that need to be synchronized on the control plane, the protocol versions of both member devices are different when the devices are upgraded one by one. Therefore, protocol compatibility must be ensured.

M-LAG ensures protocol compatibility using the following methods:

- M-LAG Hello packets carry the protocol version number, which does not change with the device upgrade, unless the functions (such as MAC address and ARP) used in the M-LAG change after the upgrade.

- If an M-LAG device finds that it runs a later version after obtaining the version number of the remote M-LAG device, the local M-LAG device processes the packets carrying the old version number and communicates with the remote M-LAG device using the method of the earlier version.

- During entry synchronization, an M-LAG device sends the original packets to the remote device, ensuring protocol compatibility. Take ARP entries as an example. ARP is a stable protocol. The processing of ARP packets in different versions is almost the same. Therefore, when the local device sends the original ARP packets to the remote device, processing of the ARP packets rarely fail due to version inconsistency.

3. M-LAG traffic model

a. Model of unicast traffic from CEs to network-side devices

As shown in Figure 3.28, when CE2 sends unicast traffic to the network, the traffic is hashed to the two member links of the link aggregation group to M-LAG member devices.

b. Model of unicast traffic exchanged between CEs

As shown in Figure 3.29, when CE1 sends traffic to CE2, Switch A learns the MAC address entry of CE2 and forwards the traffic to CE2.

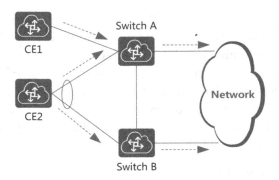

FIGURE 3.28   Unicast traffic from CEs to network-side devices.

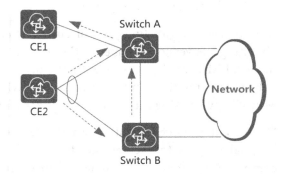

FIGURE 3.29   Unicast traffic exchanged between CEs.

When CE2 sends traffic to CE1, CE2 performs hash calculation first. If the traffic is hashed to Switch A, Switch A can forward the traffic to CE1 because it has learned the MAC address
entry of CE1. Because Switch A synchronizes the MAC address
entry of CE1 to Switch B, Switch B learns the entry from the
peer-link interface (that is, the outbound interface in the MAC
address entry of CE1 on Switch B is the peer-link interface). If the
traffic is hashed to Switch B, Switch B sends the traffic to Switch
A through the peer-link interface. Similarly, Switch A forwards
the traffic to CE1.

   c. BUM traffic model

In Figure 3.30, CE2 sends broadcast traffic over the two
M-LAG links. After reaching Switch A and Switch B, the traffic continues to be broadcast to other CEs and the network side.

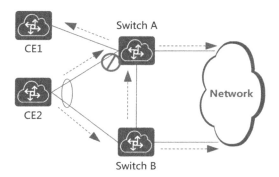

FIGURE 3.30   CE2 sending broadcast traffic.

Note that peer-link interfaces are added to all VLANs by default. Therefore, the broadcast traffic is also broadcast on the peer-link. After Switch B sends the broadcast traffic to Switch A through the peer-link, the broadcast traffic is not sent to CE2 through the M-LAG member interface of Switch A to prevent loops. (Switch A also sends the broadcast traffic to Switch B and traffic processing is similar, which is not mentioned here.) The peer-link has an interface isolation mechanism. All traffic received on the peer-link interface of an M-LAG device is not sent through the M-LAG member interface.

d. Network-side traffic model

M-LAG processing varies according to the network-side protocol.

– xSTP network: M-LAG member devices negotiate to be the root bridge of an xSTP network.

– TRILL or VXLAN network: M-LAG member devices use the same identifier (virtual nickname or virtual VTEP) to negotiate with other devices. A network-side device considers the M-LAG member devices as one device for negotiation. Traffic is forwarded based on the mechanism of TRILL or VXLAN, implementing load balancing.

– IP network: Normal route negotiation is performed. If M-LAG member devices have the same link status, the links function as ECMP links to load balance traffic.

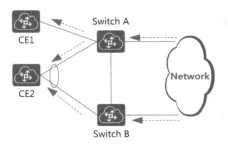

FIGURE 3.31   Network-side traffic model.

When a network-side device sends unicast or broadcast traffic to either member device in an M-LAG, the M-LAG member device processes the traffic in the same way as it processes traffic from CEs. That is, unicast traffic is forwarded normally, and interface isolation is performed for broadcast traffic, as shown in Figure 3.31.

4. M-LAG failure scenarios

1. M-LAG member interface failure

As shown in Figure 3.32, if an M-LAG member interface on one member device fails, network-side traffic received by the device is sent to the other member device through the peer-link, and all traffic is forwarded by the other member device. The process is as follows:

– If the M-LAG member interface on Switch B fails, the connected network-side device is unaware of the failure and still sends traffic to Switch B.

– Because the M-LAG member interface fails, Switch B clears the MAC address entry of the CE. In this case, the M-LAG system triggers a MAC address entry synchronization process. Switch A synchronizes the MAC address entry of the CE to Switch B, and the outbound interface in the entry is the peer-link interface.

– Because the M-LAG member interface fails, dual-homing of the CE to the M-LAG changes to single-homing, and the interface isolation mechanism is disabled.

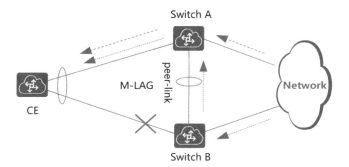

FIGURE 3.32   M-LAG member interface failure.

- After receiving traffic from a network-side device to the CE, Switch B searches for the MAC address entry of the CE and sends the traffic to Switch A through the peer-link interface accordingly. Switch A then forwards the traffic to the CE.

After the faulty M-LAG member interface recovers and goes Up, MAC address entry synchronization is triggered in the M-LAG system. After the synchronization, the outbound interface in the MAC address entry of the CE on Switch B is restored to the M-LAG member interface, and Switch B forwards traffic to the CE through the interface.

2. Peer-link failure

As shown in Figure 3.33, if the peer-link fails, the two devices in the M-LAG cannot forward traffic at the same time. If they forward traffic simultaneously, problems such as broadcast storm and MAC address flapping occur. Therefore, only one device can forward traffic in this scenario. The process is as follows:

- If an M-LAG member device detects that the peer-link interface fails, it immediately initiates a DAD process to perform DAD through the keepalive link. If the local device does not receive any keepalive packet from the remote device within a specified period, the local device considers that the remote device fails. If the local device receives keepalive packets from the remote device, the local device considers that the peer-link fails.

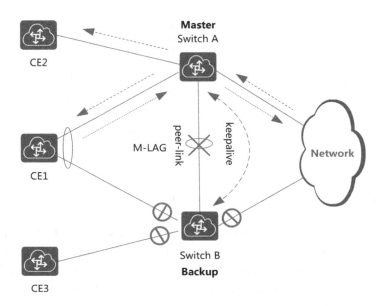

FIGURE 3.33   Peer-link failure.

- If the peer-link fails, the M-LAG backup device sets all physical interfaces except the peer-link interface, stack interface, and management interface to Error-Down state. In this case, all traffic is forwarded only through the M-LAG master device.

- After the faulty peer-link recovers, peer-link interfaces go Up, and M-LAG member devices renegotiate with each other. After the negotiation and network-side protocol convergence are complete, the interfaces in Error-Down state on the backup device go Up after a delay of 2 minutes instead of being restored immediately, to ensure that the M-LAG interface isolation mechanism takes effect.

When the peer-link fails in the preceding scenario, a CE (for example, CE3) that is single-homed to the backup device cannot access the network side or receive network-side traffic because interfaces on the backup device are in Error-Down state. Therefore, the single-homing networking is not recommended.

If modular switches are used to establish an M-LAG, it is recommended that the peer-link interface and M-LAG member interface be deployed on different cards. If the peer-link interface and M-LAG member interface are deployed on the same card, both interfaces fail when the card fails. If the fault occurs on the master device, the backup device automatically sets its physical interfaces to Error-Down state. In this case, traffic sent to the M-LAG is discarded.

3. Device failure

In case of a device failure, traffic is forwarded in the same way as in case of a peer-link failure. As shown in Figure 3.34, a device detects that the peer-link interface is Down, performs DAD, and determines that the other device fails. If the master device fails, the backup device becomes the master device and forwards traffic. If the backup device fails, the master device continues to forward traffic. After the faulty member device recovers, the normal member device detects that the peer-link interface is Up and performs M-LAG negotiation. After the negotiation and network-side protocol convergence are complete, the interfaces in Error-Down state on the backup device go Up after a delay of 2 minutes instead of being restored immediately, to ensure that the M-LAG interface isolation mechanism takes effect.

4. Uplink failure

In most cases, an uplink failure does not affect traffic forwarding of an M-LAG. As shown in Figure 3.35, although the uplink of Switch A fails, the entries about traffic forwarding to the network

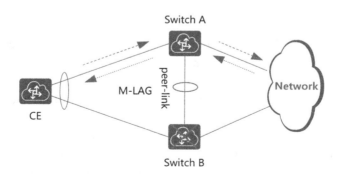

FIGURE 3.34 Device failure.

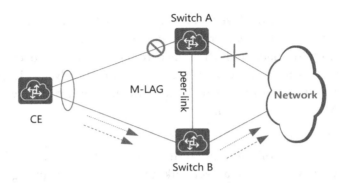

FIGURE 3.35   Uplink failure.

side are synchronized by Switch B to Switch A through the peer-link, and Switch A can send network-side traffic to Switch B for forwarding. The traffic sent from a network-side device to the CE is not sent to Switch A for processing because the uplink interface of Switch A fails.

However, if an interface on the keepalive link fails, both member devices consider that the remote device fails after DAD. In this case, two master devices exist, and the traffic sent from the CE to Switch A is discarded because the uplink outbound interface of Switch A fails.

To solve this problem, use either of the following methods:

– Use the link between management interfaces as the keepalive link.

– Configure the Monitor Link function to associate the M-LAG member interface with the uplink interface. If the uplink fails, the M-LAG member interface is triggered to go Down, preventing traffic loss.

5. A server dual-homed to a network running a tunneling protocol through an M-LAG

When a server is dual-homed to a network running tunneling protocols such as TRILL and VXLAN through an M-LAG, the two M-LAG member devices are virtualized into one device to communicate with network-side devices, as shown in Figure 3.36.

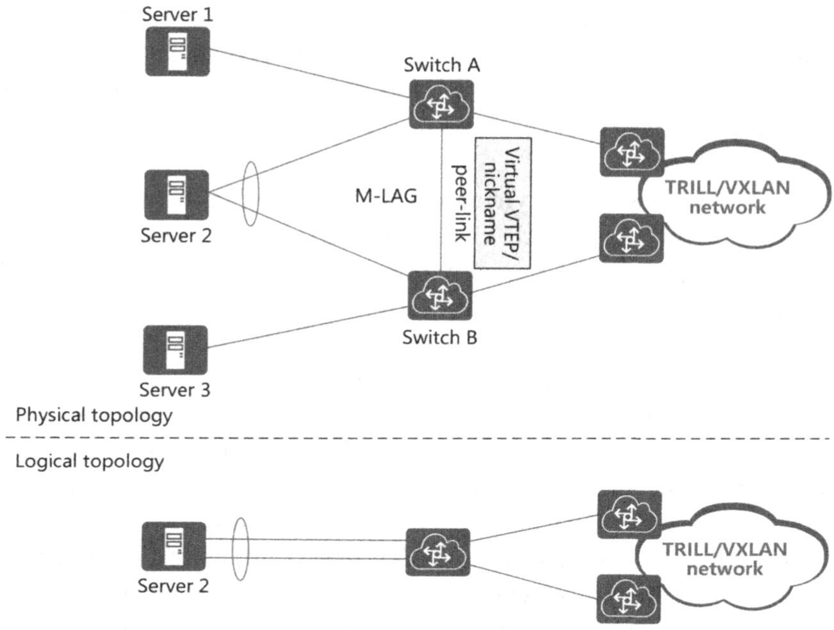

FIGURE 3.36 A server dual-homed to a network running a tunneling protocol through an M-LAG.

When an M-LAG is connected to a TRILL network, the two member devices negotiate a virtual nickname through the peer-link or be manually configured with a virtual nickname. When sending Hello packets, the two devices encapsulate the virtual nickname in the packets. In this way, network-side devices consider that they establish TRILL neighbor relationships with the same device, and all routes to servers point to the virtual nickname.

When an M-LAG is connected to a VXLAN network, the two member devices virtualize a VXLAN tunnel endpoint (VTEP). Regardless of whether a VXLAN tunnel is manually established or automatically established using Border Gateway Protocol Ethernet Virtual Private Network (BGP EVPN), Switch A and Switch B use the virtual VTEP IP address to establish VXLAN tunnels with other devices.

3.1.2.5 NVO3 Technologies

The preceding sections describe virtual chassis, L2MP, and multi-chassis link aggregation technologies that can solve problems of xSTP technologies.

However, these technologies are fundamentally traditional network technologies and are still hardware-centric. Different from these technologies, NVO3 technologies are overlay network technologies driven by IT vendors and aim to get rid of the dependency on the traditional physical network architecture.

An overlay network is a virtual network topology constructed on top of a physical network. Each virtual network instance is implemented as an overlay, and an original frame is encapsulated on a Network Virtualization Edge (NVE). The encapsulation identifies the device that will perform decapsulation. Before sending the frame to the destination endpoint, the device decapsulates the frame to obtain the original frame. Intermediate network devices forward the encapsulated frame based on the outer encapsulation header and are oblivious to the original frame carried in the encapsulated frame. The NVE can be a traditional switch or router, or a virtual switch in a hypervisor. The endpoint can be a VM or physical server. A VXLAN network identifier (VNI) can be encapsulated into an overlay header to identify a virtual network to which a data frame belongs. Because a virtual DC supports both routing and bridging, the original frame in the overlay header can be a complete Ethernet frame containing a MAC address or an IP packet. Figure 3.37 shows the model of NVO3 technologies.

The sender in the figure is an endpoint, which may be a VM or physical server. An NVE may be a physical switch or a virtual switch on a hypervisor. The sender can be connected to an NVE directly or through a switching network. NVEs are connected through a tunnel. A tunnel is used to encapsulate a packet of one protocol into a packet of another protocol. The ingress NVE of a tunnel encapsulates a packet of a protocol into a packet of the encapsulation protocol, and the egress NVE of the tunnel

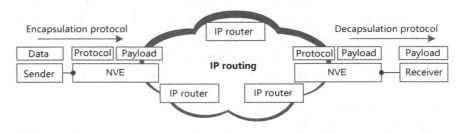

FIGURE 3.37   NVO3 technology model.

decapsulates the encapsulated packet to obtain the original packet. When the encapsulated packet is transmitted over the tunnel, the original packet is the payload of the encapsulated packet. NVEs perform network virtualization functions to encapsulate and decapsulate packets. In this way, nodes on the Layer 3 network only need to forward packets based on outer headers and do not need to be aware of tenant information.

To some extent, NVO3 and L2MP technologies are similar. They both build an overlay network on the physical network. The difference is that L2MP technologies add a new forwarding identifier to the original Layer 2 network, thereby requiring that chips on hardware devices support L2MP technologies. In contrast, NVO3 technologies reuse the current IP forwarding mechanism and only add a new logical network that does not depend on the physical network environment on the traditional IP network. The logical network is not perceived by physical devices, and its forwarding mechanism is the same as the IP forwarding mechanism. In this way, the threshold of NVO3 technologies is greatly lowered, and this is why NVO3 technologies become popular on DCNs in a few years.

Typical NVO3 technologies include VXLAN, Network Virtualization Using Generic Routing Encapsulation (NVGRE), and Stateless Transport Tunneling (STT), among which VXLAN is the most popular one. Huawei's CloudFabric DCN solution also uses VXLAN as the overlay network technology. VXLAN is described in more detail in Chapter 6.

3.2 ARCHITECTURE AND SOLUTION EVOLUTION OF DCNs FOR FINANCIAL SERVICES COMPANIES

3.2.1 Architecture of Financial Services Companies' Networks

1. Architecture

The architecture of a financial services company's network consists of the service domain, channel domain, and user domain, as shown in Figure 3.38. Bidirectional arrows indicate dependencies, and unidirectional arrows show composition.

The user domain contains internal users and external users. Internal users include branch users, DC users, and headquarters (HQ) users. External users include Internet users and extranet third-party users.

The local user access zone in the DCN zone is used to provide access to internal users. Branch users access the DCN through the

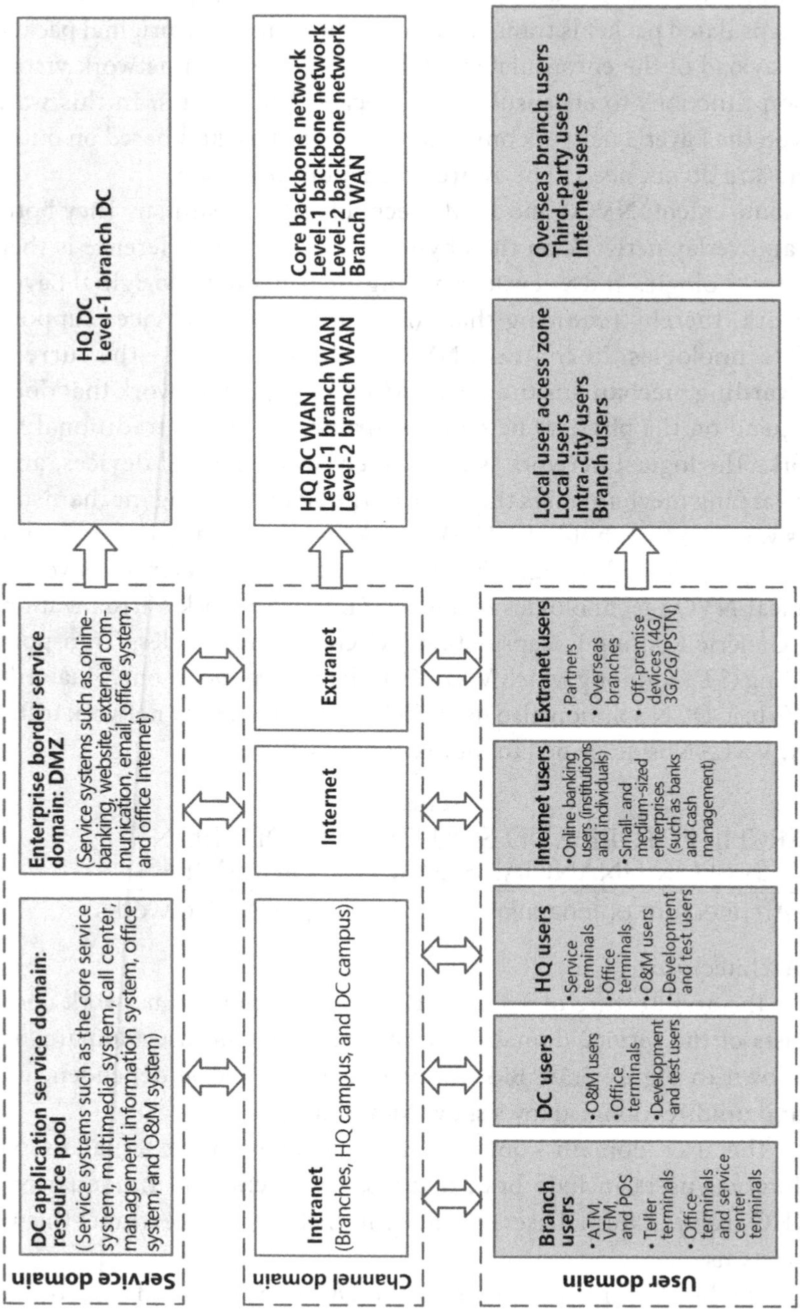

FIGURE 3.38 Architecture of a financial services company's network.

intranet channel domain. HQ users access the intranet channel domain through a metropolitan area network (MAN) and then access the wide area network (WAN) access zone of the DCN. Internet users access the Internet access zone of the DCN through the Internet channel domain. Extranet users access the extranet access zone of the DCN through the extranet channel domain (mainly through leased lines).

In the channel domain, intranet channels are created on the core backbone network, level-1 backbone network, level-2 backbone network, and branch WAN of the financial services company. Among these,

- Branches are connected to the branch WAN.

- The branch WAN is connected to the level-2 backbone network.

- The level-2 branch DCN is connected to the level-2 backbone network.

- The level-2 backbone network is connected to the level-1 backbone network.

- The level-1 branch DCN is connected to the level-1 backbone network.

- The level-1 backbone network is connected to the core backbone network.

- The HQ DCN is connected to the core backbone network. Banks are currently reconstructing their networks with flat architectures; therefore, the number of level-2 backbone networks will gradually decrease.

The service domain provides various financial services, covering all host zones and server zones in the HQ DCN and level-1 branch DCN of the financial services company. The provided financial services include core services (such as accounting, customer information, background services, treasury services, and settlement services), intermediate services (such as bank card services, international services, agent services, external services, and credit services), channel services (such as online banking, telephone banking, mobile banking, self-service banking, and integrated teller services), and management support systems (such as operation management, risk management, management platforms, and office systems).

2. DCN architecture of financial services companies

On a financial services company's DCN, switching core devices aggregate traffic from physical zones. These physical zones are not large; there are currently about 1000 physical servers in the largest physical zone. Most new servers use 10GE NICs, and some use GE NICs for out-of-band management.

Firewalls are deployed at the border of each physical zone. In most cases, the number of firewall policies is greater than 50,000. Changing these policies takes a lot of work, and they may need to be changed 10,000 times per year. Firewall configuration changes account for more than 60% of all network configuration changes. Load balancers (LBs) are deployed in each physical zone to serve multiple application systems. A distributed Domain Name Service (DNS) system is deployed in the campus. Branch clients access applications in the DC and level-5 DR applications access the database through domain names.

Figure 3.39 shows the first example of common DCN zones.

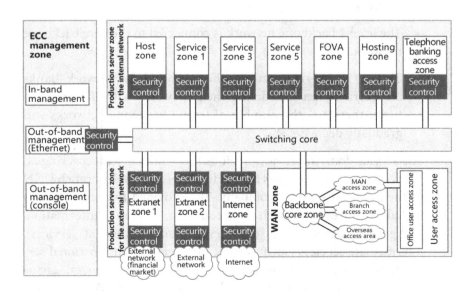

FIGURE 3.39   DCN zone example 1: bank DCN architecture.

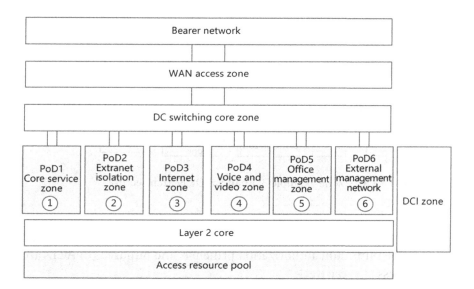

FIGURE 3.40   DCN zone example 2: financial DCN architecture.

In the preceding zones, service zone 1 is the counter service zone, service zone 3 is the online banking service zone, and service zone 5 is the office and email service zone. The overseas service zone (FOVA zone) is divided into three sub-zones by time zone, and the hosting zone is used to process web access for subsidiary services, the Internet zone, and the online banking zone.

Figure 3.40 shows the second example of common DCN zones. The DCN is divided into the core service zone, extranet isolation zone, Internet zone, voice and video zone, office management zone, and external management network based on characteristics and importance of applications, security isolation, and O&M management.

Figure 3.41 shows the third example of common DCN zones. The DC adopts a traditional zone design. That is, the Layer 2 network uses STP networking, and the Layer 3 network uses an OSPF routing design. Firewalls are deployed to isolate the production network from the office network, making the network more secure. Most service zones in the user domain do not have a firewall deployed. Heterogeneous firewalls and anti-DDoS devices are deployed in the online banking zone (Internet zone). The main pain point in network O&M is that a large number of security ACLs are changed daily.

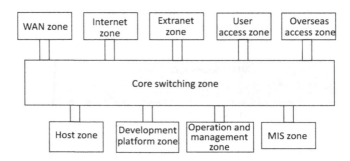

FIGURE 3.41   DCN zone example 3: financial DCN architecture.

The security authorization in each zone of a DC is processed based on the minimum authorization principle, and hundreds of ACLs are processed every week.

The preceding example architectures are applicable to different service scales and requirements. All examples adopt a horizontally partitioned, vertically tiered design. The network in each zone uses an "aggregation + distribution + access" architecture, and the gateway is deployed at the aggregation layer.

3. Network architecture in the server zone

Figure 3.42 shows the mainstream server zone network architecture in a financial DC. Core devices of the DC are deployed independently. The "aggregation + distribution + access" architecture is used in the service zone. vPC, stacking, or M-LAG technology is used at each layer to prevent loops and reuse links.

Aggregation devices include routing core and switching core devices. High-end modular switches function as aggregation devices. Two modular switches are virtualized into four devices using virtual system (VS) technology. Two VS1 devices function as routing core devices, and two VS0 devices function as Layer 3 switching core devices. Firewalls are deployed between the switching core and routing core devices. Static routes are configured between firewalls and switching core devices and between firewalls and routing core devices. Server gateways are deployed on switching core devices. Access switches are deployed in Top of Rack (TOR) mode, and distribution switches are deployed in End of Row (EOR) mode.

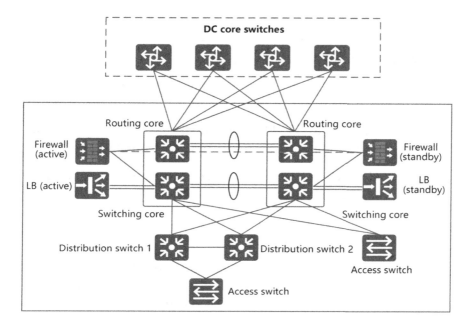

FIGURE 3.42   Server zone network architecture.

If source network address translation (SNAT) is not performed for servers in a zone, gateways of the servers are deployed on LBs. If SNAT is performed for servers in a zone, gateways of the servers are deployed on core switches.

Except the online banking zone and extranet zone, the web, application, and database of other functional zones are deployed in the same zone. Access traffic exchanged between them does not pass through firewalls.

Servers are connected to TOR switches using Ethernet cables. Eight or twelve servers are deployed in a rack.

TOR switches are connected to the zone aggregation switches in the network cabinet through jumpers. The access switch used for out-of-band management of network devices is deployed in the network cabinet. The distribution switch in each zone has four gigabit uplinks. Routing core devices in each zone are connected to core switches of the DC through 40GE ports.

A common pain point of a DC LAN is that servers are deployed at fixed locations. Servers in different service zones can only connect to corresponding access switches in fixed network zones.

This is not flexible. In the long run, it will lead to resource shortage in general and abundant resources in some zones. However, the pain points do not end there.

- Network configurations are performed manually or using scripts, which is inefficient and error-prone.

- The maintenance workload of firewall policies is heavy. It is difficult to determine whether firewall policies are useful or redundant.

- IP addresses are assigned manually, which is time-consuming and inefficient.

- Some IP addresses are public IP addresses, which may conflict with other IP addresses.

- Some applications are accessed through IP addresses and do not support SNAT. As a result, the Layer 2 network is large, gateways are deployed on LBs, performance is poor, and the fault domain is large.

4. Geo-redundancy network architecture with two active-active sites and a DR site

Financial DCs generally use a geo-redundancy network architecture with two active-active sites and a DR site. The network architecture of some banks is evolving to the geo-redundancy network architecture with multiple sites at multiple cities. The primary DC, intra-city DR DC, and remote DR DC are connected through the WAN, and application systems can be deployed across campuses. Production Internet egresses are deployed at the primary DC and remote DR DC. The total bandwidth ranges from 10 to 20 Gbit/s, and the total number of servers is about 7000. To prevent Layer 2 broadcast storms, large banks use Layer 3 IP interconnection between DCs in the same city. Layer 2 interconnection is used by some small- and medium-sized enterprises or just for temporary service migration.

In typical Layer 2 interconnection between intra-city DCs, zone switching core devices (gateway switches) are directly connected to Dense Wavelength Division Multiplexing (DWDM) devices. Eth-Trunks are configured to allow packets from corresponding VLANs to pass through to implement Layer 2 extension, as shown in Figure 3.43.

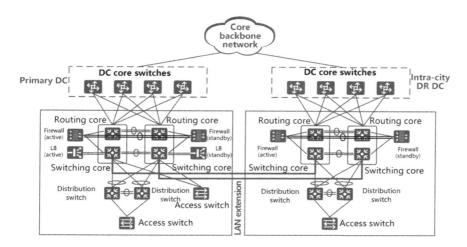

FIGURE 3.43   Network architecture for interconnection between intra-city DCs.

Layer 3 interconnection can be implemented in two ways. Core switches can be either connected directly to DWDM devices, or connected to customer-premises equipment (CPE) and multiple custom edge (MCE) devices that are connected to DWDM devices.

3.2.2 Financial Service Development Requirements and DCN Evolution

1. Financial service development requirements

Financial services are developing in three trends: digitization, mobilization, and democratization. Digitization helps banks better manage risks and be more customer-centric through data analysis. Mobilization uses mobile Internet technologies to meet customer requirements anytime and anywhere. Democratization allows more and more people to enjoy financial and banking services.

Digitalization requires big data and AI as technical support to implement precision marketing and risk management. Mobilization requires banks to ensure good service experience of a large number of users, including providing services anytime and anywhere, meeting their access requirements, handling access requests from hundreds of millions of users with each user accessing the network dozens of times in average, and requiring a network that supports fast-growing services, a large number of users, and massive amounts of data. The network needs to be resilient to ensure consistent user experience when coping with traffic surges by dozens of times. Banks need to provide personalized services, innovate financial services, respond to customer requirements, and update products quickly. The time to market (TTM) is shortened from months to days, requiring the network to quickly respond to changing customer needs and support rapid product launches. Channel innovation, 4K facial recognition, and fingerprint identification require the network to provide large-scale computing capabilities. In addition, the network must provide 24/7 services. A large number of users must be able to access the network at the same time. Network faults have a greater impact, and therefore, a more reliable network is required.

Banks urgently need to accelerate service innovation to improve competitiveness in the face of Internet finance. They need to improve customer acquisition capabilities, accurately analyze customer behavior, and provide competitive products and services. They need to improve customer experience, handle issues in real time, and implement in-event risk control. They need to provide better services and a platform for querying petabytes of historical data with real-time and long-term queries. They also need to collect comprehensive customer data especially unstructured data, analyze and mine customer habits accordingly, predict customer behavior, and segment customers effectively to improve the effectiveness and relevance of services marketing and risk control.

Financial services companies generally have their own business development strategies. Common trends include groupization, industrialization, mobility (mobile Internet), digitization (big data analytics), and intelligence. Groupization requires that unified network access be provided for each subsidiary, some branch services be processed by the HQ, and branch cloud and branch hosting services be provided. Industrialization requires that financial IaaS, SaaS, and PaaS services be provided for customers outside the group, and an industry cloud be constructed.

To address these challenges, banks require their IT department to meet the following requirements:

- Supports cloud-based key services to cope with surging Internet financial access traffic. Accelerates the launch of financial products to improve user experience. Simplifies traditional IT O&M and focus on service innovation.

- Adopts big data analytics to support precision marketing and real-time risk control. Supports query for large amounts of historical data in real time to improve efficiency, enhance competitiveness, and cope with Internet finance challenges.

- Deploys key service systems in active-active mode to ensure zero service interruption and zero data loss, meeting increasing regulatory requirements.

The requirements on the network are as follows:

1. The rapid release of innovative Internet applications requires the network to provide flexible resource allocation and agile application deployment capabilities.

2. To cope with the uncertain and explosive growth of users, the network must be able to handle high concurrency, large bursts, and flexible on-demand expansion.

3. The heavy integration of new technologies represented by big data and cloud computing in financial services requires a high-performance network with high bandwidth and low latency. The network architecture must be scalable, open, and compatible.

4. Risk prevention and control, and the prevalence of severe cyber security events require comprehensive improvement of network security assurance. A lot of pressure is received from the public when a fault occurs. A complete security protection system is required and the network architecture has to be highly available to ensure business continuity. In addition, applications are visualized and management is automated, facilitating rapid troubleshooting.

5. Internet finance poses many requirements on network availability, high performance, flexibility and elasticity, agility and automation, and security and controllability, as shown in Figure 3.44.

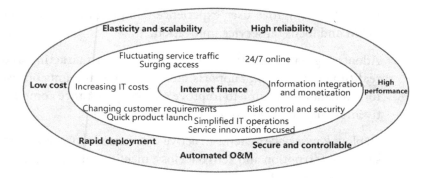

FIGURE 3.44 Network requirements of Internet finance.

6. Cloud DCs have a large number of VMs, rapid growth, and a large migration scope. Therefore, cloud DCs have the following new requirements on the network:

   – As east-west traffic increases and the Layer 2 network topology becomes larger, higher bandwidth for network interconnection is required. The network evolves from "GE access+10GE uplink" to "10GE access+40GE uplink," and then to the popular "25GE access+100GE uplink."

   – VM deployment and migration within a DC equipment room module, between multiple DC equipment room modules, and between DCs in the same city must be flexible. To meet this requirement, a large Layer 2 network without broadcast storms is required.

   – Multiple sites on a DCN must be planned based on the multi-active or active/standby deployment of application systems in the cloud environment.

   – Access switches must support larger MAC address tables and more host routing entries.

7. Big data applications have high data volume, many different data types, and high processing speed. Scenarios with many-to-one or many-to-many traffic models pose the following new requirements on the network:

- The network must be highly available and scalable, and support ECMP.

- Big data applications also generate burst traffic. Therefore, network devices must have strong buffer and queue functions to mitigate the impact of burst traffic.

- Big data applications require a good oversubscription ratio. In most cases, the oversubscription ratio of servers to the access layer is about 3:1, and the oversubscription ratio of the access layer to the aggregation layer and of the aggregation layer to the core layer is about 2:1.

- Big data applications require the network to provide sufficient access bandwidth and low latency.

- AI technologies require the network to achieve high performance, low latency, and zero packet loss.

In conclusion, the development of financial services requires resource pooling, flexibility, elasticity, automation, and servitization of the network architecture.

In addition, financial technology (FinTech) is widely used in traditional financial services companies and Internet finance companies. FinTech development also poses a series of requirements on the network.

2. FinTech development requirements

FinTech has evolved from FinTech 1.0 to FinTech 2.0 and is now evolving to FinTech 3.0. As shown in Figure 3.45, FinTech brings business transformation and corresponding IT technology transformation in each phase.

In the FinTech 1.0 phase, the core of IT transformation is electrization and automation. Key features include replacing manual operations with computers and building an electronic funds transfer system. A chimney-type system architecture is used, where data flows only within a city, and networks are constructed in a distributed manner.

In the FinTech 2.0 phase, the core of IT transformation is network-based and centralized. A geo-redundancy network architecture with

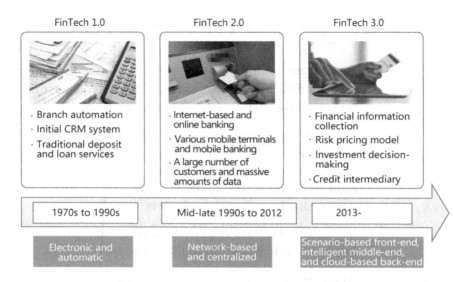

FIGURE 3.45   FinTech development trend.

two active-active sites and a DR site is constructed. The DCN adopts a horizontally partitioned, vertically tiered modular design to adapt to the rapid development of services.

In the FinTech 3.0 phase, the core of IT transformation is to use the front-end, middle-end, and back-end system design ideas and leverage big data, cloud computing, and artificial intelligence to implement scenario-specific front-end, intelligent middle-end, and cloud-based back-end. The DCN uses the resource pool, large Layer 2 and fat-tree architecture and automation design to support the forwarding of east-west traffic that dominates in DCs and meet requirements for flexible service deployment and elastic scaling.

To sum up, the development of FinTech requires the network to support cloudification, resource pooling, and automation.

In addition, IT requirements on the network need to be analyzed in financial DCN planning and design, and mainstream ICT technologies are used for planning and design.

3. Network function requirement analysis

DCN function requirements in the financial services industry mainly include VM and bare metal (BM) access, Docker-based container access, data replication, data backup, unified communications,

IP voice system, call center, video conference system, and video sur-veillance system.

1. VM and BM access

Access switches that support 10GE access and 40GE uplink provide network access for VMs and BMs. A server is dual-homed to two switches, and NICs of the server work in active/standby or active-active mode. The mainstream virtualization manage-ment platforms for virtual servers include KVM, VMware ESXi, and Microsoft Hyper-V, and their cloud management platforms are OpenStack, vCenter, and System Center (integrated in Azure Pack), respectively.

The type and number of server NICs and the virtualization ratio of virtual servers affect the selection and number of access switches, and need to be investigated clearly.

In addition, the web front-end processor, encryptor, Linux server, midrange computer, and MySQL database server must also be connected to the network, and thereby need to be considered.

2. Docker-based container access

Docker container access is similar to VM or BM access. Docker containers can be assigned IP addresses independently or use host IP addresses through NAT. In addition, Docker con-tainers can access the network in VLAN-based Layer 2 mode or routing mode. In practice, Docker containers on multiple net-work segments must be deployed on one host, and IP addresses of Docker containers remain unchanged when the containers are restarted. When Docker containers connect to the network, a plug-in is required to provide connection with the container platform (such as Kubernetes) so that the container platform can automatically deliver network configurations.

3. Data replication

There are three data replication scenarios.

   - Data storage in a DC: construct a storage area network (SAN) in the DC for SAN storage. Use the service IP network in the DC for network-attached storage (NAS) and General Parallel File System (GPFS) storage.

- Synchronous data replication between DCs in the same city: For fiber channel (FC), the network needs to provide a logical channel for synchronous data replication. For NAS/GPFS replication between intra-city DCs, use the service IP network between intra-city DCs.

- Asynchronous data replication between DCs in different cities: For FC over IP, the network needs to provide IP connections for asynchronous data replication. For NAS/GPFS replication between DCs in different cities, use the service IP network between the DCs when the traffic is light, and plan an independent logical channel between the DCs when the traffic is heavy.

4. Data backup

DC data backup has the following requirements on the network:

- Data backup in a DC: If the LAN-free mode is used, construct a SAN in the DC. If the LAN-based mode is used, use the service IP network in the DC. In LAN-free mode, data is replicated from a fast random storage device such as a disk array or a server hard disk to a backup storage device such as a tape library or a tape drive.

- Data backup between DCs in the same city requires Layer 3 connections and an IP network between the DCs.

- Data backup between DCs in different cities also requires Layer 3 connections and an IP network between the DCs.

- NAS deployment must be considered in IP network design.

- Data backup flows should not pass through firewalls, unless otherwise required.

5. Unified communications

The unified communications system has the following requirements on the network:

- Branches are connected to DCs through Layer 3 networks.

- The network transmission latency must be less than 150 ms, the packet loss rate less than 1%, and the jitter less than 20 ms.

6. Voice over Internet Protocol (VoIP) system

The VoIP system has the following requirements on the network:

- DCs in the same city are connected through the Layer 3 network.

- Branches are connected to DCs through Layer 3 networks.

- The network transmission latency must be less than 150 ms, the packet loss rate less than 1%, and the jitter less than 20 ms.

7. Call center

The call center has the following requirements on the network:

- Call center agents are connected to DCs through Layer 3 networks.

- The network transmission latency must be less than 100 ms, the packet loss rate less than 0.1%, and the jitter less than 10 ms.

- A session border controller (SBC) should be deployed in the Internet demilitarized zone (DMZ) for multimedia users to access the call center.

8. Video conference system

The video conference system has the following requirements on the network:

- DCs in the same city are connected through the Layer 3 network.

- Branches are connected to DCs through Layer 3 networks.

- The network transmission latency must be less than 200 ms, the packet loss rate less than 1%, and the jitter less than 50 ms.

9. Video surveillance system

The video surveillance system has the following requirements on the network:

- Branches are connected to DCs through Layer 3 networks.

- Each video stream occupies 4–6 Mbit/s bandwidth. The network transmission latency must be less than 300 ms and the packet loss rate less than 0.5%.

4. Network attribute requirement analysis

In addition to network function requirements, network quality-related requirements also include network attribute requirements that are critical to the stable and healthy running of the network, as shown in Table 3.1.

TABLE 3.1 Network Attribute Requirement Analysis

Dimension	Requirement
High availability requirements	Important systems are deployed in two equipment rooms in two regions. Applications or databases in the same city are deployed in active-active mode. Applications in different cities are deployed in triple-active mode. Faults that affect a large number of devices on the network are not allowed. Automatic switchover is supported when a fault occurs, ensuring that the RPO is 0 and the recovery time objective (RTO) is less than 20 minutes. Broadcast storms are not allowed A failure of the controller does not affect traffic forwarding on the network. A failure of a single node or link and active/standby deployment of NICs do not affect service traffic forwarding Version upgrade and spare part replacement do not affect service traffic forwarding
Flexible expansion requirements	The network architecture and cabling for server access are standardized Services can be flexibly deployed on servers in different DCs and equipment rooms. The access locations of servers are irrelevant to services VASs can be flexibly added in the DMZ using Service Function Chaining (SFC) Layer 2 and Layer 3 networks are decoupled from Layer 4 to Layer 7 VASs, facilitating flexible expansion and deployment of VASs, as well as fabric network construction and capacity expansion
High security requirements	Isolation between tenants and security isolation within tenants enhance the security in cloud DCs. The network supports firewall virtualization, software-based deployment, and resource pooling The network supports microsegmentation security for PMs on the same network segment
Automation requirements	IP addresses are automatically allocated The controller can connect to the cloud platform to automatically deliver Layer 2 and Layer 3 network configurations during VM and BM provisioning. Devices at Layer 4 to Layer 7 are managed in a unified manner and their configurations are automatically delivered
O&M requirements	Services are recovered within 15 minutes The network topology, forwarding quality, and forwarding path are visualized. Lightweight fault locating and monitoring tools are provided Analysis of associations between services and networks is intelligent

5. DCN technical architecture evolution

The DCN technical architecture evolves in three phases based on different service requirements and technology development, as shown in Figure 3.46.

1. DCN 1.0

The DCN 1.0 phase is also called the modular and hierarchical phase. In this phase, the network is partitioned based on horizontal partitioning and vertical tiering principles. The logical structure is tightly coupled with the physical layout. Horizontal partitioning allows network devices that carry similar services and have similar security levels to be grouped into a network zone, facilitating security policy implementation and optimal data exchange. Vertical tiering allows network devices with different network functions to work at different independent layers. The logical architecture is clear, facilitating function expansion, adjustment, and management. The network consists of the core layer, aggregation layer, and access layer. DCN 1.0 has the following key features:

– Most resources are limited to one functional zone, and a small amount of resources can be shared between zones.

– Functional zones and physical locations are bound.

– Functional zones are connected to Layer 2 core devices based on requirements.

– Functional zones cannot be flexibly added or deleted.

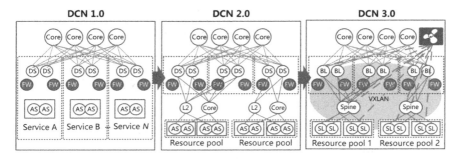

FIGURE 3.46   Evolution of the DCN technical architecture.

- Multiple Spanning Tree Protocol (MSTP) and Virtual Router Redundancy Protocol (VRRP) technologies are used, and the link utilization is only 50%.

- Most servers use GE NICs.

2. DCN 2.0

The DCN 2.0 phase is also called the resource pool phase and supports virtualization. The logical architecture is irrelevant to physical locations.

In this phase, physical network technologies and computing technologies are integrated. DCN 2.0 has the following key features:

- (Optional) Layer 2 core devices implement network-wide resource sharing within a DC.

- Access devices constitute resource pools.

- Layer 2 connectivity is implemented as required, and M-LAG or vPC technology is used to prevent loops.

- Functional zones can be flexibly added or deleted.

- The link utilization reaches 100%.

- Most servers use GE NICs.

- The network provides GE access and 10GE interconnection.

3. DCN 3.0

The DCN 3.0 phase is also called the cloud-based phase. In this phase, the overlay network architecture is used, and the SDN controller connects to the cloud platform to implement end-to-end automation. DCN 3.0 has the following key features:

- Network-wide resources can be shared within a DC.

- Server leaf nodes connecting to a VXLAN network constitute resource pools.

- Layer 2 connectivity is implemented as required.

- The link utilization reaches 100%.

- Functional zones can be flexibly added or deleted.

- The spine-leaf network has strong horizontal scalability.

- Most servers use 10GE NICs.

- The network provides 10GE access and 40GE uplink.

It is recommended that the DCN 3.0 technical architecture be used for financial DCN planning and design.

3.2.3 Target Architecture and Design Principles of Financial Cloud DCs

1. Target architecture of financial cloud DCs

The target architecture of financial cloud DCs consists of five parts, as shown in Figure 3.47. The five functional parts collaborate with each other to implement technology standardization, capability servitization, fast supply, resource elasticity, and management automation of cloud DCs.

1. Infrastructure: It consists of servers, storage devices, network and security devices, and L1 infrastructure in the equipment room, including air conditioning and cooling, PoE power supply, and cabling. Infrastructures of different DCs are connected through a Data Center Interconnect (DCI) network.

2. Resource pools: The pools include the production resource pool, R&D and test resource pool, DMZ resource pool, hosting resource pool, and DR resource pool that use the OpenStack cloud platform as the management domain. Each service resource pool consists of the storage resource pool (centralized or distributed storage), compute resource pool (VM and PM resource pools), database resource pool, and network resource pool.

3. Service domain: The service platform integrates and provides the following types of services in a unified manner:

   - Computing services: elastic IP address (EIP) and BM services.

   - Storage services: elastic block storage and distributed block storage services.

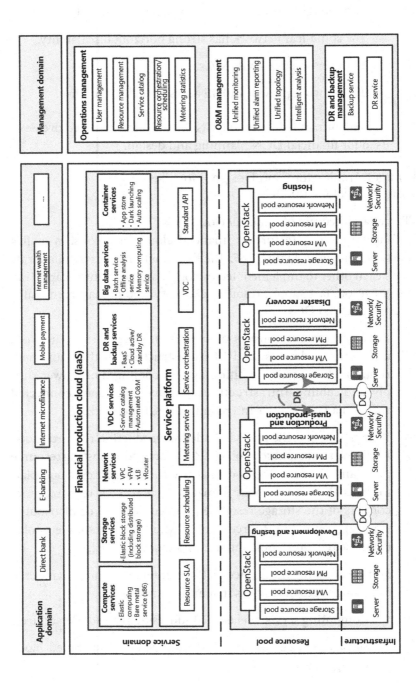

FIGURE 3.47 Target architecture of financial cloud DCs.

- Network services: Virtual Private Cloud (VPC), virtual firewall (vFW), virtual load balancer (vLB), and virtual router (vRouter) services.

- Other services: virtual data center (VDC) services (service catalog management and self-O&M), DR services (backup as a service and cloud active/standby DR), big data services (batch processing, offline analysis, and memory computing services), and container services (application store, gray release, and elastic scaling).

- The service platform provides various basic assurance capabilities for services, including resource Service-Level Agreement (SLA), resource scheduling, metering and charging, service orchestration, VDC, and standard application programming interface (API).

4. Application domain: It includes various applications built based on standard APIs provided by the service domain for specific business domains. The applications include direct banking, e-banking, Internet micro-loan, mobile payment, and Internet wealth management.

5. Management domain: It provides operation management, O&M management, and DR management. Operation management includes user management, resource management, service catalog, resource orchestration and scheduling, and metering statistics. O&M management includes unified monitoring, unified alarm, unified topology, and intelligent analysis. DR management includes backup as a service (BaaS) and DR services.

2. Network planning and design objectives of financial cloud DCs

The ultimate goal of network planning and design is to support service development and meet the requirements of elastic service provisioning, fast application deployment, information sharing, distributed system expansion, and flexible load scheduling. To meet these requirements, the SDN design objectives of cloud DCs are proposed. These objectives are summarized as ICS DCN, as shown in Figure 3.48.

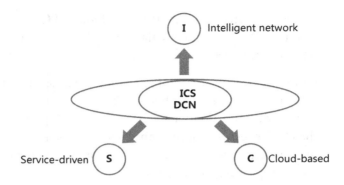

FIGURE 3.48   SDN design objectives of cloud DCs

The I objective refers to intelligent network, including AI Fabric with features such as high performance, low latency, and zero packet loss, Artificial Intelligence for IT Operations (AIOps), and Artificial Intelligence and Security (AISec). The C objective refers to cloud-based architecture. The DCN connects to the cloud platform through SDN to implement an automated, software-based, and resource-based network. In addition, it is required that the cloud DCN provides network access for non-cloud servers and can connect to non-cloud networks. The S objective refers to service-driven, which is oriented to applications. It focuses on services, provides network functions as services, and is a higher layer of cloudification.

The SDN design objectives of cloud DCs are clarified as follows:

1. High availability

    –   99.999% availability: The network architecture has several layers, and fault domains are isolated.

    –   Network faults can be rectified automatically.

    –   The fault domain is reduced, and Layer 2 loops are eliminated.

    –   Broadcast storms on the LAN are eliminated, and abnormal traffic is detected and blocked.

    –   The underlay network is a pure IP network, and VLANs are terminated on edge devices.

    –   The architecture is hierarchical and distributed.

Key technologies include spine-leaf, VXLAN, EVPN, and M-LAG.

2. High security

   - End-to-end and multi-dimensional security protection

   - Security groups, microsegmentation security, and distributed software firewalls provide more refined security.

   - SFC makes the security manager more flexible and convenient.

   - The security manager manages firewall policies in a unified manner.

   - The big data security, security manager, and network controller are associated to implement intelligent security.

3. High performance

   - Low latency (ns-level) and zero packet loss

   - The network supports 10GE/25GE access and 40GE/100GE uplink.

4. Flexible and elastic scaling and large-scale pooling of computing and storage resources

   - Improve the access capability of physical zones. A single physical zone can accommodate 7000 servers.

   - Physical zones support cross-equipment room module deployment, and server access is appropriately decoupled from physical locations.

   - Optimize the division of physical zones in a DC to ensure physical zones can be properly combined to improve the sharing capability of computing and storage resource pools.

   Key technology: EVPN-based VXLAN resource pools enable flexible and elastic server deployment.

5. Automation and servitization: Interconnection with the cloud platform implements on-demand, self-service, agile, and automatic delivery of network services, thereby supporting fast service rollout.

- Automatic deployment and reclamation of Layer 2 and Layer 3 network configurations, load balancing policies, and firewall policies are implemented.

- IP address resources can be automatically allocated and reclaimed.

- The dynamic binding of network services in the compute node provisioning process is implemented.

- The SDN controller based on the service-oriented architecture provides northbound REST APIs.

6. Visualized and intelligent O&M

- Topology visualization, resource visualization, traffic visualization, and forwarding path visualization

- Traffic hotspot map and full-flow visualization

- Hardware fault prediction

- VM profile

Intelligent O&M implements collaboration among monitoring, analysis, and control, analyzes data obtained from monitoring, and automatically reports analysis results to the network through the controller.

3. Design principles

Depending on the importance, the network architecture design principles are as follows:

- High reliability: The network architecture design must ensure continuous reliability to support 24/7 network running. Meet high reliability requirements of systems and applications in the same DC and across different DCs, and support service continuity of intra-city active-active DCs and remote DR DCs. Consider the scale of network fault domains, prevent architecture risks caused by a large network fault domain, and control the scope of broadcast and multicast traffic to prevent loops.

- High security: Based on classification of application systems, determine access control rules between different service systems and deploy policies based on the rules. Deploy firewalls based on physical zones to improve resource pooling capabilities based on the principle of "on-demand deployment, nearby protection, and single-side execution." Optimize security policy management by improving security zone planning and policy automation management capabilities.

- Easy management: Use standardized architectures and deployment solutions to gradually improve the automation capability and reduce the workload of service deployment. Build an all-in-one network O&M system, and use visualization and AIOps to simplify O&M and quickly rectify faults.

- High performance: The network supports large capacity and low latency, meeting the connection at any time, real-timeness, interaction, and intelligence requirements of service applications. Meet the network bandwidth requirements of heavy-load applications such as distributed storage, distributed database, and big data analytics platform. Meet high scalability requirements of large-scale x86 server clusters and distributed storage.

- Future-oriented: The network supports high bandwidth, low latency, large buffer, 10GE/25GE access, and 40GE/100GE uplink. The network supports network SDN and cloud computing platforms to implement network automation.

- Evolution: The architecture can be gradually evolved, taking into account the coexistence of traditional and new network architectures. The network design can meet development requirements of computing, storage, and database, and support technology evolution in a certain period of time. The network architecture supports evolution to distributed and cloud-based DCs.

3.3 ARCHITECTURE AND SOLUTION EVOLUTION OF DCNs FOR CARRIERS

3.3.1 Architecture of Carriers' Networks

Carriers' networks have a hierarchical architecture, which consists of the access layer, aggregation layer, and core layer. Network and telecommunications devices are deployed in network equipment rooms of corresponding levels. With the reconstruction of DCs in mainstream carriers' network equipment rooms around the world, traditional end offices (switching centers) of carriers are transformed into DCs similar to those of cloud service providers, as shown in Figure 3.49. Carriers' DCNs use a three-level architecture consisting of backbone DCs, central DCs, and edge DCs, which vary in quantity, coverage, and scale. Typically, three to ten backbone DCs are constructed to provide services for a large province or a small- and medium-sized country. A single backbone DC is large and contains more than 2000 servers. The number of central DCs ranges from ten to hundreds. They provide services for large- and medium-sized cities or small- and medium-sized provinces. A single central DC contains 500–2000 servers. The number of edge DCs ranges from hundreds to thousands. They mainly provide services for districts and counties. A single edge DC contains dozens to hundreds of servers.

Carriers' services are classified into two types of services (external and internal) containing seven services.

External services include hosting, public cloud, content delivery network (CDN), IPTV, and MAN pipe services. The hosting service is the IaaS infrastructure leasing or hosting service for enterprises or individual customers. For the convenience of enterprise or individual customers, the hosting service is deployed in edge or central DCs. The public cloud service is the standard IaaS/PaaS/SaaS cloud service provided for enterprises or individual customers. Customers can access the resources through the Internet or leased lines. The public cloud service is generally deployed in backbone or central DCs to fully utilize these large DCs and reduce costs. Multimedia applications have high application latency requirements and consume a large amount of bandwidth. Carriers provide CDN and IPTV services for customers using multimedia applications. CDN and IPTV services are generally deployed in central or backbone DCs. To enable enterprises or individual customers to access the network and implement intra-enterprise communication, carriers also provide the MAN pipe service.

Internal services refer to the voice and data communication service and IT system service required for internal operation of carriers. They are usually called the telco cloud service and enterprise data center (EDC) service. The IT system service for enterprises is generally deployed in backbone DCs. The telco cloud service system is distributed in backbone, central, and edge DCs based on its service characteristics, as shown in Figure 3.50.

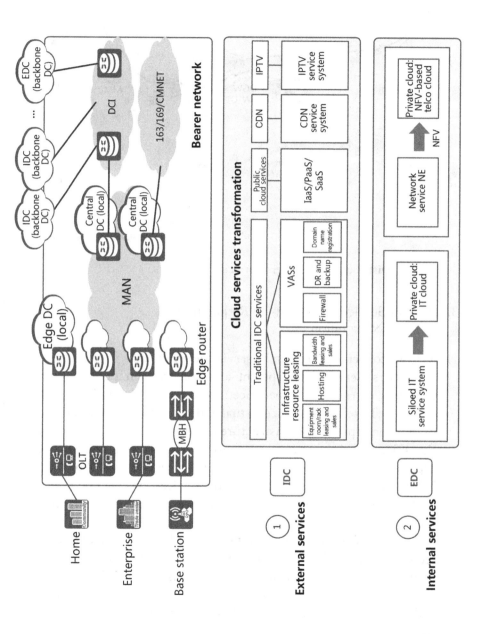

FIGURE 3.49 Architecture of carriers' DCNs.

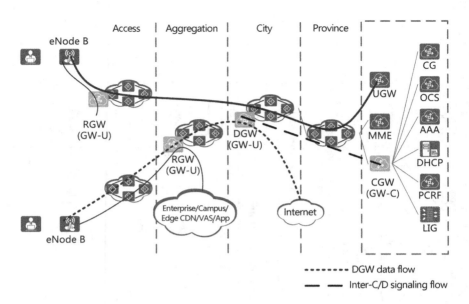

FIGURE 3.50   Distribution of carriers' services.

3.3.2 Carrier Service Development Requirements and DCN Evolution

1. ICT DC convergence driven by service development

Carriers' service development and construction periods vary. Currently, carriers have experienced the following phases: DC-based enterprise IT services, service virtualization, external service cloudification, enterprise IT infrastructure cloudification, and CT service cloudification. As carriers' key network requirements and service experience indicators vary depending on services, as shown in Table 3.2, most DCs currently adopt the vertical construction mode. In this mode, DC resources are seriously wasted, and the O&M cost keeps increasing.

As carriers' commercial, operation, and R&D modes change, they require an open, always-online, and automated ICT infrastructure to meet service development requirements, as shown in Figure 3.51. Therefore, global tier-1 carriers have started network transformation and released the 2020 transformation strategy. Their network reconstruction approaches are gradually unified.

TABLE 3.2   Service Requirements and Experience Indicators in Different Phases

Service Type	Service Requirement	Experience Indicator
IT cloud	Unified management, improved efficiency, reduced OPEX, and compatibility with existing devices and systems	90% resource utilization and minute-level troubleshooting
Telco cloud	Shortened TTM of services, accelerated service innovation, carrier-class SLA, and unified management of multi-level DCs	VM HA switchover:<90 seconds; active/standby switchover: 5 minutes; hardware switches: line-rate forwarding; software switches: 40 Gbit/s forwarding
Hosting	Integration of scattered resources, improved resource utilization, and fast network service provisioning	Service provisioning time shortened from one month to several minutes, and unified management of resources in multiple DCs
Public cloud	Support for a large number of tenants, openness, and standardization	Number of tenants increased from 4000 to 16,000; service provisioning time shortened from one month to several minutes
DC-based MAN	Cost reduction, simplified network, openness, and standardization	Service provisioning time shortened from one month to several minutes, and CAPEX reduced by 80%

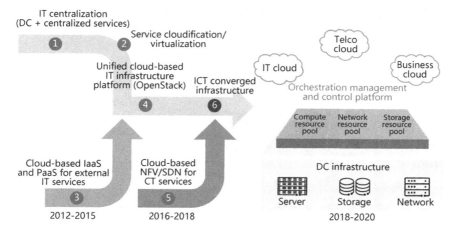

FIGURE 3.51   ICT converged network.

2. DC service and traffic transformation driven by 5G services

The commercial deployment of 5G services poses higher requirements on user experience and network quality. Take 4K videos as an example. Carriers need to reduce the round-trip latency by 10–15 ms and improve the video mean opinion score (vMOS) by 0.2. Heavy-traffic services, such as CDN, need to be forwarded along the shortest path to reduce the impact on carriers' MAN. In addition, centralized control and unified management are required to reduce the operation cost. 5G 3GPP specifications are also promoting the separation of the control plane and user plane of core NEs.

As services on the user plane are gradually migrated down to central and edge DCs (shown in Figure 3.52), carriers' DC architecture faces the following challenges:

- Scalability: Services are deployed in edge DCs. Region DCs and POP DCs in a region need to be deployed and managed in a unified manner. In the future, telco cloud services will be further deployed in edge DCs, and hundreds of DCs need to be managed in a unified manner.

- Reliability: The cloud core mobile service requires that the network service traffic interruption time cannot exceed 5 seconds. Once the interruption time exceeds 5 seconds, calls of mobile phone users will be interrupted. A network upgrade interrupts services for less than 5 seconds, ensuring zero service interruption.

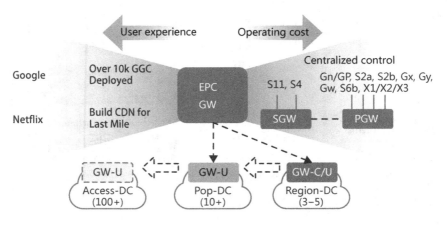

FIGURE 3.52 Challenges faced by carriers' DCs.

- Openness: Components such as the VNF, network, and cloud OS are purchased by layer. The components must support a rich ecosystem.

- High quality: Heavy traffic, low latency, and high concurrency are the top concerns for 4K and VR live broadcast services. In addition, the CDN is deployed closer to users, and traffic needs to be forwarded locally, reducing network overheads and achieving optimal user experience.

3.3.3 Target Architecture and Design Principles of Carrier Cloud DCs

Figure 3.53 shows the target architecture of carrier cloud DCs.

- Management components, such as management and orchestration (MANO), OpenStack, and network controllers are deployed in region DCs and remotely manage virtualized network functions

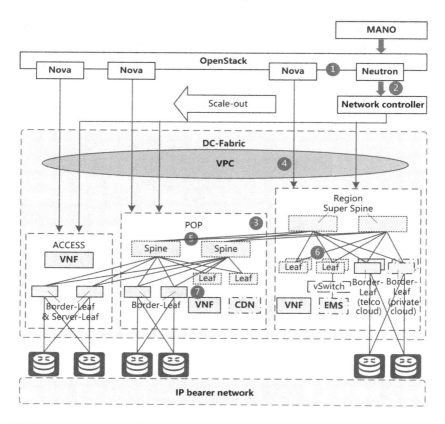

FIGURE 3.53   Target architecture of carrier cloud DCs.

(VNFs) and switches, reducing server costs and management overheads.

- As a de facto standard, OpenStack becomes a unified platform for open interconnection with the IaaS layer.

- The multi-level spine-leaf architecture and multi-level DCs support horizontal scaling.

- On the large-scale SDN, the network controller can manage more than 1000 devices, implementing cross-DC VPC automation.

- Switches on the entire network are not stacked to prevent single points of failure. M-LAG and ECMP technologies are used to implement fast link switching. The traffic interruption time during a network upgrade is less than 5 seconds, ensuring that mobile services are not interrupted.

- The networking of DCs is evolving to distributed hybrid overlay networking, allowing simultaneous access of VNFs and physical network functions (PNFs). Traffic does not need to be diverted and is forwarded with low latency.

- The data network supports IPv6 and connects to IPv4 and IPv6 networks through dynamic routing.

CHAPTER 4

Functional Components and Service Models of Cloud DCNs

Servers, network devices, storage devices, LBs, and security devices form the core of traditional DCs. As IT and enterprise services are not interconnected, independent and large-scale systems cannot quickly respond to fast-growing service requirements. This has led to the emergence of cloud DCs. A cloud DC is a service-oriented architecture where all devices, systems, and functions are considered services. In order to manage these services, a new architecture (cloud platform or SDN controller) is needed to rapidly respond to fast-growing service requirements.

4.1 SERVICE MODELS OF CLOUD DCNs

While service models of cloud DCNs provided by different vendors may not be identical, they often have common or similar features. The network service model of OpenStack is a typical example.

OpenStack, an open-source project, is the most popular open-source cloud OS framework. As a cloud OS, the main task of OpenStack is to manage resources. OpenStack aims to offer a cloud management platform that allows for simple operation, high scalability, and unified standards.

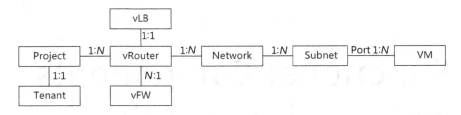

FIGURE 4.1   Neutron service model.

It manages compute, storage, and network resources. All resources are managed by multiple components, including Nova, Cinder, Neutron, Swift, Keystone, Glance, Horizon, and Ceilometer. Of these, Neutron is responsible for network service provisioning, and the basic service model described in this section refers to the network service model defined by Neutron, which is shown in Figure 4.1.

The service model involves the following key elements:

Tenant: indicates the person or enterprise that applies for compute, storage, and network resources. When a tenant applies for resources from OpenStack, the tenant can only perform actions within these resources. For example, when a VM is created, the resources used by the VM are those applied for by the tenant.

Project: In Neutron, one project corresponds to one tenant. Starting from Keystone V3, OpenStack recommends that a project be used in Neutron to uniquely identify a tenant.

vRouter: If a logical network created by a tenant has multiple network segments and Layer 3 communication is required, or an internal network needs to communicate with an external network, you need to create a vRouter on the logical network. On the other hand, if a tenant requires only a Layer 2 network, you do not need to create a vRouter. The vRouter has the following functions:

- Provides Layer 3 communication between network segments of the logical intranet.

- Connects an internal network to an external network.

- Provides the NAT capability to allow access from the internal network to the external network.

Network: indicates an isolated Layer 2 broadcast domain and can contain one or more subnets.

Subnet: specifies an IPv4 or IPv6 address range. A subnet contains multiple VMs, with each VM's IP address assigned within the subnet. When creating a subnet, you need to define the IP address range and mask, and define a gateway IP address for the network segment to allow communication between VMs and the external network.

Port: identifies an object connected to a logical network in the Neutron network model. A port can be a vNIC of a VM or a physical NIC of a bare metal (BM).

vLB: provides L4 load balancing services for tenant services and provides health check services for LB services.

vFW: indicates Firewall as a Service (FWaaS) V2.0 defined in Neutron, which is an advanced service of Neutron. Similar to traditional firewalls, vFWs use firewall rules on vRouters to control network data of tenants.

4.1.1 Typical OpenStack Service Model

This section describes typical OpenStack network service models:

- Compute resources only need to communicate with each other at Layer 2.

- Compute resources need to communicate with each other at Layer 3, but do not need to connect to an external network.

- Compute resources need to communicate with each other at Layer 3 and connect to an external network.

1. **Service model 1: Compute resources need to communicate with each other at Layer 2 only.**

   If compute nodes need to communicate with each other at Layer 2, you only need to set up a temporary test environment to verify service functions or performance. The test nodes must be on the same network segment. In this case, you can orchestrate the service model shown in Figure 4.2, create one or more subnets, and mount interfaces of compute node NICs on the same network segment to the same network.

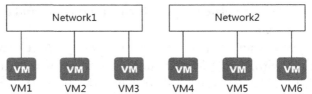

FIGURE 4.2   Compute resources only need to communicate with each other at Layer 2.

2. **Service model 2: Compute resources need to communicate with each other at Layer 3, but do not need to connect to an external network.**

You need to divide the network into different network segments and then deploy/configure security isolation for security purposes or in either of the following situations:

- User services place high requirements on the network. For example, they need to be deployed by layer (such as by web, application, or database layer), and each layer needs to be isolated by network segment.

- Multiple services need to be isolated from each other to reduce BUM packet interference.

In this case, you can achieve Layer 3 isolation by dividing the compute resources to be isolated into different subnets, with Layer 2 isolation achieved by using a different network for each subnet, as shown in Figure 4.3.

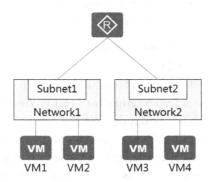

FIGURE 4.3   Compute resources need to communicate with each other at Layer 3 but do not need to access the external network.

If different subnets need to communicate with each other, a router needs to be deployed in the service model to provide Layer 3 forwarding. For comparison between the service model and traditional physical networking, the router is equivalent to a Layer 3 switch, the network is equivalent to a VLAN, subnets are equivalent to network segments in a VLAN, and the gateway IP address of the network segment is equivalent to the IP address of a VLAN interface (for example, the IP address of a VLANIF interface) on a Layer 3 switch.

3. **Service model 3: Compute resources need to communicate with each other at Layer 3 and connect to an external network.**

Based on service model 2, if the user network needs to access an external network (for example, the Internet or private network), it needs to do so through the router based on service model 2. The system administrator can create an external network during system initialization in OpenStack. When configuring service networks, tenants are only able to select such external networks. OpenStack allows a router to connect to only one external network, as shown in Figure 4.4.

In this service model, as well as providing the Layer 3 forwarding capability for packets between different subnets, the router also provides the NAT capability for internal and external networks. This capability includes SNAT for access from the internal network to the external network and destination NAT (DNAT) for access from the external network to the internal network.

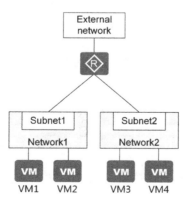

FIGURE 4.4 Compute resources need to communicate with each other at Layer 3 and connect to an external network.

4.1.2 FusionSphere Service Model

FusionSphere is a cloud platform developed by Huawei based on open-source OpenStack and enhanced for commercial use. Therefore, its basic features are similar to those of OpenStack.

In addition to basic components defined by OpenStack, the FusionSphere service model maps physical resources to logical NEs from the perspective of DC resource deployment. Before introducing the FusionSphere service model, let's learn about the basic principles and concepts of Huawei DC networking.

In Figure 4.5, physical resource management in DCs is generally planned by Point of Delivery (PoD).

As a physical concept, a DC implements centralized data processing, storage, transmission, switching, and management in a physical space such as an equipment room. Key devices in a DC include servers, storage devices, and network devices. Infrastructure such as the power supplies, cooling systems, fire control systems, and monitoring systems are also key components of a DC.

To facilitate resource pooling in a DC, it is divided into one or more physical partitions, with each partition called a PoD. It therefore follows that a PoD is a basic deployment unit of a DC, with one physical device belonging to only one PoD.

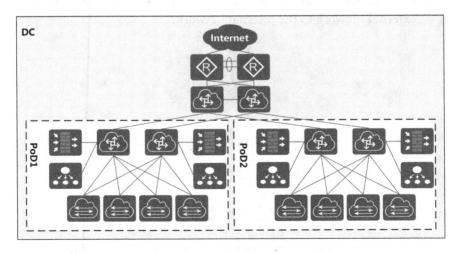

FIGURE 4.5　PoD-based physical resource management.

The following describes the FusionSphere service model, as shown in Figure 4.6.

- An availability zone (AZ) is a logical concept that represents a fault isolation area. For example, hosts in an AZ share the same power supply system and network facility. If the facility fails, these resources are unavailable for all hosts in the AZ. AZs can be flexibly mapped to DCs according to the actual deployment. For example, in a large-scale public cloud, an AZ can contain multiple DCs; whereas in a small- or medium-sized private cloud, multiple independent AZs can be configured in a single DC. You can also plan a DC to be an AZ.

- A VPC is a basic service unit of FusionSphere, and one VPC can be deployed across several PoDs. Alternatively, VPCs of the same tenant can be deployed in different PoDs. One VPC corresponds to one vRouter, and one vRouter is represented as a virtual routing forwarding (VRF) on a switch.

- In FusionSphere, a VDC is a resource unit and corresponds to a tenant. A VDC can be deployed across PoDs, and tenants rent resources by VDC. Note that one VDC may contain multiple VPCs.

The logical NEs in a VPC are the same as those in OpenStack. A vRouter is a logical unit that provides the routing function in a VPC, with one vRouter corresponding to one VPC. The vLB and vFW are responsible for load balancing and firewall services in the FusionSphere service model, with one vLB or vFW corresponding to one vRouter. The service models of the two types of NEs are described in the next section. In addition, a vRouter can connect to multiple subnets, each of which can connect to multiple VMs.

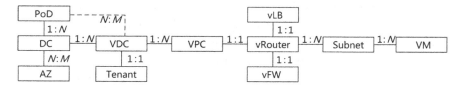

FIGURE 4.6   FusionSphere service model.

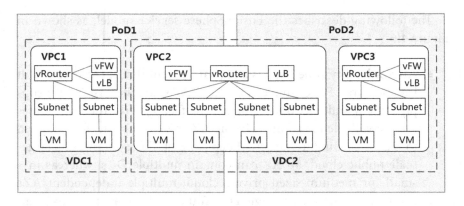

FIGURE 4.7 Deployment relationship of the FusionSphere service model.

Figure 4.7 shows the deployment relationship of the FusionSphere service model.

The deployment principles of the FusionSphere service model are as follows:

- A PoD is also a basic deployment unit for carrying services. A set of resources can be deployed in one or more PoDs, each of which can carry multiple sets of resources.

- A VDC can be deployed across PoDs, and tenants rent resources by VDC.

- As mentioned already, a VPC can be deployed across several PoDs, and different VDCs of the same tenant can be deployed in different PoDs. When creating a VDC, the tenant's VDC must have been deployed across several PoDs.

4.1.3 iMaster NCE-Fabric Service Model

iMaster NCE-Fabric is the SDN controller in Huawei's DCN solution. It is used to automatically deliver network services orchestrated by users or the cloud platform. Compared with manual configuration of network devices, iMaster NCE-Fabric greatly improves service provisioning efficiency, shortening the time taken to just minutes.

iMaster NCE-Fabric also provides a network service orchestration model. In addition, the service model construction on iMaster

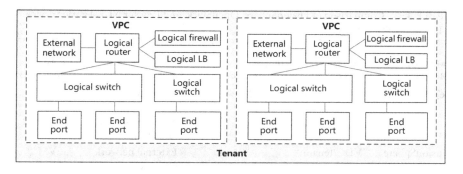

FIGURE 4.8   iMaster NCE-Fabric service model.

NCE-Fabric is similar to that of OpenStack and FusionSphere. As shown in Figure 4.8, the basic service model of iMaster NCE-Fabric includes the tenant, VPC, logical router, logical switch, logical firewall, logical LB, and end port.

On iMaster NCE-Fabric, administrators can authorize tenants to use a certain number of VPCs, and configure usage quotas of logical routers, logical switches, logical firewalls, and logical LBs. The logical router, logical switch, logical firewall, and logical LB provide the FaaS. In other words, the network is abstracted into multiple services for tenants.

- The logical NEs in the iMaster NCE-Fabric service model map to components of the OpenStack and FusionSphere service models.

- The logical router corresponds to the vRouter of the cloud platform.

- The logical switch corresponds to the network or subnet of the cloud platform.

- The logical firewall and logical LB correspond to the vFW and vLB of the cloud platform, respectively.

- An end port is a logical access point that is connected to a logical switch. It can be a VM, a physical server, or a third-party device.

Table 4.1 compares the FusionSphere, OpenStack, and iMaster NCE-Fabric service models at the resource management layer, logical organization layer, network entity layer, and access entity layer.

TABLE 4.1 Comparison of FusionSphere, OpenStack, and iMaster NCE-Fabric Service Models at Different Layers

(a) Service Model	Resource Management Layer	(b) Logical Organization Layer	(c) Network Entity Layer	(d) Access Entity Layer
(e) OpenStack	Project/tenant	None	External network vRouter Network/Subnet vFW/vLB	VM
FusionSphere	VDC/tenant	VPC	External network vRouter Subnet vFW/vLB	VM
iMaster NCE-Fabric	Tenant	VPC	External network Logical router Logical switch Logical firewall/LB	End port

- Resource management layer: is a basic resource unit. At this layer, DC resources are allocated by tenant, and corresponding tenant administrators are specified.

- Logical organization layer: is a basic service unit and defines the logical organization of networks and compute entities. Networks between service units are securely isolated.

- Network entity layer: includes various network entities of a service unit.

- Access entity layer: includes all access entities contained in a service unit. The access entity may be a compute node or a network node outside the management scope of iMaster NCE-Fabric.

In addition to network service orchestration, iMaster NCE-Fabric also provides security service orchestration. iMaster NCE-Fabric uses the microsegmentation security control model to orchestrate security access control within a service or between services, providing refined security management capabilities for users.

Microsegmentation, known as refined group-based security isolation, groups services on the network based on certain rules (such as IP addresses, IP network segments, MAC addresses, VM names, containers, and OS). Then, the orchestrator is used to deploy access control policies based on groups to implement security management of network services. Microsegmentation aims to implement refined security management within a DCN while ensuring efficiency. iMaster NCE-Fabric uses the group-based policy (GBP) model for orchestration.

The security group model of Neutron is complex. This means that application developers (that is, users of the real cloud infrastructure) are required to consider the interconnection between development systems at different layers, causing unnecessarily complex operations for application developers. This is a major problem for application developers, who want to describe their application network and security requirements in simple terms. The GBP model solves this problem.

It does this by providing a declarative, user intent-oriented architecture, as shown in Figure 4.9. In this model, the user plane corresponds to the application architecture, as opposed to various network elements in Neutron. When using the model, you need to first define various groups, followed by network features between the groups, including security, performance, and network services.

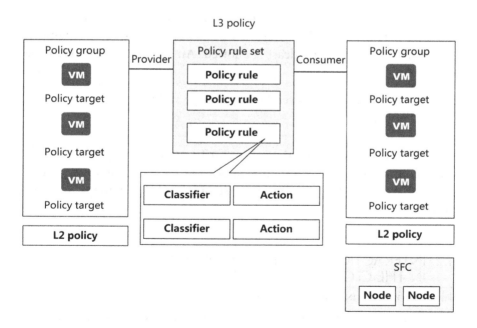

FIGURE 4.9   GBP model.

- Policy target: is an object used to define policies. Typically, a policy target is an object that can be located, such as an NIC or an IP address.

- Classifier: is a measure taken to classify network traffic based on IP addresses, MAC addresses, traffic directions, and other conditions.

- Action: indicates the action taken by the classifier for a certain type of traffic after the classifier classifies the traffic. The value can be Allow, Redirect, or Drop.

- Policy rule: is a group of classifiers and actions.

- Policy rule set: is a group of policy rules.

- Policy group: is a group of policy targets that have the same attributes. The policy group can provide consumption rule sets.

In addition to the preceding entities, the GBP model defines the following network policies:

- L2 policy: is a set of policy groups, which identifies a Layer 2 switching domain. Layer 2 network attributes can be defined, for example, whether Layer 2 broadcast is allowed. An L2 policy must reference an L3 policy.

- L3 policy: defines a Layer 3 routing space and can reference multiple L2 policies.

The GBP orchestration model is a security policy management model centered on application flows. It is applicable to scenarios where enterprise private cloud system administrators centrally manage security policies.

4.2 INTERACTION BETWEEN COMPONENTS IN THE CLOUD DCN SOLUTION

4.2.1 Cloud DCN Solution Architecture

Before describing interaction between components, we must introduce the architecture of the cloud DCN solution. It includes the components and interfaces between them, and varies depending on the solution scenario. The following factors determine the solution architecture:

- Whether the overall solution contains the cloud platform and whether the SDN controller needs to interconnect with the cloud platform.

- Whether the network overlay or hybrid overlay needs to be used. The architecture of the cloud DCN solution in different scenarios is divided into the following four layers:

  1. Service presentation or orchestration layer

     The service presentation layer is oriented toward DC users. The cloud platform at this layer provides GUIs for service, network, and tenant administrators, implementing service management, automatic service provisioning, as well as resource and service guarantee.

     The service orchestration layer consists of Nova, Neutron, and Cinder components of the cloud platform. These components interoperate to realize virtualization, pooling, and collaborative scheduling of compute, storage, and network resources in DCs.

  2. Network control layer

     This layer is at the core of the solution. Here, the SDN controller implements network modeling and instantiation, works with physical and virtual networks to provide network resource pooling and automation, and provides the network-wide view to uniformly control and deliver service flow tables. The SDN controller is the key component for separating SDN network control and forwarding. Huawei SecoManager is used as an example. The SecoManager can model and instantiate L4–L7 VASs, pool and automate security resources, as well as deliver L4–L7 VAS configurations.

  3. Network service layer

     This layer is the infrastructure of a DCN that provides high-speed channels for services, for example, L2–L3 basic network services and L4–L7 VASs. The distributed network overlay is recommended for the cloud DCN solution, which also supports centralized network overlay and distributed hybrid overlay.

4. Computing access layer

The cloud DCN solution supports the access of virtual servers, physical servers, and bare metal servers (BMSs).

Virtual server: Virtualization technology virtualizes one physical server into multiple VMs and vSwitches; VMs connect to fabric networks through vSwitches.

Physical server: Physical servers connect to fabric networks through logical ports. The cloud platform directly interacts with BMS to manage them as instances.

4.2.2 Interaction between Components during Service Provisioning

This section uses a typical service provisioning process as an example to describe the interaction between components in the cloud DCN solution.

4.2.2.1 Service Provisioning Scenario

The following example uses the cloud platform and network overlay. The cloud platform functions as the service orchestration entry.

In Figure 4.10, the cloud DCN solution uses the SDN controller to map logical networks dynamically constructed on the cloud platform to physical networks through network modeling and to complete automatic configuration delivery. Service provisioning involves both network and compute services:

- Network service provisioning: Administrators allocate network resources to specified services or applications (including L4–L7 VASs) through the cloud platform.

- Compute service provisioning: Administrators create, delete, and migrate compute and storage resources on the cloud platform.

Before service provisioning, the system administrator creates a project, binds the project to a tenant administrator, and sets resource quotas for the project. The quotas include:

- Number of vCPUs: When creating a VM, you can allocate a specified number of vCPUs to the VM. If the number of vCPUs is large, the VM computing capability is strong. This parameter specifies the number of vCPUs that can be allocated to all VMs in the project.

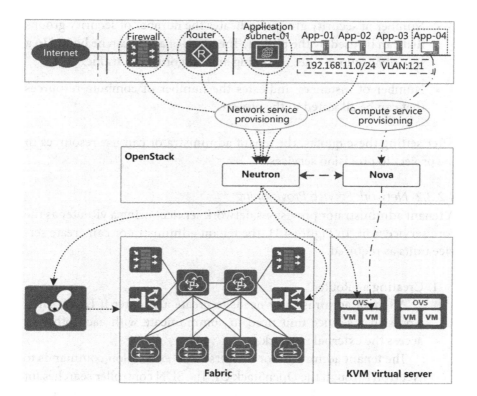

FIGURE 4.10   Unified service provisioning on the cloud platform.

- Memory: When creating a VM, you can allocate the memory of a specified size to the VM. This parameter specifies the memory size that can be allocated to all VMs in the project.

- Storage: When creating a VM, you can allocate a specified number of disks to the VM. The disks can be either local disks on a server or remote disks mounted to a network.

- Number of floating IP addresses: indicates the number of public IP addresses that can be used to access an external network in the project.

- Number of vRouters: indicates the number of vRouters that can be created in the project.

- Number of networks or subnets: indicates the number of Layer 2 networks that can be created in the project.

- Number of security groups: indicates the number of security groups that can be used by the project. The vNIC of a VM can be bound to a security group to filter incoming and outgoing VM traffic.

- Number of instances: indicates the number of compute resources that can be allocated to the project.

After setting these quotas, the tenant administrator can use resources in the project to provision services.

4.2.2.2 Network Service Provisioning

A tenant administrator provisions network services using a vRouter as the core service unit. In Figure 4.11, the tenant administrator can create service units as required.

1. Creating a vRouter

   A tenant administrator needs to create a vRouter if Layer 2 networks in a service unit need to communicate with each other or access the external network.

   The tenant administrator delivers vRouter creation commands to Neutron through the OpenStack UI. The SDN controller searches for the AZ corresponding to OpenStack and creates a VRF on a gateway of a fabric network corresponding to the AZ to implement Layer 3 isolation. After a vRouter is created, the VXLAN gateway generates a default route that points to the gateway added when a subnet of an external network is created.

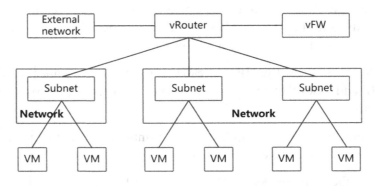

FIGURE 4.11   Service units among which a vRouter is the core.

2. Creating a network or subnet

The network refers to a Layer 2 network, where a tenant administrator delivers a command for creating a subnet to Neutron through the OpenStack UI. After receiving the subnet creation message, the SDN controller obtains the VLAN ID, queries the database to find the corresponding VNI, delivers a VBDIF interface to the VXLAN gateway, creates an interface of the Layer 3 VXLAN gateway, and binds the interface to the VRF corresponding to the service unit.

The tenant administrator can provision a network as either a routed network or an intranet. A routed network must be associated with a vRouter. If the network is an intranet, the network is used only for Layer 2 communication. If you need to access an external network in the future, you can associate a vRouter with the network at any time.

3. Creating a vFW

When Layer 2 networks in a service unit communicate with each other or access the external network, security access control is required. A tenant administrator needs to create and configure access rules on vFWs, and associate them with vRouters.

4. Associating an external network

An external network can be created only by system administrators and is used to connect to the exterior of a DC, such as the Internet. Tenant administrators can view all external networks in the system and associate vRouters with external networks so that service units can communicate with external networks.

4.2.2.3 Compute Service Provisioning

Compute service provisioning is a process of allocating compute and network resources in a unified manner. The provisioning process requires coordination and interaction between cloud platforms, SDN controllers, network devices, and hypervisors, as shown in Figure 4.12.

The following describes the compute service provisioning process from the three aspects: VM login, VM logout, and VM migration.

1. VM login

- The tenant administrator creates a VM on the cloud platform portal and sends a VM creation request to Nova through Horizon.

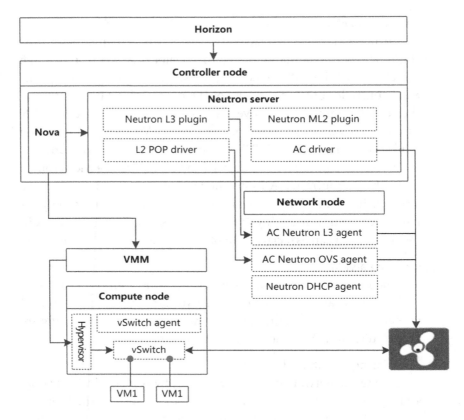

FIGURE 4.12   Compute service provisioning process.

- Nova invokes an interface of Neutron to apply for creating an interface that corresponds to a VM.

- Neutron broadcasts CreatePort messages through Neutron L2 Plugin.

- The SDN controller obtains the VM host information through the vSwitch agent.

- Nova works with the VMM to create a VM and associate it with the created interface.

- The vSwitch detects that the interface is connected to the VM and reports the interface login message (carrying the interface ID) to the SDN controller.

- The SDN controller compares interface IDs to obtain reported VXLAN information on VM access.

- The SDN controller uses NETCONF interfaces to deliver configurations to the TOR switches connected to hosts and connects VMs to the specified VXLAN network.

2. VM logout

- A tenant administrator deletes a VM on the cloud platform portal. Horizon sends a VM deletion request to Nova.

- Nova invokes an interface of Neutron to apply to delete an interface corresponding to the VM.

- Neutron broadcasts the DeletePort message through Neutron L2 Plugin.

- The SDN controller obtains interface information of the deleted VM through the vSwitch agent.

- Nova works with the VMM component to delete VMs.

- The vSwitch detects that the interface goes offline and sends the interface offline message to the SDN controller.

- The SDN controller deletes VM forwarding entries on the network device and reclaims related resources.

3. VM migration

VM migration is the migration of VMs in a Layer 2 domain, and the process of reclaiming and provisioning service resources (mapping to VM login and logout processes).

A tenant administrator performs VM migration on the portal of the cloud platform and notifies Nova and Neutron of the VM migration event and related parameters through southbound RESTful interfaces. The SDN controller then updates the routing table based on the topology relationship, instructs the destination VTEP to update the MAC address, routing table, and traffic classification table, and also instructs the source VTEP to delete the VNI, MAC address, route, and traffic classification table.

4.3 INTERACTION TECHNOLOGIES BETWEEN CLOUD DCN COMPONENTS

This section describes key technologies used in the cloud DCN solution, including OpenFlow, NETCONF, OVSDB, and Yet Another Next Generation (YANG), which is a data modeling language used for NETCONF.

As described above, the cloud DCN solution includes four layers: service presentation/collaboration layer, network control layer, network service layer, and computing access layer. The service presentation or orchestration layer consists of the cloud platform and analyzer, and the network control layer mainly includes the controller. The network service layer mainly includes network devices, and the computing access layer consists of virtual servers, physical servers, and BMSs.

Service interaction between these layers depends on different technologies. For example, the network control layer interconnects with the service presentation or orchestration layer through RESTful and Remote Procedure Call (RPC) APIs, interconnects with the network service layer through protocols such as NETCONF and SNMP, and interconnects with the computing access layer through protocols such as OpenFlow and Open vSwitch Database (OVSDB).

4.3.1 OpenFlow

4.3.1.1 Introduction to OpenFlow

The development of cloud computing is based on virtualization technology, which abstracts various physical resources into logical resources and decouples physical resources from physical networks. According to the on-demand allocation principle, IT resources such as compute, storage, and network resources can be easily provided for users.

Network virtualization plays an important role in this process. As a new network architecture, SDN separates the control plane from the data forwarding plane. The software platform in the centralized controller implements programmability to control underlying hardware and flexibly allocates network resources on demand, implementing network virtualization.

To implement SDN, the SDN controller needs to use interfaces to manage and configure various network devices, thereby requiring a protocol to implement communication between the control layer and the data forwarding layer. OpenFlow is a standard interface protocol that meets this requirement.

The following describes an OpenFlow switch based on the OpenFlow protocol.

4.3.1.2 Components of an OpenFlow Switch

In Figure 4.13, an OpenFlow switch consists of one or more flow tables, one group table, one meter table, one or more OpenFlow channels, and OpenFlow ports. The flow table and group table are used to perform lookup and forwarding of data packets, while the meter table is used to collect statistics on packets. An OpenFlow switch communicates with an external controller using an OpenFlow channel with the controller through OpenFlow. The controller also manages the OpenFlow switch through OpenFlow.

1. Flow table

   A flow table contains flow entries and is delivered by the controller to an OpenFlow switch through OpenFlow to guide data packet forwarding.

   Using OpenFlow, the controller can add, delete, and update flow entries in flow tables reactively (in response to packets) and proactively. Each flow table of an OpenFlow switch contains a set of flow entries, each of which is composed of match fields, counters, and a set of instructions to apply to matching data packets.

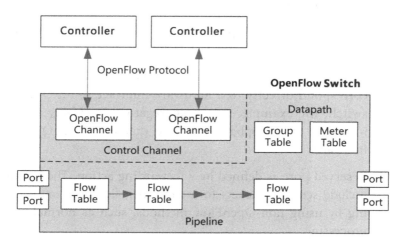

FIGURE 4.13   Components of an OpenFlow switch.

2. Pipeline

A pipeline is a set of associated flow tables that provide packet matching, forwarding, and modification in an OpenFlow switch.

3. Group table

A group table contains a series of flooding action sets and complex forwarding processing, and may contain common actions of multiple flow entries. A group table consists of group entries, with each including a series of specific action buckets. Actions may exist in multiple action buckets, of which one or more are selected for each data packet by the group table.

4. OpenFlow channel

An OpenFlow channel is an interface between an OpenFlow switch and an OpenFlow controller, and is used by the OpenFlow controller to manage the OpenFlow switch.

5. Meter table

A meter table includes multiple meter entries and determines the number of flows. It is mainly used for a QoS operation taken on a data packet, for example, rate limiting.

6. OpenFlow port

OpenFlow ports are classified into physical ports, logical ports, and reserved ports.

- Physical port: corresponds to a hardware interface of an OpenFlow switch.

- Logical port: is defined by an OpenFlow switch, but does not correspond directly to a hardware interface of an OpenFlow switch, for example, a link aggregation interface, a tunnel interface, or a loopback interface that is not an OpenFlow interface.

- Reserved port: is defined by a forwarding action. These actions include sending packets to a controller, flooding, or forwarding by using non-OpenFlow methods, such as normal switch processing.

4.3.1.3 Working Modes of an OpenFlow Switch

OpenFlow switches come in two types: OpenFlow-only and OpenFlow-hybrid switches. OpenFlow-hybrid switches support both OpenFlow operation (using the pipeline to look up a table for received data packets) and normal Ethernet switching operation. In addition to flow table lookup, OpenFlow-hybrid switches support traditional Layer 2 Ethernet forwarding, VLAN isolation, Layer 3 routing, ACL, and QoS processing. They must also provide a classification mechanism before flow table lookup and determine whether to route traffic to either the OpenFlow pipeline or the normal pipeline. OpenFlow-hybrid switches allow packets to go from the OpenFlow pipeline to the normal pipeline. In Figure 4.14, pipeline processing includes both inbound and outbound processing. The outbound processing is optional and is used to specify the processing on an output port.

Pipeline processing involves one or more flow tables, which are sequentially numbered, starting at 0. The ID of the upper-level flow table must be smaller than that of the lower-level flow table. The Instructions parameter in the upper-level flow table provides the ID of the lower-level flow table to be queried.

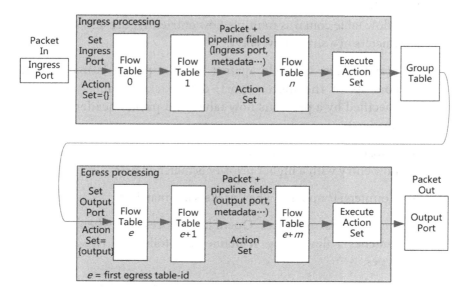

FIGURE 4.14   Pipeline processing.

When a packet enters an OpenFlow switch, the OpenFlow switch parses the packet. Pipeline processing always starts at the first flow table: the packet is first matched against flow entries of flow table 0. Flow entries match packets in priority order, and if a matching entry is found, the OpenFlow switch executes the instructions associated with the specific flow entry. However, if no match is found in a flow table, the OpenFlow switch checks whether the flow table contains a table-miss entry. If so, then the OpenFlow switch takes the action defined in the table-miss flow entry (either discarding the packet, continuing to the subsequent flow table, or forwarding the packet to the controller). If no table-miss entry is found, the OpenFlow switch discards the packet.

If the packet is forwarded to an output port, the OpenFlow switch checks whether there is a flow table available for outbound processing. If so, the OpenFlow switch processes the packet in the outbound direction. If not, the OpenFlow switch directly forwards the packet.

4.3.1.4 OpenFlow Table

1. Flow table

    A flow table contains multiple flow entries. Figure 4.15 shows the parameters of a single flow entry.

    - Match fields: is a field against which a packet is matched. It consists of the ingress port, data packet header, metadata specified by a previous flow table, and packet headers (tunnel headers).

    - Priority: indicates the matching sequence of flow entries. The flow entry with a higher priority is matched first.

    - Counters: is updated when packets are matched.

    - Instructions: indicates OpenFlow processing when a packet matches a flow entry. It defines the following two execution types:

Match Fields	Priority	Counters	Instructions	Timeouts	Cookie	Flags

FIGURE 4.15   Parameters of a flow entry.

TABLE 4.2   Actions

(f) Action	Type	(g) Description
(h) Write-Actions action(s)	Required	Adds a specific action to the running action set. If the specific action exists, the original action is overwritten
Goto-Table next-table-id	Required	Specifies the ID of the subsequent flow table in pipeline processing
Apply-Actions	Optional	Applies the specific action(s) immediately without changing the action set
Clear-Actions	Optional	Deletes all actions in the action set immediately
Write-Metadata metadata/mask	Optional	Writes the masked metadata value into the metadata field
Stat-Trigger stat thresholds	Optional	Generates a flow statistics overflow event and sends it to the controller

- Action set: refers to a set of actions that are executed in a specific sequence. An OpenFlow switch does not immediately modify the packet content until the packet no longer needs to match the lower-level flow table. There is only one action in an action set, and the actions are executed in a specified sequence.

- List of action: refers to a series of actions that need to be executed immediately. The list of action is similar to the action set. In this case, an OpenFlow switch immediately modifies the packet content and carries out the specified action.

  The OpenFlow protocol defines actions of required and optional types (Table 4.2).

- Timeouts: indicates the aging time of flow entries.

- Cookie: indicates the opaque data value selected by the controller. It may be used by the controller to send information such as flow modification and deletion to an OpenFlow switch. This parameter is not used when packets are forwarded.

- Flags: is used to change the management mode of flow entries.

A flow table entry is identified by its match fields and priority. The table-miss flow entry wildcards all match fields (all fields omitted) and has the lowest priority (0). Each flow table does not need to support all entry content, and the features of a flow table are configured and managed by the controller.

Group Identifier	Group Type	Counters	Action Buckets

FIGURE 4.16   Parameters of a group entry.

2. Group table

A group table consists of one or more group entries, which are referenced by flow entries. This means the actions defined in a flow entry can also direct packets to the associated group entry and provide packet forwarding for all flow entries that reference this group entry. Figure 4.16 shows the parameters of a group entry.

- Group identifier: indicates the group ID, which uniquely identifies a group entry on an OpenFlow switch.

- Group type: determines group semantics.

- Counters: is updated when packets are processed by a group.

- Action buckets: refers to an ordered list of action buckets, with each containing a set of actions.

3. Meter table

A meter table is composed of meter entries, which are referenced by flow entries. Specifically, an action related to a flow entry may also direct a packet to a meter entry to measure and control the rate of the packet for all flow entries that reference it. Per-flow meters enable OpenFlow to implement various simple QoS operations such as rate limiting. They can be used to implement complex QoS policies, such as DSCP priority marking, which can classify a group of packets into multiple categories based on the flow rate. Meters are independent of per-port queues, and they can be combined to implement complex QoS frameworks such as Differentiated Service (DiffServ) based on flow operations and classification.

Figure 4.17 shows the parameters of a meter entry.

- Meter identifier: indicates the meter ID, which uniquely identifies a meter entry on an OpenFlow switch.

Meter Identifier	Meter Band	Counters

FIGURE 4.17   Parameters of a meter entry.

Band Type	Rate	Burst	Counters	Type Specific Arguments

FIGURE 4.18   Sub-parameters of the Meter Band parameter.

- Meter band: refers to a list of meter bands, where each specifies the rate of the band and the way to process the packet.

- Counters: is updated when packets are processed by meters. Figure 4.18 shows the sub-parameters of the meter band parameter.

- Band type: defines how packets are processed.

- Rate: defines the lowest rate at which the band can apply.

- Burst: defines the granularity of Meter Band.

- Counters: is updated when packets are processed by meter bands.

- Type-specific arguments: Some band types have optional arguments.

4.3.1.5 Information Exchange on an OpenFlow Channel

An OpenFlow channel transmits the protocol packets of an OpenFlow switch, which supports three types of messages: Controller-to-Switch, Asynchronous, and Symmetric. A Controller-to-Switch message is initialized by the controller and sent to an OpenFlow switch. An Asynchronous message is used by an OpenFlow switch to asynchronously report the network status and change to the controller, whereas a Symmetric message can be sent by both the controller and OpenFlow switch.

1. Controller-to-Switch message

    The Controller-to-Switch message is initiated by the controller to manage or obtain the OpenFlow switch status. The following types of messages are available:

    - Features: The controller sends this message to query the identity and supported functions of an OpenFlow switch.

    - Configuration: The controller sends this message to query and configure parameters of an OpenFlow switch.

    - Modify-State: The controller sends this message to manage the state on OpenFlow switches. The primary purpose is to add, delete, and modify flow/group entries in the OpenFlow tables and to set switch port properties.

- Read-State: The controller uses this message to collect various information from the switch.

- Packet-out: The OpenFlow switch uses this message to send a data packet matching a flow entry.

- Barrier: The controller uses this message to ensure message dependencies have been met or to receive notifications for completed operations.

- Role-Request: The controller uses this message to set its ID and the role of its OpenFlow channel, or query that role. This is mostly useful when the OpenFlow switch connects to multiple controllers.

- Asynchronous-Configuration: The controller uses this message to set an additional filter on the Asynchronous messages that it wants to receive on its OpenFlow channel, or to query that filter.

2. Asynchronous message

- Asynchronous: An OpenFlow switch initiates this message to denote a packet arrival or switch state change. Asynchronous messages are classified into the following types:

- Packet-in: Upon receipt of a data packet, an OpenFlow switch sends a Packet-in message to the controller and sends the data packet to the controller if no matching entry exists in the flow table.

- Flow-Removed: This message is triggered when a flow entry on an OpenFlow switch is deleted due to a timeout, modification, or another reason, or when a response to a flow entry deletion request from the controller is received.

- Port-Status: This message is sent when the port status of an OpenFlow switch changes (for example, the port is Down).

- Role-Status: When an OpenFlow switch is managed by another controller, the switch sends a Role-Status message to the original controller.

- Controller-Status: An OpenFlow switch sends a Controller-Status message to report the status change of an OpenFlow channel to all controllers.

- Flow-monitor: An OpenFlow switch sends a Flow-Monitor message to notify the controller of changes in the flow table.

3. Symmetric message

Symmetric messages can be initiated by the controller and OpenFlow switch. Symmetric messages fall into the following types:

- Hello: Hello messages are exchanged between the OpenFlow switch and controller upon connection startup.

- Echo: Echo messages can be sent from either the OpenFlow switch or the controller. They can be used to measure the latency or determine the liveness of a controller-switch connection.

- Error: This message is sent to report an error to the remote end.

- Experimenter: Experimenter messages provide a standard way for OpenFlow switches to offer additional functionality.

4.3.2 NETCONF

4.3.2.1 Introduction to NETCONF

NETCONF provides a mechanism to install, maintain, and delete configurations of network devices. You can use NETCONF to obtain configurations and status of the network devices. NETCONF enables network devices to provide a set of standard APIs. Applications can use these APIs to deliver configurations to network devices and obtain them.

NETCONF is the basic module of the automatic configuration system. XML is a universal language for NETCONF communications and provides a flexible and comprehensive encoding mechanism for hierarchical data contents.

NETCONF can be used together with XML-based data conversion technologies such as Extensible Stylesheet Language Transformations (XSLT) to provide a tool for automatic generation of configuration data. This tool can query various configuration data from one or more databases and convert the data into a specified configuration data format by using the XSLT script based on the requirements of different application scenarios. The configuration data is then uploaded to devices through NETCONF.

NETCONF classifies data on network devices into configuration and status data. Configuration data can be modified. NETCONF operations can be performed to change the status of network devices to the expected status. The status data cannot be modified, and mainly includes the running status and statistics about network devices. NETCONF distinguishes configuration data from status data, reducing the size of configuration data and facilitating management of configuration data.

4.3.2.2 NETCONF Network Architecture

Figure 4.19 illustrates the typical NETCONF network architecture. The architecture requires at least one network management system (NMS) running on a server to manage devices. The NETCONF network architecture consists of the following components:

1. NETCONF client

    The NETCONF client uses NETCONF to manage network devices.

    • The client sends <rpc> elements to a NETCONF server to query or modify configuration data.

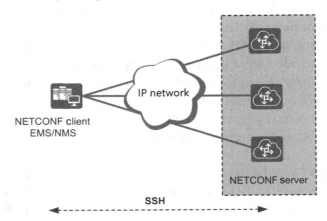

FIGURE 4.19   NETCONF network architecture.

- The client learns the status of a managed device based on the generated alarms and events reported by the NETCONF server of the managed device.

2. NETCONF server

A server maintains the configuration data of managed devices, responds to the <rpc> elements sent by clients, and sends requested management data to the clients.

- After receiving a request from a client, the server parses the request, processes it based on the configuration management framework (CMF), and then returns a response to the client.

- If an alarm is generated or an event occurs on a managed device, the NETCONF server of the managed device reports the alarm or event to a NETCONF client to notify the client of the status of the managed device.

A NETCONF session is a logical connection between a client and a server. A network device must support at least one NETCONF session. The data that a NETCONF client obtains from a NETCONF server includes configuration and status data.

- The NETCONF client can modify and operate the configuration data so that the status of the NETCONF server migrates to a user-expected status.

- The NETCONF client cannot modify status data. Status data includes the running status of the NETCONF server and other statistics.

4.3.2.3 NETCONF Framework

Similar to the International Organization for Standardization (ISO) or OSI model, the NETCONF framework uses a hierarchical structure. A lower layer provides services for the upper layer. The hierarchical structure enables each layer to focus only on a single aspect of NETCONF and reduces the dependencies on different layers.

NETCONF is partitioned into four layers, as illustrated in the following table.

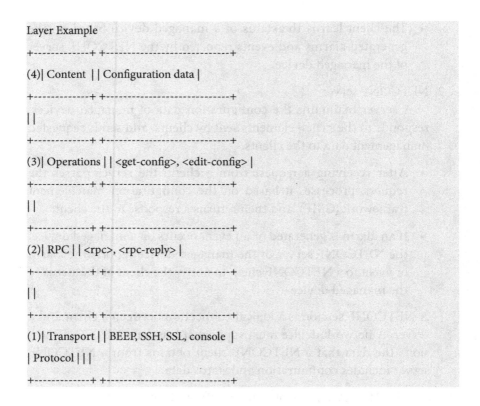

Table 4.3 describes the NETCONF framework.

4.3.2.4 NETCONF Capabilities

A NETCONF capability is a set of functionality that supplements basic NETCONF functionality. A network device can add protocol operations through the capability set to extend the operation scope of existing configuration objects.

Each capability is identified by a unique Uniform Resource Identifier (URI). The URI format is as follows:

urn:ietf:params:xml:ns:netconf:capability:{name}:{version}

In the preceding URI, the name refers to the capability name, and the version refers to the capability version.

The capability definition may name on one or more dependent capabilities. The NETCONF server must support any capabilities upon which it depends. In addition, NETCONF provides semantic specifications for defining capabilities. Device vendors can define proprietary capabilities as required.

TABLE 4.3   NETCONF Framework

(i) Layer	Example	(j) Description
(k) Transport layer	BEEP, SSH, SSL, console	The transport layer provides a communication path for interaction between the NETCONF client and server NETCONF can be carried on any transport protocol that meets all of the following requirements: • The transport protocol is connection-oriented. A permanent link is established between the NETCONF client and server, and then, data is transmitted reliably and sequentially • NETCONF user authentication, data integrity, and security depend on the transport layer • The transport protocol provides a mechanism to distinguish whether the session type is client or server for NETCONF
RPC layer	<rpc>, <rpc-reply>	The RPC layer provides a simple RPC request and response mechanism independent of transport protocols. The client uses the <rpc> element to encapsulate RPC request information and sends the RPC request information to the server. The server uses the <rpc-reply> element to encapsulate RPC response information (content at the operation and content layers) and sends the RPC response information to the client
Operations layer	<get-config>, <edit-config>	The operations layer defines a series of basic operations used in RPC, which constitute the basic capabilities of NETCONF
Content layer	Configuration	The content layer consists of configuration data involved in network management. The configuration data depends on vendors' devices All the layers except the content layer have been standardized for NETCONF. The content layer has no standard NETCONF data modeling language or data model. Common NETCONF data modeling languages include Schema and YANG. YANG is a widely used data modeling language designed for NETCONF

The NETCONF client and server exchange capabilities to notify each other of their supported capabilities. The client can send operation requests only within the capabilities supported by the server.

4.3.2.5 NETCONF Configuration Datastore

A configuration datastore is a collection of complete configuration parameters for a device. Table 4.4 describes NETCONF-defined configuration datastores.

TABLE 4.4   NETCONF-Defined Configuration Datastores

(I) Configuration Datastore	Description
(m) <running/>	Stores the effective configuration, status information, and statistics on the current device Unless the NETCONF server supports the candidate capability, this configuration datastore is the only mandatory standard datastore To support modification of the <running/> configuration datastore, a device must have the writable-running capability
<candidate/>	Stores the configuration data to be run by a device An administrator can perform operations on the <candidate/> configuration datastore. Any change to the <candidate/> datastore does not affect the current device To support the <candidate/> configuration datastore, the current device must have the candidate capability
<startup/>	Stores the configuration data loaded for device startup, which is similar to the saved configuration file. To support the <startup/> configuration datastore, the current device must have the Distinct Startup capability

4.3.2.6 XML Encoding

The NETCONF client and server communicate through the RPC mechanism. They must establish a secure and connection-oriented session for communication. The client sends an RPC request message to the server. After processing the request message, the server sends a response to the client. The RPC request from the client and the response from the server are encoded in XML format.

XML is a NETCONF encoding format, allowing complex hierarchical data to be expressed in a text format that can be read, saved, and manipulated with both traditional text tools and tools specific to XML.

XML-based network management uses XML's powerful data presentation capabilities to describe managed data and management operations so that computers can easily parse management information. XML-based network management helps computers efficiently process network management data, improving network management capabilities.

The XML encoding format file header is **<?xml version= "1.0" encoding= "UTF-8" ?>**, where

- **<?** is the start of an instruction.

- **xml** identifies an XML file.

- **version** indicates the NETCONF version. "1.0" indicates that the XML1.0 standard version is used.

- **encoding** is the character set encoding format. Only UTF-8 encoding is supported.

- **?>:** is the end of an instruction.

4.3.2.7 RPC Mode

NETCONF uses an RPC-based communication model. NETCONF uses XML-encoded <rpc> and <rpc-reply> elements to provide the transport-protocol-independent framework of NETCONF request and response messages. Table 4.5 covers some commonly used RPC elements.

4.3.3 OVSDB

Network devices at the SDN infrastructure layer may have multiple implementation forms, such as hardware and software. With the improvement of the performance of common processors, software-based network devices can meet network transmission requirements in multiple scenarios. In particular, software-based network devices have higher flexibility and better integration with virtualization software, to enable software-based switches to operate well in the SDN field.

An open vSwitch (OVS) is a software-based open-source virtual switch. The OVS complies with Apache 2.0 and supports multiple standard management interfaces and protocols, such as NetFlow, sFlow, Switched Port

TABLE 4.5   RPC Elements

(n) Element	Description
(o) <rpc>	Encapsulates a request that the client sends to the server
<rpc-reply>	Encapsulates a response message for an <rpc> request message. The server returns a response message, which is encapsulated in the <rpc-reply>element, for each <rpc> request message
<rpc-error>	Notifies a client of an error occurring during <rpc> request processing. The server encapsulates the <rpc-error> element in the <rpc-reply> element and sends the <rpc-reply> element to the client
<ok>	Notifies a client that no errors occurred during <rpc> request processing. The server encapsulates the <ok> element in the <rpc-reply> element and sends the <rpc-reply> element to the client

Analyzer (SPAN), Remote SPAN (RSPAN), command line interface (CLI), LACP, and 802.1ag. It can be deployed across physical servers. The Open vSwitch Database (OVSDB) is a lightweight database developed for the OVS. It stores various configuration information, such as bridges and interfaces, on the OVS.

An OVS is divided into the user and kernel spaces. The OVS has a plurality of components in the user space, and the components are mainly responsible for implementing data exchange and OpenFlow flow table functions. Therefore, they are the core of the OVS. The OVS also provides tools for switch management, database setup, and interaction with kernel components. Figure 4.20 illustrates the position of the OVSDB in the OVS. The main components are as follows.

- ovs-vswitchd: indicates a daemon that works with the Linux kernel-compatible module to implement flow-based switching.

- ovsdb-server: indicates the OVSDB server, which provides lightweight database services and stores OVS configuration information, including interfaces, switching content, and VLANs. The OVS works based on the configuration information in the database.

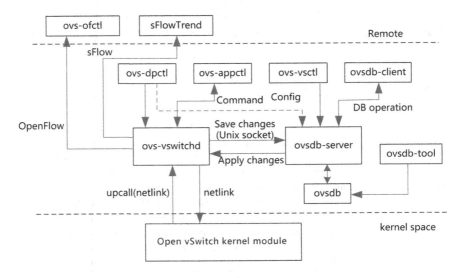

FIGURE 4.20   OVS core components and their relationships.

- ovsdb-client: indicates the OVSDB client, which is used to configure the server.

- ovs-dpctl: is a tool used to configure the kernel module of the OVS to control forwarding rules.

- ovs-vsctl: obtains or modifies the ovs-vswitchd configuration. This tool updates the database of the OVSDB server.

- ovs-appctl: sends commands to ovs-vswitchd.

- ovs-ofctl: controls the flow table content when the OVS functions as an OpenFlow switch.

The OVSDB consists of ovsdb-server and ovsdb-client. ovsdb-server is located on the OVS. ovsdb-client sends database configuration and query commands (ovs-vsctl commands) to ovsdb-server through the OVSDB management protocol.

The OVSDB is a lightweight database. It is a JSON file that records network configurations such as bridges, interfaces, and QoS. The ovsdb/spe file defines OVSDB table specifications. Before creating a database, you need to prepare a schema file. The file is a character string in JSON format and defines the database name, and all tables contained in the database. Each table contains a JSON array of columns. A DB file is created using this schema file.

4.3.4 YANG

YANG is a data modeling language used to model the operations and content layers of NETCONF, as well as configuration and status data manipulated by NETCONF, RPC, and NETCONF notifications.

YANG is proposed by the NETMOD working group and released in IETF RFC 6020. This language is a modular language between UML (advanced) and implementation, and is similar to the ASN.1 language in that it describes any object in a tree. This is similar to the Management Information Base (MIB) of SNMP (MIB is described in ASN.1). However, YANG is more flexible than SNMP (SNMP defines the entire tree hierarchy too rigidly, so its application scope is limited). YANG claims that it is compatible with SNMP. It defines the following four types of nodes to model configuration data and state: leaf nodes, leaf-list nodes, container nodes, and list nodes.

4.3.4.1 Function Description

YANG is a language used to model data for the NETCONF protocol. A YANG model defines a data hierarchy that can be used for NETCONF-based operations, including configuration, state data, RPCs, and notifications. This allows a complete description of all data transmitted between the NETCONF client and server.

YANG models the hierarchical organization of data as a tree in which each node has a name, and either a value or a set of child nodes. YANG provides clear and concise descriptions of the nodes, as well as the interaction between those nodes.

YANG structures data models into modules and submodules. A module can import data from other external modules. The hierarchy can be augmented, allowing one module to add data nodes to the hierarchy defined in another module. This augmentation can be conditional, with new nodes appearing only if certain conditions are met.

YANG models can describe constraints to be enforced on data, restricting the appearance or value of nodes based on the presence or value of other nodes in the hierarchy. These constraints can be executed by either the client or the server, and valid content must comply with them.

YANG defines a set of built-in types and has a type mechanism that defines additional types. Derivative types can restrict the basic type set of valid values. They use the constraint mechanism similar to range or pattern. These constraints can be executed by the client and server.

YANG allows reusable node groups (grouping). The instantiation of these groups can refine or extend nodes, which can be tailored to specific needs. Derivative types and groups (grouping) may be defined in one module or submodule, and used in the module in which they are located or another module or submodule that introduces or contains the module.

The YANG data hierarchy contains the definition of a list, in which list entries are identified by keywords, which distinguish list entries. Such lists can be sorted by users or automatically by the system. The user sorting list defines the operations on the order of the items in the operation list.

YANG modules can be translated into an equivalent XML syntax called YANG Independent Notation (YIN), allowing applications using XML parsers and XSLT scripts to operate on the models. The conversion from YANG to YIN is lossless. Therefore, the content of YIN can be rolled back to YANG.

YANG fairly processes the relationship between high-level data modeling and low-level bit stream encoding. You can view the advanced view of the data model of the YANG module and understand how data is encoded into NETCONF operations.

YANG is an extensible language that can be extended by standard organizations, vendors, and individuals. The declaration syntax allows these extensions to coexist with the standard YANG declaration in a natural manner, and the extensions in the YANG module are prominent to be noticed by readers.

YANG maintains compatibility with SNMP SMIv2 for scalability. SMIv2-based MIB modules can be automatically converted to YANG modules that allow read-only access. However, YANG does not care about reverse YANG-SMIv2 conversion.

Similar to NETCONF, YANG aims to smoothly integrate the local management architecture of devices. This protects or exposes elements in your data model using its existing access control mechanisms.

4.3.4.2 YANG Development

The current YANG version is 1.0. Currently, the IETF NETMOD working group is designing YANG1.1. IETF working groups are also proactively promoting the standardization of the YANG model. Currently, the IETF has released some standard YANG models (IP, interface, system management, and SNMP configuration), and a large number of drafts are being evolved.

RFC 6991: Common YANG Data Types

RFC 7223: A YANG Data Model for Interface Management RFC 7224: IANA Interface Type YANG Module

RFC 7277: A YANG Data Model for IP Management

RFC 7317: A YANG Data Model for System Management RFC 7407: A YANG Data Model for SNMP Configuration

The IETF and ONF have started to require the submission of models written in YANG. Especially for IETF, all model drafts are written using YANG.

More and more vendors require YANG adaptation. Products from vendors such as Huawei, Cisco, and Juniper have provided NETCONF and YANG functions. Currently, Huawei's unified controller supports the YANG model, which promotes development. In addition, more and more YANG-based tools are launched in the industry, such as YangTools (with new functions added) of OpenDaylight (ODL), pyang of Google, and Yang Designer.

CHAPTER 5

Constructing a Physical Network (Underlay Network) on a DCN

THIS CHAPTER DESCRIBES THE key points in the physical network design of a DCN. A DCN's physical network, also known as the underlay network, usually adopts the spine-leaf architecture, in which leaf nodes are classified into server leaf nodes, service leaf nodes, and border leaf nodes. The chapter will elaborate on protocols used on such a network, including OSPF or External Border Gateway Protocol (EBGP). It will also detail how servers can connect to server leaf nodes in M-LAG, stack, or standalone mode. Finally, the chapter will outline how service leaf nodes and border leaf nodes can be deployed independently or combined, as well as how border leaf nodes can be connected to external PEs in multiple networking modes or through route advertisement.

5.1 PHYSICAL NETWORK AND NETWORK INFRASTRUCTURE

DCNs typically adopt the spine-leaf architecture for their physical network. Table 5.1 describes the roles and their functions in physical networking of the cloud DCN solution, while Figure 5.1 illustrates the recommended networking mode in the industry.

TABLE 5.1   Roles and their Functions in Physical Networking

(a) Role	Function
(b) Fabric	Network failure domain that is managed by an SDN controller. It contains one or more spine-leaf architectures
Spine	Core node on a VXLAN fabric network. It provides high-speed IP forwarding and connects to leaf nodes through high-speed interfaces
Leaf	Access node on a VXLAN fabric network. It connects various network devices to the VXLAN fabric network
Service leaf	Functional node that connects VAS devices, such as firewalls and LBs, to a VXLAN fabric network
Server leaf	Functional node that connects virtual and physical servers to a VXLAN fabric network
Border leaf	Functional node that connects to routers or transmission devices outside a DC to forward traffic from an external network to a VXLAN fabric network in a DC

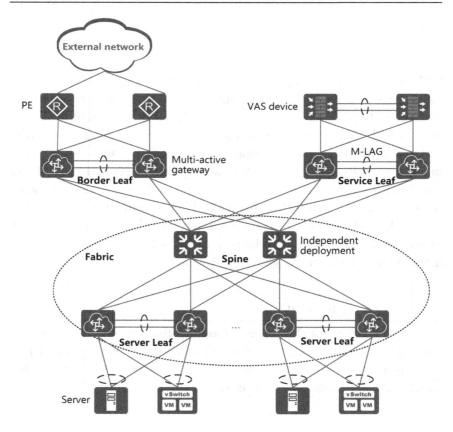

FIGURE 5.1   Recommended physical networking.

A well-designed fabric network can provide consistent access across access nodes. Fabric networks, which feature high bandwidth, large capacity, and low network latency, contain one or more spine-leaf architectures. And such an architecture contains three types of leaf nodes — server leaf nodes, service leaf nodes, and border leaf nodes — which are essentially the same at the forwarding plane. Where they differ is in terms of access devices. Because the spine-leaf architecture is used, the network is flattened, ensuring the east-west traffic forwarding path across the entire network is short and the forwarding efficiency high.

Another key advantage of the fabric network is that it can be elastically scaled. That is, when the number of servers increases, you only need to add leaf nodes. Then, if the spine forwarding bandwidth is insufficient for the increasing number of leaf nodes, you can add spine nodes.

For the spine-leaf architecture, the recommended configuration of spine and leaf nodes is as follows:

- Spine node: Spine nodes forward traffic between leaf nodes at a high speed. It is recommended that spine nodes be deployed independently, with the specific number of spine nodes depending on the oversubscription ratio of leaf nodes. While different industries and customers have varying requirements on the oversubscription ratio, an oversubscription ratio of 1:9 to 1:2 is generally used.

- Leaf node: A leaf node connects to service servers and VAS servers, and functions as a north-south gateway. While leaf nodes can be deployed flexibly, M-LAG active-active is the recommended deployment mode. If the requirements for reliability and packet loss are not high, virtual chassis technologies such as CSS/iStack can also be used. Each leaf node is connected to all spine nodes, forming a full-mesh topology.

Leaf nodes and spine nodes are connected through Layer 3 routed interfaces, and they communicate at Layer 3 by configuring a dynamic routing protocol. OSPF or BGP is recommended. For details about routing protocol selection, see the following sections in this chapter.

ECMP is recommended for implementing load balancing and link backup, as shown in Figure 5.2. In this case, leaf nodes forward data traffic to spine nodes through multiple ECMP paths, guaranteeing reliability

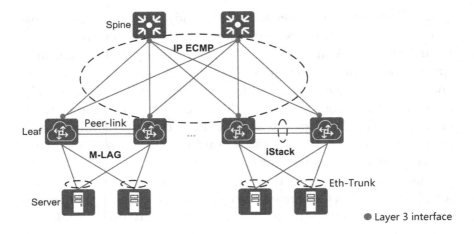

FIGURE 5.2   Using ECMP on the fabric network.

while ensuring that network bandwidth improves. It is important to be aware that ECMP links need to use the load balancing algorithm based on the Layer 4 source port number at the transport layer. Because VXLAN uses User Datagram Protocol (UDP) encapsulation, the destination port number of a VXLAN packet is always 4789, while its source port number is variable.

5.2 PHYSICAL NETWORK DESIGN ON A DCN

The last section of this chapter outlines the basic concepts about the physical network of a DCN. Designing a physical network is the first step in DCN design. You should now have a grasp of the basics of physical networking in the cloud DCN solution and be ready to start to design the physical network.

This section covers the following key points in the physical network design, which can help you design a DCN:

• Routing protocol selection

• Server access mode selection

• Design and principles of border and service leaf nodes

• Egress network design

5.2.1 Routing Protocol Selection

In most cases, either OSPF or EBGP can be used on an underlay network. OSPF is preferred for most networks. If the scale of the network is large, EBGP is recommended as the underlay network needs to be partitioned into areas, and flexible control of BGP is required.

1. OSPF deployment on the underlay network

    When the number of leaf nodes is less than 100, OSPF is recommended on the underlay network. The route planning is as follows:

    On a single fabric network, OSPF is deployed on all physical switches where spine and leaf nodes are configured. Only OSPF area 0, in which all physical switches are deployed, is planned. OSPF neighbor relationships are established using addresses of Layer 3 routed interfaces to implement connectivity on the underlay network. It is recommended that the network type be P2P, as shown in Figure 5.3.

    In a scenario where multiple fabric networks are deployed and constitute a VXLAN domain on the overlay network (that is, a DCN is divided into two fabric networks and is managed through a set of management interfaces), it is recommended that only one OSPF process be deployed on all devices, interconnected devices between

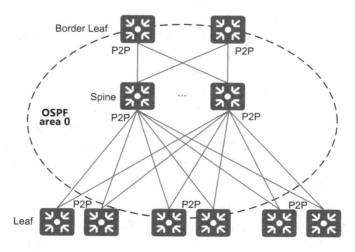

FIGURE 5.3   Recommended OSPF planning for a single fabric network.

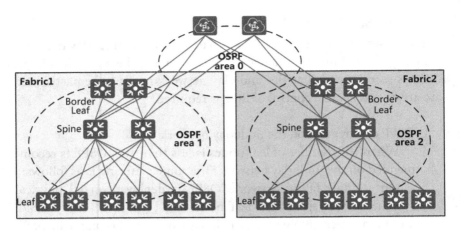

FIGURE 5.4 Recommended OSPF planning for multi-fabric deployment (single VXLAN domain).

multiple fabric networks be deployed in OSPF area 0, and Fabric1 and Fabric2 be deployed in OSPF areas 1 and 2 respectively. In this case, there is only one OSPF process, and connectivity is achieved on the underlay network, as shown in Figure 5.4.

If multiple fabric networks form two VXLAN domains on the overlay network (that is, two DCNs are managed through two sets of management interfaces, but they need to be interconnected), it is recommended that OSPF be deployed on each fabric network, and interconnected devices between fabric networks exchange routes through BGP.

OSPF is easy to deploy and provides fast convergence for smaller-scale networks; therefore, it is preferred on the underlay network of a small- or medium-sized DCN. Most enterprises deploy BGP EVPN on the control plane of the overlay network. When OSPF is selected for the underlay network, OSPF and BGP packets are placed in different queues, and VRFs and routing entries are isolated from each other, which, in turn, keeps fault domains on the underlay and overlay networks isolated.

2. EBGP deployment on the underlay network

When there are more than 100 leaf nodes on a large-scale network, EBGP is recommended on the underlay network as using OSPF would slow down protocol and fault convergence. The route planning is as follows.

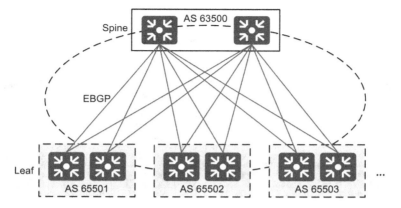

FIGURE 5.5   Recommended EBGP planning for a single fabric network.

On a single fabric network, spine nodes are added to the same autonomous system (AS), and each group of leaf nodes is added to an AS. EBGP peer relationships are established between leaf nodes and all spine nodes. As shown in Figure 5.5, EBGP is used to implement Layer 3 interconnection on the entire network.

In a scenario where multiple fabric networks are deployed, the processing method is similar to that for a single fabric network. All spine nodes are deployed in the same AS, as are each group of leaf nodes, and EBGP runs between spine and leaf nodes. Each fabric network is connected to the peer AS through interconnected leaf nodes on the DCN. The interconnected leaf nodes are deployed in an AS and establish peer relationships with spine nodes through EBGP, as shown in Figure 5.6.

The EBGP configuration on the underlay network is complex as groups of leaf nodes, and spine nodes need to be allocated to different ASs, and full-mesh EBGP connections need to be established between spine and leaf nodes. What's more, TCP-based BGP connections need to be manually specified, which requires heavy configuration workload and the maintenance is complex. However, OSPF processes only need to be enabled on corresponding interfaces when these OSPF processes are being used.

BGP provides an independent routing domain for each area. The fault domain of BGP is smaller than that of OSPF. In addition, BGP provides various route control methods, which increases flexibility

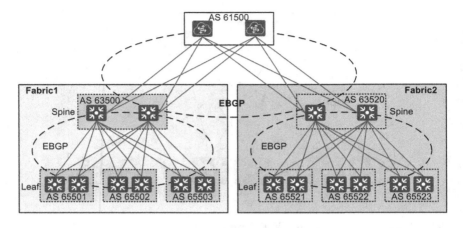

FIGURE 5.6   Recommended EBGP planning for multiple fabric networks.

and facilitates expansion of route control compared with OSPF during network deployment. You can select either of the two protocols to deploy a DCN based on the actual situation.

5.2.2 Server Access Mode Selection

Typically, the server access mode is selected based on leaf node model selection and deployment mode.

- To ensure reliability, leaf nodes usually use the virtual chassis technology, such as iStack, or inter-device link aggregation technology, such as M-LAG, to connect to service servers, which are dual-homed to leaf nodes. The preceding section compares iStack and M-LAG. M-LAG is recommended because its loose coupling of the control plane facilitates upgrade and increases reliability.

- Server leaf nodes are selected based on the server access bandwidth and the oversubscription ratio of leaf nodes. Hardware models of server leaf nodes are selected mainly based on the access bandwidth, oversubscription ratio, and special service requirements. Server access bandwidth shows the rate of an interface through which a server is connected. Specifically, 10GE or 25GE is often used, whereas GE access and 100GE access are less frequently used. The oversubscription ratio can be designed based on customer requirements and is usually 1:9 to 1:2. However, it is not only the requirements of

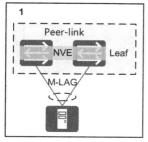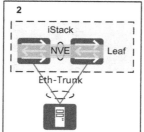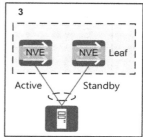

FIGURE 5.7   Three server access modes.

customers that need to be considered, but also those of some special services. For example, if servers require IPv4/IPv6 dual-stack access or support of chip-based features such as microsegmentation and AI Fabric, you need to select the appropriate hardware models based on the specific requirements.

There are three server access modes:

- (Recommended) Use an Eth-Trunk to connect a server to the M-LAG composed of leaf nodes. See No. 1 in Figure 5.7.

- Use an Eth-Trunk to connect a server to the iStack system composed of server leaf nodes. See No. 2 in Figure 5.7.

- Connect a server to server leaf nodes in active/standby mode. See No. 3 in Figure 5.7.

Table 5.2 compares the preceding three access modes.

5.2.3 Design and Principles of Border and Service Leaf Nodes

Border and service leaf nodes are not used to connect to service servers. Rather, border leaf nodes function as north-south gateways of a DCN to send north-south traffic to the peer PEs and receive traffic sent to the DCN from PEs, whereas service leaf nodes are used to connect to VAS devices, such as firewalls and LBs.

The forwarding model and deployment mode of border and service leaf nodes are similar to those of server leaf nodes connected to service servers. Therefore, M-LAG or iStack is used to ensure reliability.

TABLE 5.2 Comparison between the Server Access Modes

(c) Access Mode	Characteristics	(d) Manageability	(e) Reliability	(f) Cost
(Recommended) Using an Eth-Trunk to connect a server to the M-LAG composed of leaf nodes	Two leaf nodes are connected through a peer-link and form a DFS group. They function as a single logical device while having their own independent control planes. A server is connected to two leaf nodes in load balancing mode Device upgrade and maintenance are simple to implement, and the running reliability is high. The downlink interfaces of leaf nodes are configured with M-LAG, and the server is dual-homed to the leaf nodes. Dual NICs of the server work in active/standby or load balancing mode. The configuration, however, is complex because each server leaf node has an independent control plane	High	High	Medium
Using an Eth-Trunk to connect a server to the iStack system composed of server leaf nodes	iStack technology virtualizes two server leaf nodes into one logical switch with a single control plane, simplifying device management. Dual NICs of the server bond together and work in either active/standby or load balancing mode to improve bandwidth efficiency. However, the upgrade and maintenance operations for the logical device are complex	Low	Medium	Medium
Connecting a server to server leaf nodes in active/standby mode	Two leaf nodes are deployed independently, and dual NICs of the server are connected to the two leaf nodes in active/standby mode. Only one of the NICs sends and receives packets simultaneously; therefore, the bandwidth efficiency is low. The VTEP IP address of received traffic will change upon an active/standby switchover of NICs, upon which the nearby server sends gratuitous ARP packets to enable traffic to be transmitted through the new VTEP	Medium	High	Medium

Hardware models of border and service leaf nodes are also selected based on the bandwidth, oversubscription ratio, and special services. The following section compares the two leaf nodes.

The following factors also need to be taken into account during the design of border and service leaf nodes:

- Border leaf nodes can connect to PEs through one, two, or four Layer 3 interfaces. Select the appropriate interconnection mode based on the PE model and capability. It is recommended that active-active border leaf nodes and static routes or BGP ECMP be deployed. For details about the interconnection between border leaf nodes and PEs, see Section 5.2.4.

- Border and service leaf nodes can be combined (leaf nodes are used as both service leaf nodes and border leaf nodes) or separately deployed. Co-deployment or separate deployment is determined based on the number of services and routes on the network.

- There may be more than one group of service leaf nodes, the number of which is determined based on the requirements of VAS devices.

The networking also involves co-deployment or separate deployment.

1. Using one device as both the service and border leaf node

   Figure 5.8 shows the scenario where service and border leaf nodes are deployed on the same group of devices. VAS devices such as L4–L7 devices are connected to service or border leaf nodes in bypass mode and are dual-homed to two switches. The topology in which firewalls are connected to gateways is simple, which, in turn, simplifies configuration and deployment. Firewalls can be added accordingly as gateways are added.

   In this scenario, north-south traffic in a DC is transmitted along the path: server -> server leaf -> spine -> service leaf (border leaf) -> VAS -> service leaf (border leaf) -> PE -> external network. When service traffic reaches the spine node, the traffic is redirected to the VAS device. After being processed by the VAS device, the traffic is sent back to the spine node and finally sent to an external network outside the DC. Traffic from an external network to the DC is transmitted along the same path but in an opposite direction.

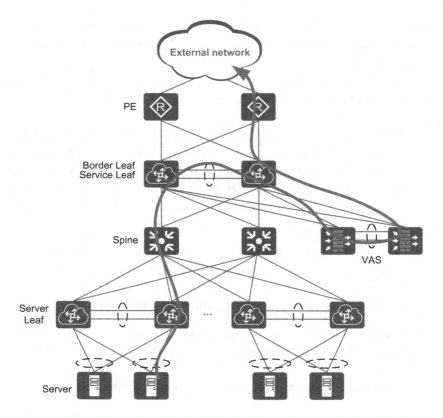

FIGURE 5.8   Using one device as both the border and service leaf node.

This networking mode requires fewer devices and lower network-ing cost. The traffic model is simple, and it is easy to maintain devices. However, its scalability is limited by the specifications of border and service leaf nodes; therefore, this mode is more applicable to small- and medium-sized DCs that focus on easy deployment, lower costs, and higher maintainability and do not have high requirements on network scalability.

2. Using different devices as service and border leaf nodes

Figure 5.9 shows the scenario where different devices are used as service and border leaf nodes, and they can be expanded independently.

In this scenario, access traffic in a DC is transmitted along the path: server -> server leaf -> spine -> service leaf -> VAS -> service

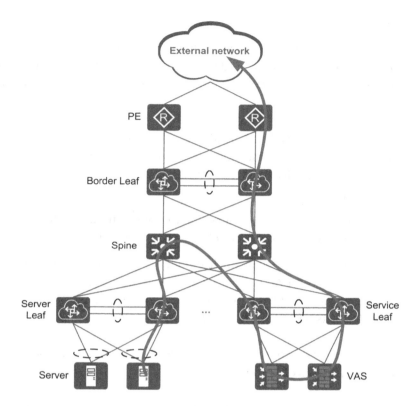

FIGURE 5.9   Using different devices as border and service leaf nodes.

leaf -> spine -> border leaf -> PE -> external network. When service traffic reaches the spine node, it is sent to the independently deployed service leaf node. The service leaf node sends the traffic to the VAS device for processing, then to the spine node, and finally to the external network outside the DC. The traffic from the external network to the internal network of the DC is processed in the same way.

This networking mode involves a large number of devices, high networking cost, and a complex traffic model and maintenance. However, this mode has high scalability and a flexible architecture, which are not supported by the co-deployment mode. The number of border and service leaf nodes can be flexibly adjusted based on the service scale. This mode is applicable to enterprises with strong technical capabilities.

5.2.4 Egress Network Design

The egress network design of a DC includes connections and configurations between border leaf nodes and PEs (egress routers). Border leaf nodes can be connected to PEs in various modes. Take the following factors into account during design:

- PEs can be deployed independently. However, it is recommended that PEs be deployed in stacking mode or by using mechanisms such as E-Trunk and VRRP to ensure reliability.

- PEs and border leaf nodes can be connected through a single Layer 3 interface, two Layer 3 interfaces (square-looped mode), or four Layer 3 interfaces (dual-homed mode), as shown in Figure 5.10.

- It is recommended that border leaf nodes be deployed in active-active mode (such as M-LAG) or stacking mode. Border leaf nodes and PEs can be connected through Layer 3 routed interfaces or virtual interfaces (such as VBDIF and VLANIF interfaces).

  You can select one of the preceding interconnection modes based on the actual status of border devices.

- For each VRF, determine the number of Layer 3 interfaces of PEs connected to border leaf nodes in a DC based on PEs. One, two, or

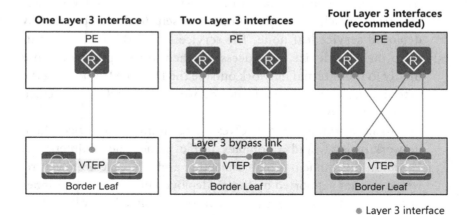

FIGURE 5.10   Three logical interconnection modes of an egress network.

four Layer 3 interfaces can be used to connect to border leaf nodes; however, four interfaces are recommended.

- Based on the specifications of physical devices, you can select the dual-homed mode (recommended mode, in which two PEs require at least four interfaces), or you can select the square-looped mode (two PEs require at least two interfaces) or direct connection mode (one interface), depending on the number of Layer 3 interfaces provided by the customer's existing border routers.

If PEs are connected to border leaf nodes through a single interface, they can be connected to border leaf nodes through stacking, inter-device link aggregation technology, or VRRP, depending on the capabilities of PEs.

If two interfaces are used for interconnection, PEs are deployed independently and form a square-looped network with active-active border leaf nodes. Two independent Layer 3 interfaces are configured on PEs, in which case, a bypass link needs to be deployed between border leaf nodes. Therefore, if an interconnection link between border leaf nodes fails, traffic can be forwarded through the bypass link.

If four interfaces are used for interconnection, PEs and border leaf nodes subsequently have four interconnection links to form a dual-homed network. The four independent interfaces are configured on PEs and support each other. Therefore, the bypass link between border leaf nodes is not mandatory. However, to ensure network robustness, you are advised to add the bypass link when there are sufficient interfaces on border leaf nodes.

The following describes route planning on the egress network.

Static or dynamic routes can be deployed between border leaf nodes and PEs. Generally, when border leaf nodes and PEs are connected through a single Layer 3 interface, you can configure static routes pointing to their peer ends on border leaf nodes and PEs to implement communication between border leaf nodes and PEs. Interconnection in square-looped networking (two Layer 3 interfaces) or dual-homed networking (four Layer 3 interfaces) is more complex. Therefore, you need to take into account the selection of static routes and dynamic routes and the route planning of the bypass link during interconnection.

The dual-homed networking is used as an example in the following section to describe route planning on the egress network.

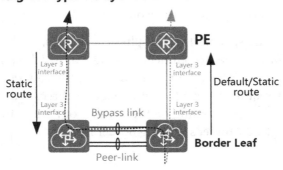

FIGURE 5.11   Static route planning in square-looped networking.

Figure 5.11 illustrates interconnection using static routes in square-looped networking. Two border leaf nodes are connected to PEs through Layer 3 interfaces, and an Eth-Trunk link is deployed between border leaf nodes as the bypass link. If border leaf nodes are configured with M-LAG, it is recommended that the bypass link and the M-LAG peer-link be deployed separately and different Eth-Trunk links are used for the former two links.

Configure a default static route on each border leaf node. The next hop of the route is the IP address of the Layer 3 interface on the peer PE. Configure a specific static route on each PE. The destination address of the route is the intranet service network segment, and the next hop of the route is the IP address of the Layer 3 interface on the peer border leaf node. It is recommended that static routes from PEs to the intranet be summarized to reduce the number of routes and facilitate O&M.

For the bypass link, you can configure a low-priority static default route on each border leaf node. The next hop of the route is the Layer 3 interconnection interface of the peer border leaf node. This way, when the static route to the PE becomes invalid, the default route with a lower priority becomes available to enable traffic to be forwarded through the bypass link.

Figure 5.12 illustrates interconnection using dynamic routes in square-looped networking. Similar to using static routes, two border leaf nodes are connected to PEs through Layer 3 interfaces, and an Eth-Trunk link is deployed between border leaf nodes as the bypass link. If border leaf nodes are configured with M-LAG, it is recommended that the bypass link

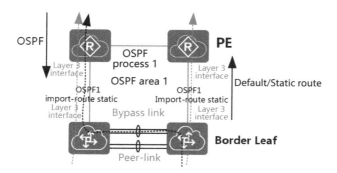

FIGURE 5.12   Dynamic route planning in square-looped networking.

and the M-LAG peer-link be deployed separately and different Eth-Trunk links are used for the former two links.

In the interconnection solution using dynamic routes, it is recommended that northbound traffic from the intranet to the extranet be transmitted to a PE through a static default route. A low-priority static default route to the peer border leaf node is configured on the border leaf node as the route of the bypass link.

OSPF is used for transmitting traffic from a PE to an intranet network segment. OSPF neighbor relationships are established between border leaf nodes and PEs, as well as between PEs through a Layer 3 link, which can be used as the bypass link. In addition, static routes are imported to OSPF on border leaf nodes. This is because border leaf nodes have static routes to the intranet network segment. After these static routes are sent to a PE through OSPF, the PE can access the intranet through these routes.

CHAPTER 6

Constructing a Logical Network (Overlay Network) in a DC

T HIS CHAPTER WILL ELABORATE on basic concepts of the most widely used VXLAN protocol in NVO3. The chapter will additionally focus on describing overlay network classification, VXLAN control plane protocol EVPN, and forwarding process of various data packets. An overlay network is a logical network constructed on an underlay network in order to meet requirements of a large Layer 2 network in a DC. Currently, NVO3 is the mainstream technology used to build an overlay network. Furthermore, this chapter will aim to describe both the construction of logical models by the controller during VXLAN configuration and the positions on the network.

6.1 OVERLAY NETWORK

In Figure 6.1, the overlay network is the software-defined logical network that is built over the existing underlay network, implementing Layer 2 communication between large-scale VMs on a DCN.

The overlay network is completely decoupled from the underlay network. Furthermore, the overlay network is a virtualized, logical network able to adapt to a variety of applications. In this manner, the underlay network can be flexibly expanded, and additionally, IP addresses are not

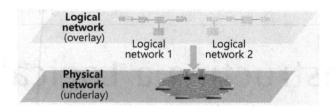

FIGURE 6.1   Overlay network.

bound to locations, resulting in flexible service deployment. This decoupling architecture facilitates SDN architecture deployment. The SDN controller is not required to consider the underlay network architecture and can flexibly deploy services on the overlay network.

The technology used to construct an overlay network is called the overlay technology, that is, NVO3 technology mentioned in Chapter 3. Overlay technology is a tunnel encapsulation technology that encapsulates Layer 2 packets over tunnels and transparently transmits the encapsulated packets. The technology finally decapsulates the packets to obtain the raw packets after the packets arrive at the destination. This is how a large Layer 2 network is built on the existing network.

Currently, mainstream NVO3 technologies include VXLAN and NVGRE. VXLAN is used by the majority of enterprises as the technical standard for constructing overlay networks. The following illustrates VXLAN technology.

6.2 VXLAN BASICS AND CONCEPTS

VXLAN is an NVO3 technology that enables Layer 2 forwarding over a Layer 3 network by using L2 over L4 (MAC-in-UDP) encapsulation. This technology, defined by the IETF, allows VMs to migrate over a large Layer 2 network and isolates tenants in a DC.

Figure 6.2 shows the VXLAN network model.

Figure 6.2 shows the new elements emerging from the VXLAN network that legacy DCNs do not have:

1. VTEP

    VTEPs are edge devices on a VXLAN network. They are a VXLAN tunnel's start and endpoints that encapsulate and decapsulate VXLAN packets. A VTEP can be an independent network device (such as Huawei CloudEngine series switch) or a server where a VM is located.

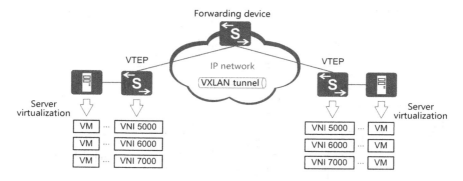

FIGURE 6.2   VXLAN network model.

2. VNI

    VLAN technology cannot provide sufficient isolation capability on DCNs, because VLAN IDs are only 12 bits long. To overcome this limitation, VXLAN uses 24-bit VNIs, which are similar to VLAN IDs but can identify approximately 16 million unique tenants. Additionally, when VXLAN packets are encapsulated, sufficient VNIs can be assigned to isolate a large number of tenants.

3. VXLAN tunnel

    A tunnel is not a new concept, and the same is true for the GRE tunnel. More specifically, the original packets are encapsulated and then transmitted over a bearer network such as an IP network. For a host, a tunnel is a direct link between the start point and end point of original packets. As the name suggests, a VXLAN tunnel is used to transmit VXLAN packets and is a virtual tunnel between two VTEPs. Figure 6.3 shows the VXLAN packet format.

A VTEP encapsulates the original Layer 2 frame sent by a VM as follows:

- VXLAN Header: It contains 8 bytes, including the 24-bit VNI field that defines different tenants on a VXLAN network. In addition, it has the VXLAN Flags (8 bits, with a value of 0000 1000) field and two reserved fields (24 bits and 8 bits, respectively).

- UDP Header: The VXLAN header and the original Ethernet frame are used as UDP data. In the UDP header, the VXLAN Port (destination port number) field has a fixed value of 4789. The value of

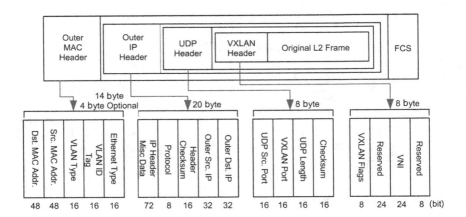

FIGURE 6.3 VXLAN packet format.

UDP Src. Port (source port number) is the hash value of the original Ethernet frame.

- Outer IP Header: This is the encapsulated outer IP header. The Outer Src. IP field (source IP address) specifies the IP address of the VTEP, where the source VM belongs. The Outer Dst. IP field (destination IP address) indicates the IP address of the VTEP where the destination VM belongs.

- Outer MAC Header: This is the encapsulated outer Ethernet header. The Src. MAC Addr. field (source MAC address) is the MAC address of the VTEP where the source VM belongs. The Dst. MAC Addr. field (destination MAC address) is the MAC address of the next-hop device on the path to the destination VTEP.

According to the VXLAN network model and packet format, VXLAN encapsulates data packets sent from VMs into UDP packets. VXLAN then encapsulates IP and MAC addresses used on the physical network, in outer headers, before sending the packets over an IP network. The egress tunnel endpoint then decapsulates and sends the packets to the destination VM. Overall, the VXLAN offers the following benefits:

- Compared with the method that uses 12-bit VLAN IDs for Layer 2 tenant isolation, VXLAN uses 24-bit VNIs to support approximately 16 million VXLAN segments, allowing DCs to accommodate a significant number of tenants. User isolation and identification are hence no longer limited.

- A VNI can be flexibly associated with other attributes. For example, if a VNI is associated with a VPN instance, complex services such as L2VPN and L3VPN are supported.

- Except the VTEP, other devices on the network are not required to identify MAC addresses of VMs, reducing the pressure of learning MAC addresses.

- VXLAN decouples physical and virtual networks by using MAC-in-UDP encapsulation to extend Layer 2 networks. Tenants can plan their own virtual networks, not limited by the physical network IP addresses or bridge domains (BDs). Overall, this greatly simplifies network management.

- The source UDP port in the VXLAN header is hashed based on inner flow information. The underlay network can implement load balancing without parsing inner packets, implementing high throughput.

6.3 VXLAN OVERLAY NETWORK

VXLAN overlay networks can be classified into network overlay, host overlay, and hybrid overlay based on the device models functioning as VTEPs.

- Network overlay: All VTEPs are deployed on physical switches.

- Host overlay: All VTEPs are deployed on vSwitches.

- Hybrid overlay: Some VTEPs are deployed on physical switches with others deployed on vSwitches.

6.3.1 VXLAN Overlay Network Types

1. Network overlay

   Network overlay is classified into centralized and distributed network overlay. The VTEPs of a VXLAN tunnel are physical switches, as shown in Figure 6.4.

   - Centralized network overlay: Leaf nodes act as Layer 2 VXLAN gateways, with spine nodes or border leaf nodes acting as Layer 3 VXLAN gateways.

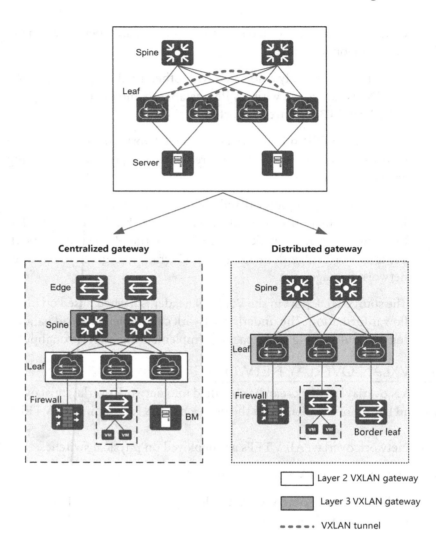

FIGURE 6.4   Centralized and distributed network overlay.

- Distributed network overlay: Leaf nodes function as Layer 2 and Layer 3 VXLAN gateways. Spine nodes are restricted to forwarding IP packets at a high speed and do not process VXLAN packets.

2. Host overlay

All VTEPs are vSwitches deployed on servers, as illustrated in Figure 6.5. East-west traffic in a DC is forwarded between vSwitches through a VXLAN tunnel, and north-south traffic is forwarded between vSwitches and vRouters. Physical switches that function as leaf and spine nodes are only able to forward IP packets at high speed and additionally do not process VXLAN packets.

3. Hybrid overlay

   VTEPs can be vSwitches or physical switches, as shown in Figure 6.6, with distributed mode frequently used.

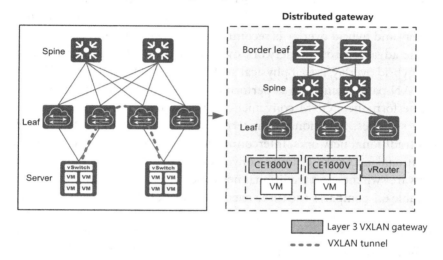

FIGURE 6.5   Host overlay.

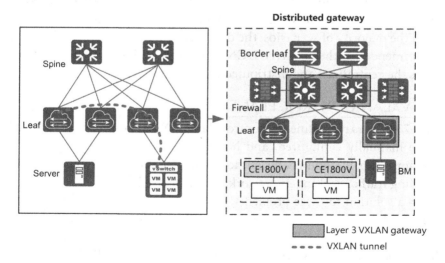

FIGURE 6.6   Hybrid overlay.

East-west traffic in a DC is forwarded through VXLAN tunnels between vSwitches and physical switches (leaf nodes). North-south traffic is forwarded through VXLAN tunnels between vSwitches or physical switches (leaf nodes) and spine nodes or border leaf nodes.

6.3.2 Comparison of VXLAN Overlay Network Types

In the preceding three VXLAN overlay network types, the network overlay is recommended for the cloud DCN solution. Comparison of the host overlay and hybrid overlay is recommended in only limited scenarios.

The advantage of the network overlay, in contrast to the host overlay and hybrid overlay, is that physical switches are able to function as VTEPs. VXLAN path calculation is performed on switches, exhibiting forwarding performance, O&M convenience, and security advantages over software switches. Additionally, in terms of interconnection between SDN and traditional networks, interconnection between physical switches and traditional networks is more convenient than interconnection between software switches and traditional networks.

Table 6.1 compares the three overlay network types.

For centralized and distributed network overlay, distributed network overlay is advantageous. In the centralized network overlay scenario, all traffic across network segments has to be forwarded through the centralized gateway, imposing substantial pressure on the centralized gateway. Therefore, the number of services supported in the centralized network overlay scenario is limited, meaning poor network scalability.

In the majority of scenarios, the distributed network overlay solution is recommended as the preferred DCN deployment solution. Table 6.2 compares the centralized and distributed network overlay.

6.4 VXLAN CONTROL PLANE

RFC 7348 does not define the control plane. VXLAN tunnels are required to be manually configured, and host addresses learned through traffic flooding. This method is efficient to implement, but it creates significant flooding traffic resulting in network expansion difficulties.

As a solution to these problems, EVPN is introduced on the VXLAN control plane. By referring to the BGP/MPLS IP VPN mechanism, EVPN defines several types of BGP EVPN routes by extending BGP. It advertises BGP routes on the network to implement automatic VTEP discovery and host address learning.

TABLE 6.1 Comparison between Network Overlay, Host Overlay, and Hybrid Overlay

(a) Item	Network Overlay	(b) Host Overlay	(c) Hybrid Overlay
(d) VTEP	Hardware switch	vSwitch	Hardware switch and vSwitch
Layer 3 VXLAN gateway	Hardware switch (distributedly deployed based on the locations where VMs go online)	vSwitch (distributedly deployed based on the locations where VMs go online)	Hardware switch and vSwitch (distributedly deployed based on the locations where VMs go online)
Access server	Virtual and physical servers	Virtual server	Virtual and physical servers
L4-L7 service type	L4-L7 services provided by hardware or software	Software L4-L7 services (vSwitch access)	L4-L7 services provided by hardware or software
Control plane	• Devices learn local Layer 2 and Layer 3 entries • Devices learn local Layer 2 entries using ingress replication of broadcast packets • The controller provides centralized control for the control plane. That is, physical network devices only forward packets based on forwarding entries delivered by the controller, and do not take any action for the control plane • Device information is synchronized through BGP EVPN	• vSwitches report ARP or ND entries to the controller through OpenFlow • The controller delivers flow tables to vSwitches for synchronization	• Hardware devices learn local Layer 2 and Layer 3 entries • Device information can be synchronized between devices through BGP EVPN • vSwitches report ARP or ND entries to the controller through OpenFlow • The controller delivers flow tables to vSwitches for synchronization • Information about hardware NVE nodes and vSwitches (software NVE nodes) is synchronized through BGP EVPN on the controller
Forwarding performance	CPU resources of servers are not occupied. Hardware devices provide high forwarding performance	VXLAN processing consumes CPU resources of servers, with the processing performance dependent on the CPU	Hardware devices do not occupy CPU resources of servers, whereas software devices occupy CPU resources for VXLAN processing

(Continued)

TABLE 6.1 (*Continued*) Comparison between Network Overlay, Host Overlay, and Hybrid Overlay

(a) Item	Network Overlay	(b) Host Overlay	(c) Hybrid Overlay
VM specifications	• The number of VPCs is restricted by VRFs and routing specifications of physical switches • The number of VMs in a VPC is restricted by entry specifications of physical switches	The number of VMs is limited by the controller capability. Compared with physical switches, VMs display multiple advantages	• The number of VPCs is limited by the controller capability. Compared with physical switches, vSwitches display multiple advantages • The number of VMs in a VPC is restricted by entry specifications of physical switches
Application scenario	• Private clouds imposing high forwarding performance, O&M, and security requirements • Access of both virtual and physical servers • Interconnection between SDN and traditional networks	• Access of virtual servers only • Networks with a large number of tenants • Networks with devices from multiple vendors, requiring decoupling VXLAN from hardware network devices	• Access of virtual servers only • Large tenant scale • Devices from different vendors are used, and VXLAN needs to be decoupled from hardware network devices

TABLE 6.2 Comparison between the Centralized and Distributed Network Overlay

Solution		Centralized Network Overlay	Distributed Network Overlay (Recommended)
Deployment	NVE node	• It is deployed on a leaf node (TOR switch) • A physical switch is used as an NVE node	• It is deployed on a leaf node (TOR switch) • A physical switch is used as an NVE node
	Layer 3 gateway	• It is centrally deployed on a spine node or border leaf node • A physical switch is used as the NVE node	• It is distributedly deployed on the nearest TOR switch based on the locations where VMs go online • A physical switch is used as the NVE node
	Border leaf node	It can be combined with a spine node and a Layer 3 gateway	It can be combined with a spine node
	L4-L7	It connects to a gateway in inline or bypass mode or a group of service leaf nodes	It connects to a border leaf node in inline or bypass mode or a group of service leaf nodes
Forwarding plane		• Inter-subnet traffic needs to be forwarded to the centralized gateway, causing a traffic detour and posing higher pressure on the centralized gateway • Layer 3 forwarding entries are centralized on the centralized gateways, posing high requirements on the centralized gateway	• Inter-subnet traffic is directly forwarded between leaf nodes, meaning traffic detours can be avoided and load balancing implemented • Layer 3 forwarding entries are distributed on leaf nodes, resulting in low gateway requirements
Control plane		BGP EVPN	BGP EVPN
Application scenario		Small-scale networks with high forwarding performance requirements and a small number of tenants	Large- and medium-scale networks with high forwarding performance requirements and a large number of tenants

Using EVPN on the control plane offers the following advantages:

- VTEPs can be automatically discovered and VXLAN tunnels can be automatically established, overall simplifying network deployment and expansion.

- EVPN can advertise Layer 2 MAC addresses and Layer 3 routing information simultaneously.

- Flooding traffic is reduced on the network.

1. EVPN principles

Traditional BGP-4 uses Update packets to exchange routing information between peers. An Update packet can advertise a type of reachable routes with the same path attributes, placed in Network Layer Reachability Information (NLRI) fields.

BGP-4 can manage only IPv4 unicast routing information. As a solution, Multiprotocol Extensions for BGP (MP-BGP) was developed as a means to support multiple network layer protocols, including IPv6 and multicast. MP-BGP extends NLRI fields based on BGP-4. After extension, the description of the address family is added to the NLRI fields to differentiate network layer protocols. These include the IPv6 unicast address family and VPN instance address family.

Similarly, EVPN defines a new sub-address family, specifically, the EVPN address family in the L2VPN address family. EVPN additionally introduces the EVPN NLRI, which defines different types of BGP EVPN routes. After the routes are advertised between EVPN peers, VXLAN tunnels can be automatically established and host addresses learned. The BGP EVPN route types are as follows:

- Type 2 route — MAC/IP route: is used to advertise the MAC addresses, ARP entries, and routing information of hosts, with Figure 6.7 showing the NLRI format in a type 2 route.

- Type 3 route — inclusive multicast route: is used for automatic discovery of VTEPs and dynamic establishment of VXLAN tunnels, with Figure 6.8 showing the NLRI format in a type 3 route.

Route Distinguisher	RD of an EVPN instance.
Ethernet Segment Identifier	Unique ID for defining the connection between local and remote devices.
Ethernet Tag ID	VLAN ID.
MAC Address Length	Length of the host MAC address carried in the route.
MAC Address	Host MAC address carried in the route.
IP Address Length	Mask length of the host IP address carried in the route.
IP Address	Host IP address carried in the route.
MPLS Label1	Layer 2 VNI carried in the route.
MPLS Label2	Layer 3 VNI carried in the route.

FIGURE 6.7   NLRI format in a type 2 route.

Prefix

Route Distinguisher	RD of an EVPN instance.
Ethernet Tag ID	VLAN ID. Here, the value is 0.
IP Address Length	Mask length of the local VTEP's IP address carried in the route.
Originating Router's IP Address	Local VTEP's IP address carried in the route.

PMSI attribute

Flags	Flag bit. This field is inapplicable in VXLAN scenarios.
Tunnel Type	Tunnel type carried in the route. The value can only be 6.
MPLS Label	Layer 2 VNI carried in the route.
Tunnel Identifier	Tunnel identifier carried in the route.

FIGURE 6.8   NLRI format in a type 3 route.

- Type 5 route — IP prefix route: is used to advertise imported external routes or advertise routing information of hosts, with Figure 6.9 showing the NLRI format in a type 5 route.

Advertised EVPN routes contain the Route Distinguisher (RD) and VPN target (known as the route target). RDs are used to distinguish different VXLAN EVPN routes. A VPN target is a BGP extended community attribute used to control the advertisement and reception of EVPN routes. Specifically, a VPN target defines the peers that can receive EVPN routes from the local end, and if the local end can receive EVPN routes from peers.

Route Distinguisher	RD of an EVPN instance.
Ethernet Segment Identifier	Unique ID for defining the connection between local and remote devices.
Ethernet Tag ID	VLAN ID.
IP Prefix Length	Length of the IP prefix carried in the route.
IP Prefix	IP prefix carried in the route.
GW IP Address	Default gateway address.
MPLS Label	Layer 3 VNI carried in the route.

FIGURE 6.9 NLRI format in a type 5 route.

There are two VPN target types:

- Export route target (ERT): The VPN target attribute is set to ERT when the local end sends EVPN routes.

- Import route target (IRT): When receiving an EVPN route from a peer, the local end compares the ERT in the received packet with its own IRT. If they are identical, the local end accepts the route. Otherwise, the local end discards the route.

2. VXLAN tunnel establishment in intra-subnet interconnection scenarios

A VXLAN tunnel is determined by a pair of VTEPs. In intra-subnet interconnection scenarios, communication is required within the same Layer 2 BD. Therefore, a VXLAN tunnel can be established providing the IP addresses of the VTEPs at both ends are reachable. EVPN can be used to dynamically establish VXLAN tunnels. This is done by establishing a BGP EVPN peer relationship between two VTEPs and using type 3 routes to transmit VNIs and VTEP IP addresses between the peers.

Figure 6.10 shows the tunnel establishment process.

- Leaf1 and Leaf2 at both ends first establish a BGP EVPN peer relationship with each other. Necessary configurations such as VTEP's IP address, VNI, and EVPN instance configurations are completed on them.

- Both Leaf1 and Leaf2 generate a type 3 EVPN route and send it to each other. The route carries the following information:

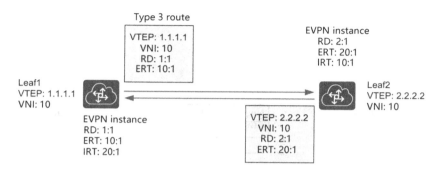

FIGURE 6.10 VXLAN tunnel establishment in intra-subnet interconnection scenarios.

local VTEP's IP address, VNI, RD of the EVPN instance, ERT, and IRT.

- After receiving the EVPN route from the peer, Leaf1 and Leaf2 check whether the ERT carried in the route matches the IRT of the local EVPN instance. If so, Leaf1 and Leaf2 accept the route. Otherwise, they discard the route.

After accepting the route, Leaf1 and Leaf2 check whether the remote VTEP's IP address and local VTEP's IP address are reachable at Layer 3. If so, Leaf1 and Leaf2 establish a VXLAN tunnel to each other. If the remote VNI matches the local VNI, Leaf1 and Leaf2 create an ingress replication list to forward broadcast, multicast, and unknown unicast packets.

3. VXLAN tunnel establishment in inter-subnet interconnection scenarios

The VXLAN tunnel establishment process in inter-subnet interconnection scenarios differs from that in intra-subnet interconnection scenarios. During VXLAN tunnel establishment using BGP EVPN, gateways (leaf nodes) are required to advertise to each other the IP routes of hosts or network segments to which they are attached. Otherwise, they cannot learn routes from each other, leading to a Layer 3 forwarding failure. In Figure 6.11, Leaf2 is required to advertise its routing information (192.168.20.1/32) to Leaf1. Otherwise, Leaf1 functioning as a Layer 3 gateway cannot learn the route to 192.168.20.1 and therefore does not know the correct method to send packets to 192.168.20.1.

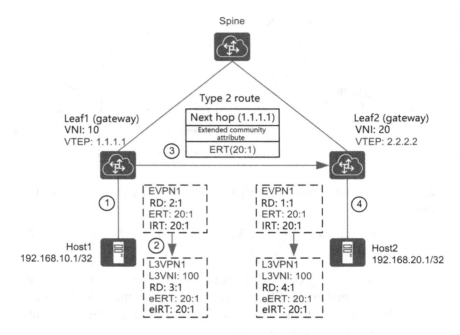

FIGURE 6.11 VXLAN tunnel establishment in inter-subnet interconnection scenarios.

In inter-subnet interconnection scenarios, tunnels are established by transmitting type 2 or type 5 routes. Type 2 and type 5 routes can carry host or network segment route information. The differences between the two routes are as follows: type 2 routes can advertise 32-bit host routes only, whilst type 5 routes can advertise both 32-bit host routes and network segment routes. Therefore, type 5 routes can enable communication between hosts on a VXLAN network and an external network. The following example uses type 2 routes.

In Figure 6.11, Leaf1 and Leaf2 transmit host routes and establish a tunnel as follows:

1. When Host1 goes online, Leaf1 can learn the ARP entry of Host1 from the ARP packet sent by Host1. In addition, Leaf1 can obtain information about the Layer 2 VNI, L3VPN instance, and Layer 3 VNI associated with the L3VPN instance based on the BD to which Host1 belongs.

Tenants on different network segments may connect to the same leaf node. To isolate different tenants, different L3VPNs can be created on the leaf node to isolate the routing tables of different tenants. As a result, different tenants' routes are stored in different private network routing tables, with Layer 3 VNIs used to identify the L3VPNs. When a leaf node receives a data packet (carrying a Layer 3 VNI) from a peer node, the leaf node finds the corresponding L3VPN based on the Layer 3 VNI. It then forwards the packet based on the routing table of an L3VPN instance.

Overall, Leaf1 obtains the following information about Host1: IP address, MAC address, Layer 2 VNI, and Layer 3 VNI of the L3VPN instance bound to the VBDIF interface. Afterward, the EVPN instance on Leaf1 can generate a type 2 route based on the preceding information.

2. The EVPN instance on Leaf1 sends the IP address, MAC address, and Layer 3 VNI of Host1 to the local L3VPN instance. This is done so the local L3VPN instance generates a route to Host1.

3. Leaf1 sends the generated EVPN route to Leaf2. In addition to the preceding host information, the route also carries the ERT of the local EVPN instance, the next hop (local VTEP's IP address) of the route, and the VTEP MAC address.

4. After Leaf2 receives the BGP EVPN route from Leaf1, Leaf2 processes the route as follows:

   - Checks whether the ERT in the route matches the IRT of the local EVPN instance. If they match, Leaf2 accepts the route. The EVPN instance extracts the host IP address and MAC address from the route to advertise the ARP entry of the host.

   - Checks whether the ERT in the route matches the IRT of the local L3VPN instance. If they match, Leaf2 accepts the route. The L3VPN instance extracts the host IP address and Layer 3 VNI from the route, and generates a route to Host1 in the routing table. The next hop of the route is set to the VXLAN tunnel interface of Leaf1.

After receiving the route, the local EVPN instance or L3VPN instance obtains the VTEP IP address of Leaf1 through the next hop. A VXLAN tunnel to Leaf1 is established, if the IP address is reachable at Layer 3.

The process of establishing a VXLAN tunnel from Leaf1 to Leaf2 matches the preceding process.

6.5 VXLAN DATA PLANE

This section describes the traffic forwarding process on the VXLAN forwarding plane.

1. Traffic model

Depending on the traffic flow direction and scope, DCN traffic can be classified into east-west traffic (transmitted within a DC) and north-south traffic (sent across the DC). The DCN traffic additionally falls into four types, with Figure 6.12 showing the distributed network overlay traffic model.

- Traffic transmitted within the same subnet of a VPC is forwarded by a TOR switch after Layer 2 VXLAN encapsulation.

- Traffic transmitted between subnets of the same VPC is forwarded by a TOR switch based on Layer 3 routes. This is done after Layer 3 VXLAN encapsulation.

- Traffic transmitted between VPCs is forwarded across subnets, and isolation for security purposes is required. Therefore, to meet this, the traffic needs to pass through a firewall and reach the Layer 3 VXLAN gateway.

- Traffic sent from a user outside the DC to a server in a VPC passes through the Intrusion Prevention System (IPS) or firewall, LB, VXLAN gateway, and TOR switch before reaching the server.

The forwarding plane forwards known unicast packets and BUM packets. This occurs on intra-subnet packets, inter-subnet packets, cross-VPC packets, and packets inside and outside a DC. The following describes each scenario.

2. Intra-subnet known unicast packet forwarding (including ARP Request/Reply packet processing)

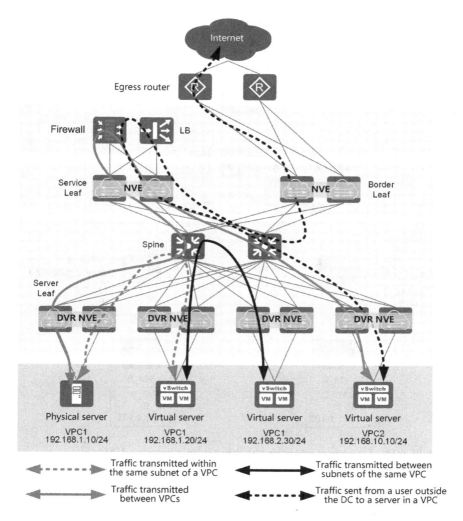

FIGURE 6.12   Four traffic models on the distributed network overlay.

In Figure 6.13, VM_A, VM_B, and VM_C are all on the network segment 10.1.1.0/24 and belong to VNI 5000. At this point, VM_A needs to communicate with VM_C.

Due to first time communication, VM_A does not have VM_C's MAC address. Therefore, VM_A broadcasts an ARP Request packet, requesting VM_C's MAC address.

1. ARP Request packet forwarding process

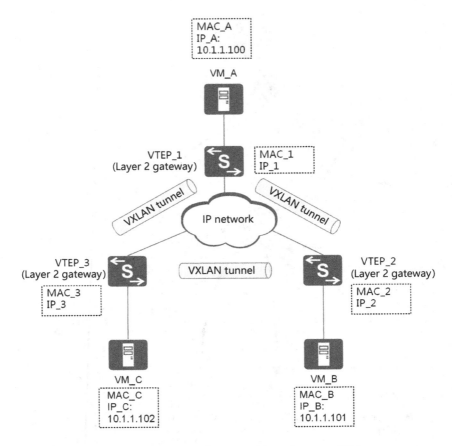

FIGURE 6.13 Communication on the same subnet between VMs.

Figure 6.14 shows the ARP Request packet forwarding process.

a. VM_A broadcasts an ARP Request packet, requesting VM_C's MAC address. The source MAC address of the packet is MAC_A, the destination MAC address is all Fs, the source IP address is IP_A, and the destination IP address is IP_C.

b. After VTEP_1 receives the ARP Request packet, it determines that the packet is required to go through the VXLAN tunnel based on the device configuration. After identifying the BD to which the packet belongs, VTEP_1 further identifies the VNI to which the packet belongs. VTEP_1 learns

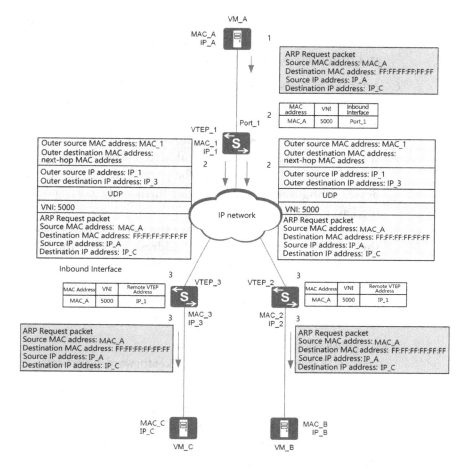

FIGURE 6.14   ARP Request packet forwarding process

the entry of MAC_A, VNI, and inbound interface (Port_1), saving the entry to the local MAC address table. Afterward, VTEP_1 replicates the packet based on the ingress replication list, carrying out packet encapsulation.

In the encapsulated packets, the following occurs: the outer source IP address is the local VTEP's (VTEP_1) IP address, the outer destination IP address is the remote VTEPs' (VTEP_2 and VTEP_3) IP addresses, and the outer source MAC address is the local VTEP's MAC address. Lastly, the outer destination MAC address is the next-hop device's MAC address, which heads toward the destination IP network.

Following the completion of encapsulation, the packet is transmitted on the IP network according to the outer MAC and IP addresses, until it arrives at the remote VTEP.

c. After the packets arrive at VTEP_2 and VTEP_3, the VTEPs decapsulate the packets to obtain the original packets sent by VM_A. VTEP_2 and VTEP_3 each learn the entry of VM_A's MAC address, VNI, and remote VTEP's IP address (IP_1), saving the entry to the local MAC address table. Next, VTEP_2 and VTEP_3 process the packets according to the device configuration and broadcast them in the corresponding Layer 2 domain.

After VM_B and VM_C receive the ARP Request packet, they check if the destination IP address matches the local host IP address. VM_B then discovers that the destination IP address is not the local IP address, discarding the packet. VM_C finds that the destination IP address is the local IP address and responds to the ARP Request packet.

2. ARP Reply packet forwarding process

Figure 6.15 shows the ARP Reply packet forwarding process.

a. Because VM_C has already learned VM_A's MAC address at this point, the ARP Reply packet is a unicast packet. The source MAC address of the packet is MAC_C, the destination MAC address is MAC_A, the source IP address is IP_C, and the destination IP address is IP_A.

b. After VTEP_3 receives the ARP Reply packet from VM_C, it identifies the VN to which the packet belongs. (The identification process is similar to step 2 of the ARP Request packet.) VTEP_3 learns the entry of MAC_C, VNI, and inbound interface (Port_3), and saves the entry to the local MAC address table. Then, VTEP_3 encapsulates the packet.

The outer source IP address is the IP address of the local VTEP (VTEP_3), and the outer destination IP address is the IP address of the remote VTEP (VTEP_1). The outer source MAC address is the MAC address of the local VTEP, and the outer destination MAC address is the MAC address of the next-hop device on the destination IP network.

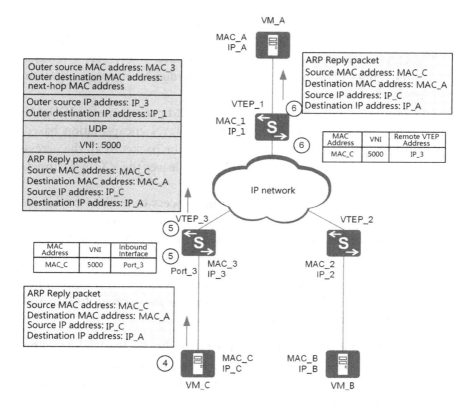

FIGURE 6.15   ARP Reply packet forwarding process.

After encapsulation, the packet is transmitted on the IP network according to the outer MAC and IP addresses, until it arrives at the remote VTEP.

c. After the packet arrives at VTEP_1, VTEP_1 decapsulates the packet to obtain the original packet sent by VM_C. VTEP_1 learns the entry of VM_C's MAC address, VNI, and remote VTEP's IP address (IP_3), and saves the entry to the local MAC address table. VTEP_1 then decapsulates the packet and sends it to VM_A.

Up until this point, VM_A and VM_C have already learned each other's MAC address. After that point, VM_A and VM_C will communicate instead in unicast mode. The encapsulation and decapsulation processes of unicast packets are similar to those shown in Figure 6.15.

3. Intra-subnet forwarding of BUM packets

Intra-subnet BUM packets are forwarded only between Layer 2 VXLAN gateways and are unknown to Layer 3 VXLAN gateways. Intra-subnet BUM packets can be forwarded in ingress replication mode.

In ingress replication mode, after a BUM packet enters a VXLAN tunnel, the ingress VTEP performs VXLAN encapsulation based on the ingress replication list and sends the packet to all the egress VTEPs in the list. When the BUM packet leaves the VXLAN tunnel, the egress VTEPs decapsulate it.

Figure 6.16 shows the forwarding process of a BUM packet in ingress replication mode. Terminal A is connected to the distributed gateway Leaf1 and is required to send BUM traffic to the VXLAN network.

- After Leaf1 receives a packet from Terminal A, it determines the Layer 2 BD of the packet, based on the access interface and VLAN ID in the packet.

- The VTEP on Leaf1 obtains the tunnel list for the VNI based on the BD and replicates the packet based on the ingress replication list. It then performs VXLAN tunnel encapsulation, before forwarding it to the outbound interface.

- After the VTEP on Leaf 2 or Leaf 3 receives the VXLAN packet, it checks the UDP destination port number, source and destination IP addresses, and VNI of the packet to determine the packet validity. Leaf2 or Leaf3 obtains the Layer 2 BD based on the VNI and performs VXLAN decapsulation to obtain the inner Layer 2 packet.

- Leaf2 or Leaf3 checks the destination MAC address of the inner Layer 2 packet and finds it is a BUM MAC address. Therefore, Leaf2 or Leaf3 broadcasts the packet onto the network connected to terminals (not the VXLAN tunnel side) in the Layer 2 BD. Specifically, Leaf2 or Leaf3 finds the outbound interfaces and encapsulation information unrelated to the VXLAN tunnel, adds VLAN tags to the packet, and finally forwards the packet to Terminal B or Terminal C.

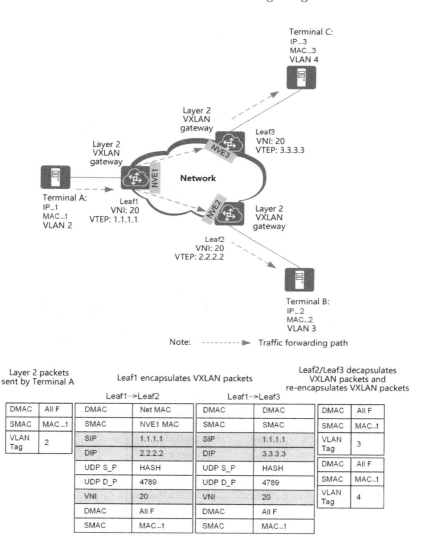

FIGURE 6.16   Forwarding process of an intra-subnet BUM packet in ingress replication mode.

4. Inter-subnet packet forwarding

Inter-subnet packets must be forwarded through a Layer 3 gateway. Figure 6.17 shows the inter-subnet packet forwarding process in distributed VXLAN gateway scenarios. Host1 and Host2 belong to different subnets and are required to communicate with each other.

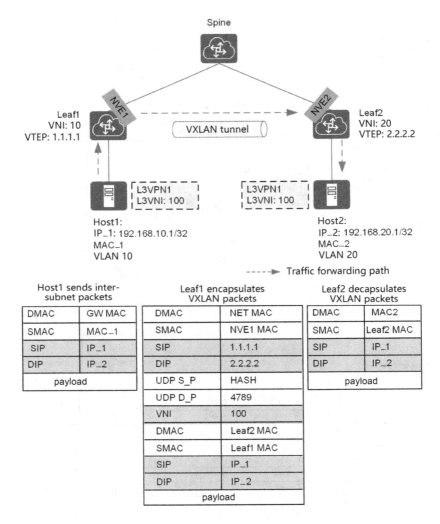

FIGURE 6.17   Inter-subnet packet forwarding.

The following describes the forwarding process. For details about the ARP forwarding process, refer to the process of forwarding known unicast packets (including the ARP Request/Reply packet processing) on the same subnet.

1. After Leaf1 receives a packet from Host1, it finds that the destination MAC address of the packet is a gateway MAC address. Therefore, this packet must be forwarded at Layer 3.

2. Leaf1 determines the Layer 2 BD of the packet based on the inbound interface and accordingly locates the L3VPN instance bound to the VBDIF interface of the Layer 2 BD. Leaf1 then searches the L3VPN routing table and locates the destination address of packet. Figure 6.18 shows the host route in the L3VPN routing table. Leaf1 obtains the Layer 3 VNI and next-hop address of the host route, and finds that the recursive outbound interface is a VXLAN tunnel interface. Therefore, Leaf1 determines that the packet must be transmitted through a VXLAN tunnel.

   – Leaf1 obtains MAC addresses based on the VXLAN tunnel's source and destination IP addresses, replacing the source and destination MAC addresses in the inner Ethernet header.

   – Leaf1 encapsulates the packet with the Layer 3 VNI.

   – Leaf1 encapsulates the VXLAN tunnel's source and destination IP addresses in the outer IP header. The MAC address of the NVE1 interface is the source MAC address, and the next-hop MAC address is the destination MAC address used in the outer Ethernet header.

1. The VXLAN packet is then transmitted over the IP network based on the IP and MAC addresses in the outer headers, and finally reaches Leaf2.

2. After Leaf2 receives the VXLAN packet, it decapsulates the packet and finds that the destination MAC address is its own MAC address. Hence, the packet must be forwarded at Layer 3.

L3VPN1

Destination Address	Layer 3 VNI	Next Hop
192.168.20.1/32	100	2.2.2.2

Recursive outbound Interface
VXLAN tunnel

FIGURE 6.18   Host route 1 in the L3VPN routing table.

L3VPN1

Destination Address	Layer 3 VNI	Next Hop	Outbound Interface
192.168.20.1/32	100	Gateway interface address	VBDIF

FIGURE 6.19   Host route 2 in the L3VPN routing table.

3. Leaf2 finds the L3VPN instance based on the Layer 3 VNI carried in the packet, searches the routing table of the L3VPN instance (as shown in Figure 6.19), and obtains the gateway interface address as the next hop of the packet. It then replaces the destination MAC address with the MAC address of Host2, replaces the source MAC address with Leaf2's MAC address, and forwards the packet to Host2.

Host2 sends packets to Host1 in the same process.

5. Inter-VPC packet forwarding

Figure 6.20 shows the access process. VM1 and VM3 are deployed on the same or different physical servers (identical process) and are connected to the same or different vSwitches. VM1 and VM3 belong to compute nodes in different VPCs. VM1 initiates access to VM3, and traffic needs to be filtered by the firewalls of the two VPCs because the traffic is transmitted across VPCs.

The following describes the forwarding process. For details related to the ARP process, refer to the process of forwarding known unicast packets on the same subnet (including the ARP Request/Reply packet processing).

- VM1 sends an ARP Request packet to request for the MAC address of the local network segment's gateway.

- After NVE1 receives the ARP Request packet, it sends an ARP Reply packet to VM1 in place of the gateway.

- VM1 sends the first data packet to VM3.

- NVE1 receives the first data packet and finds that the destination address does not belong to the network segment of VRF-A. The packet matches the default route in VRF-A and is then sent to VRF-A on the service leaf node. The two distributed

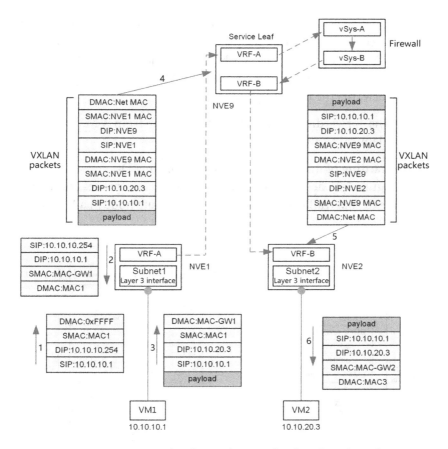

FIGURE 6.20   Cross-VPC packet forwarding on the distributed overlay.

VRFs (VRF-A) on NVE1 and the service leaf node exchange information through a VXLAN overlay tunnel (Layer 3 VNI interconnection).

- The data packet of VM1 matches the default route in VRF-A on the service leaf node and is forwarded to vSys-A of the firewall. The firewall is connected to the service leaf node through a VLAN, and hence, the packet is forwarded as a common Ethernet packet. The firewall searches for the route for VPC communication in vSys-A and then forwards the packet to vSys-B. The firewall searches for the route in vSys-B and finally forwards the packet to VRF-B of the service leaf node.

- The service leaf node searches for the host route of VM3 in VRF-B and forwards the packet to VRF-B of NVE2 through the VXLAN tunnel (Layer 3 VNI interconnection). NVE2 searches for the host route, removes the VXLAN tag from the packet, and forwards the packet to VM3.

- The forwarding process for packets sent from VM3 to VM1 is identical to above.

6. Forwarding of internal and external traffic in a DC

   Figure 6.21 depicts the forwarding and access process of internal and external traffic in a DC. VM1 is a compute node in a VPC of the

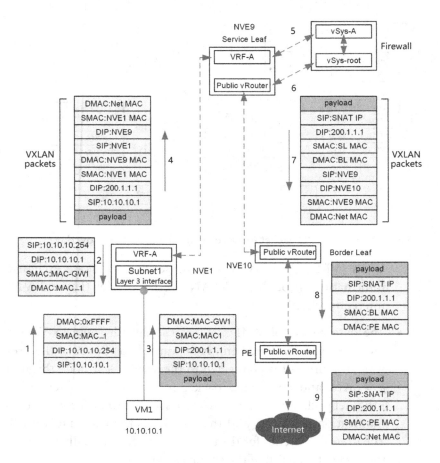

FIGURE 6.21   Distributed VXLAN traffic leaving a DC.

DCN and needs to access an IP address of a network, such as the Internet, outside the DC.

The following describes the forwarding process. For details related to the ARP process, refer to the process of forwarding known unicast packets on the same subnet (including the ARP Request/Reply packet processing).

1. VM1 sends an ARP Request packet to request the MAC address of the gateway on the local subnet.

2. After NVE1 receives the ARP Request packet, it sends an ARP Reply packet to VM1 in place of the gateway.

3. VM1 sends the first data packet to the public IP address on the Internet.

4. After receiving the first data packet, NVE1 finds that the destination address is not in the network segment of VRF-A and then sends the data packet to VRF-A on the service leaf node through the default route in VRF-A. The two distributed VRFs (VRF-A) interwork through a VXLAN overlay tunnel (Layer 3 VNI interconnection).

5. The data packet of VM1 matches the default route in VRF-A on the service leaf node and is forwarded to vSys-A of the firewall. The firewall is connected to the service leaf node through a VLAN, and hence, the packet is forwarded as a common Ethernet packet.

6. The firewall searches for the route in vSys-A and forwards the data packet to vSys-root. In addition, the firewall translates the source address of the data packet. vSys-root of the firewall forwards the data packet to the public vRouter of the service leaf node through the default route.

7. The public vRouter of the service leaf node forwards the data packet to the public vRouter of the border leaf node through the default route. The public vRouters on the service leaf node and border leaf node are connected through Layer 3 VNIs.

8. The border leaf node forwards packets through the underlay network to the VRF on the PE connecting to Internet links.

9. The PE forwards the packet from the DC.

6.6 MAPPING BETWEEN SERVICE MODELS AND NETWORKS

A logical model is formed after modeling and abstraction of a user network on the controller. Table 6.3 describes common logical entities when a VPC is created on the controller.

The controller translates the VPC logical network model designed by an administrator into configuration commands, during automatic service configuration. The controller then delivers them to forwarders through NETCONF or OpenFlow. A physical model is formed after the controller translates the logical model into configurations and delivers them to physical devices. For example, after a VPC is created and delivered to a physical device, an L3VPN instance is created on the physical device.

The distributed VXLAN overlay networking example is used to describe mapping between logical and physical models. Figure 6.22 illustrates these physical models.

1. Computing access side

   When a VM goes online, the VM has a VLAN assigned by a vSwitch. In this case, a BD must be configured on the vSwitch to map the VLAN. The BD (No. 1 in Figure 6.22) is a Layer 2 domain defined in VXLAN. On a distributed VXLAN network, an NVE node functions as a DVR. Therefore, a VRF has required to be created (No. 3 in Figure 6.22) for the VM to be able to correspond to a logical router.

TABLE 6.3 Common Logical Entities on the Controller

Logical Entity	Description
Logical router	Serves as the Layer 3 service gateway of servers
Logical switch	Corresponds to a Layer 2 subnet, equivalent to several Layer 2 switches
Logical port	Is the port on the network device side of a server connected to an access switch
End port	Corresponds to a NIC port of a server
Service function (SF)	Refers to an L4-L7 VAS device, such as a firewall
Logical router (distributed)	Connects to a Layer 3 gateway of an external network when a VPC accesses the external network. Multiple VPCs can share the same logical router
External gateway	This refers to logical modeling of external networks. If a tenant's VPC needs to access networks outside a DC, such as the Internet or remote private network, an external gateway must be bound to the VPC

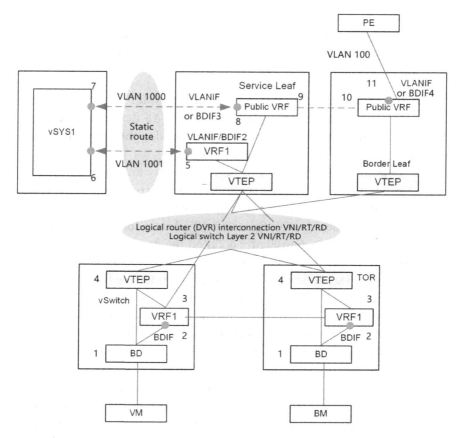

FIGURE 6.22   Physical models.

The gateway address and IP address of the online VM are obtained by binding a VBDIF interface (No. 2 in Figure 6.22) to the VRF. The IP address of the VBDIF interface is used as the service gateway address of the VM. Layer 2 broadcast of the VM is performed in the BD, and Layer 3 routes are queried based on the VRF. If traffic forwarding is required across NVE nodes, a VXLAN tunnel needs to be established between VTEPs (No. 4 in Figure 6.22) to encapsulate and decapsulate VXLAN packets.

The controller uses OpenFlow to deliver Layer 2 and Layer 3 flow tables to the vSwitch, after modeling and computing are complete. The vSwitch forwards service packets based on the flow tables.

The physical NVE node participates in BGP EVPN peer establishment and information synchronization. The physical model of the physical NVE node is similar to that of the vSwitch, and details are not provided.

2. Service leaf node, border leaf node, and firewall

A service leaf node is connected to a VAS device. A firewall is used as an example. The firewall is connected to the service leaf node through two different Layer 3 interfaces.

One Layer 3 interface can be regarded as an internal interface (No. 6 in Figure 6.22) and corresponds to the trusted zone. The interface is in the same routing domain as the service server but belongs to different subnets. Configure an IP address for the interconnection interface (No. 6 in Figure 6.22) on the firewall and connect the interface to the interconnection interface (No. 5 in Figure 6.22) on the service leaf node. The interconnection interface (No. 5 in Figure 6.22) on the service leaf node connects to the firewall through a VLANIF or VBDIF interface, and the interface is bound to the VRF of the tenant. In distributed networking, the RT of VRF1 is the same as the RT configured in BGP EVPN. Multiple distributed VTEPs can learn the host routes of the private network.

When traffic reaches the firewall, the firewall has a vSYS corresponding to the user VRF. In the vSYS, the firewall takes actions such as security filtering, SNAT, EIP, and IPsec for the traffic and sends it out through another interface (No. 7 in Figure 6.22). The interface (No. 7 in Figure 6.22) can be regarded as an external interface and corresponds to the untrusted zone. The interface is connected to another routing domain (public VRF, No. 9 in Figure 6.22) on the service leaf node.

The public VRF is shared by all user networks. The service leaf node and border leaf node are decoupled and are distributed VTEPs. Therefore, public VRFs (No. 9 and No. 10) are created on both the service leaf node and border leaf node, and routes are synchronized through BGP EVPN. The Layer 3 interface (No. 8 in Figure 6.22) connecting the public VRF on the service leaf node to the vSYS of the firewall and the Layer 3 interface (No. 11 in Figure 6.22) connecting the public VRF on the border leaf node to the PE belong to different network segments on the same private network.

3. Route advertisement and traffic forwarding

1. From the internal network to the external network

When a VM needs to access the external network, the controller configures a default route in VRF1 on the service leaf node. The next hop is the interface of the firewall (No. 6 in Figure 6.22). The service leaf node advertises the default route to BGP EVPN so that all EVPN peers (physical VTEPs) learn the default route. For the vSwitch, the controller participates in EVPN calculation, learns the default route, and delivers the flow table to the vSwitch.

When VRF1 on the access side receives a packet destined for an external network, it matches only the default route because it cannot match any specific route. Therefore, the packet is encapsulated into a VXLAN packet and forwarded to VRF1 on the service leaf node through the VXLAN tunnel.

The service leaf node searches VRF1 for a route and finds that the next hop is the interface of the firewall (No. 6 in Figure 6.22), so the service leaf node forwards the packet to the firewall. After processing the packet, the firewall matches the packet with the default route and sends it to the public VRF on the service leaf node.

The public VRF of the service leaf node matches the default route, so the packet is sent to the public VRF of the border leaf node. Then, the border leaf node sends the packet to the PE based on the default route.

2. From the external network to the internal network

In the public VRF on the service leaf node, the static route to the service network segment is configured and synchronized to the public VRF on the border leaf node through BGP EVPN.

When the border leaf node receives a packet with the destination address on the service network segment from the PE, it matches the packet with the static route in which the next hop is the service leaf node.

The service leaf node sends the packet to the firewall for processing.

The firewall sends the return packet to VRF1 on the service leaf node.

VRF1 on the service leaf node searches the table for the route synchronized through EVPN and forwards the packet to the corresponding VTEP. The VTEP decapsulates the VXLAN packet and sends it to the VM.

CHAPTER 7

Constructing a Multi-DC Network

A S SERVICES ARE DEPLOYED across regions, distributed multiple DCs become increasingly popular. Service subsystems are deployed across DCs, so the DCs support DR and load sharing, and provide the nearest services for users in different locations, improving user experience. This chapter focuses on Huawei multi-DC solution, which comprises multi-site and multi-PoD solutions. The multi-site solution is applicable to long-distance multi-DC construction and service collaboration between networks managed by multiple controllers. The multi-PoD solution is mainly developed for building multiple DCs within a short distance and using one set of controllers for unified management.

7.1 MULTI-DC SERVICE REQUIREMENTS AND SCENARIOS

The preceding sections cover the design and technical principles of the physical network (underlay network) and logical network (overlay network) on a single DCN. This section covers service scenarios and customer requirements in the multi-DC scenario.

7.1.1 Multi-DC Service Scenarios

With the development of the Internet, cloud computing, and big data, virtualization and resource pooling have become mainstream requirements. Resources cross regions and DCs need to be integrated to form a unified resource pool. In addition, deployment of service systems is distributed across multiple DCs to form the multi-active model. Services can be provided close to the user, improving user experience. Therefore, the distributed multi-DC solution has become the mainstream solution. Currently, service scenarios of multiple DCs include cross-DC service deployment, geo-redundancy, network-level DR, and distributed cloudification:

1. Cross-DC service deployment

   Typically, an application needs to be implemented by multiple or even hundreds of subsystems. Because the scale of a single DC is limited, one DC cannot accommodate all subsystems. Therefore, different subsystems of applications are deployed in different DCs. In this case, this application is deployed across multiple DCs. Additionally, different subsystems provide different functions. Some subsystems need to be deployed in multiple DCs in distributed mode, and some need to be deployed in centralized mode. As such, the entire service system is deployed across DCs. For example, in the following scenario, the web, app, and database subsystems are deployed in DC1, DC2, and DC3, respectively. The web subsystem invokes the app subsystem, and the app subsystem invokes the database subsystem. Different subsystems need to communicate with each other at Layers 2 and 3 across DCs to ensure applications are operating normally. Figure 7.1 shows the cross-DC service deployment scenario.

   In this case, the network needs to provide interworking capabilities between DCs to ensure smooth interaction at the service layer.

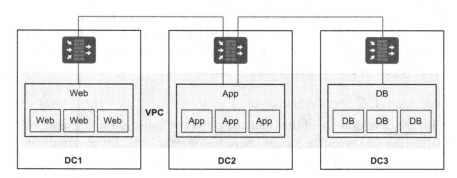

FIGURE 7.1   Cross-DC service deployment.

2. Geo-redundancy

In most cases, the geo-redundancy DC solution indicates that a remote DR DC is added to synchronize data with two active-active (primary) DCs in the same city.

Two active-active DCs in the same city: The same two service systems are deployed in two DCs in the same city. Traffic is routed to application servers in different DCs through load balancing. The two service systems run in the two DCs in the same city and provide services for users, doubling service capabilities and performing DR in real time. If the service system in a DC fails, the service system in the other DC continues to provide services, greatly improving the continuity and reliability of services. In this situation, users are unaware of faults. Subsystems in different DCs need to communicate with each other at Layer 2 and Layer 3, and security policies of the same subsystem must be consistent. The DCs provide the same services in active-active mode.

The remote DR DC is the backup of the two active-active DCs in the same city and is used to back up the data, configurations, and services of the two active-active DCs. If a fault occurs in the two active-active DCs due to natural disasters, the remote DR DC can quickly recover data and applications to ensure normal service operation and reduce loss.

In Figure 7.2, multiple VPCs are deployed in primary DCs A and B to carry the same services. In addition, subsystems in primary DCs A and B communicate with each other at Layer 2 and Layer 3, and the two DCs provide the same services externally. In this case, the network needs to meet requirements of DCI. The DR DC needs to synchronize its status with the primary DCs in real time. Therefore, the network needs to satisfy requirements of interconnection between the DR DC and primary DCs.

3. Network-level DR

A large number of applications provide services using cluster software. Multiple servers on a network are associated by the cluster software and appear as a logical server to the rest of the network to provide consistent services. In a cluster, multiple servers operate in load balancing mode to improve the overall service processing capability of the cluster. In addition, these servers support each other to improve

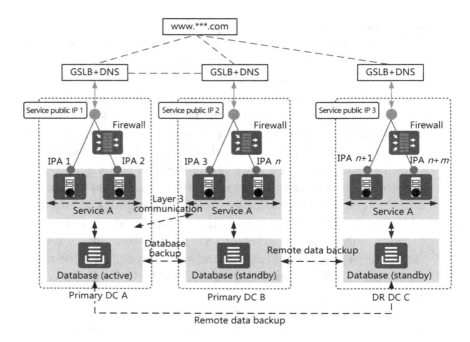

FIGURE 7.2   Geo-redundancy scenario.

reliability in the system. When servers in a cluster are deployed in different DCs, if a DC is faulty, servers in other DCs in the cluster can provide services to perform DR on application systems across DCs.

The cluster software of most vendors requires Layer 2 interconnection between servers; therefore, deployment of server clusters across DCs requires the network to provide large Layer 2 network capabilities across DCs. In addition, the cluster uses a virtual IP address (VIP) to provide external services, and the VIP is externally advertised through the frontend network of the DC. For this reason, the network needs to provide cross-DC gateways for the VIP of the cluster. These cross-DC gateways can work in active/standby or active-active mode, with the active and standby gateways advertising active and standby routes. Normally, north-south traffic is transmitted through the active gateway of the active DC based on the active route. However, if the active DC is faulty, services are switched to the standby route and traffic is forwarded through the standby gateway in the backup DC. The active-active gateways advertise ECMP routes externally. In addition, north-south traffic is load-balanced to two

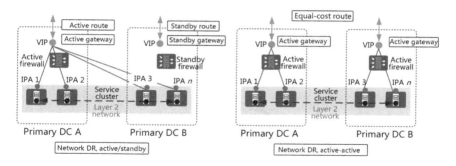

FIGURE 7.3   Network-level DR scenario.

DCs based on ECMP routes. If a DC is faulty, traffic is switched to the gateway of the other DC. Firewalls must protect the north-south traffic of the cluster and can be deployed in active/standby or active-active mode, as shown in Figure 7.3.

4. Distributed cloudification architecture

Distributed cloudification involves services that are deployed in multiple DCs in a distributed manner. Each DC can bear traffic in real time and provide services. Multiple DCs are interconnected through the DCI backbone network to form a unified resource pool for real-time synchronization of data, and services can be directly switched if any site fails. Multiple sites work in active-active mode, and edge DCs provide services close to users, ensuring a short latency and good user experience.

The central DC is the main source of data and pushes the website content to edge DCs through the backbone network. Subsequently, the edge DCs send the content to users. During this process, multiple DCs need to communicate with each other at both Layer 2 and Layer 3, as shown in Figure 7.4.

7.1.2 Multi-DC SDN Network Requirements

In the cloud DC, network resources are virtualized into resource pools to decouple services from physical networks. In addition to implementing on-demand self-service and automatic deployment of service networks, SDN technology also supports multi-tenant, elastic scaling, and fast deployment. In multi-DC scenarios, issues such as cross-DC service deployment, service interworking, automatic deployment through SDN, and cross-DC service DR and multi-active need to be resolved.

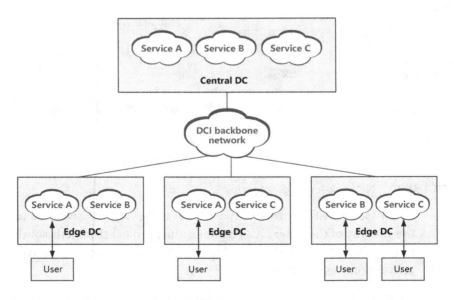

FIGURE 7.4 Distributed cloudification scenario.

TABLE 7.1 Main Multi-DC Service Requirements

Main Service Requirement	Requirement Analysis	Solution
Cross-DC service deployment	Large VPC	Layer 2 and Layer 3 communication across fabric networks in a VPC
	Security isolation	Routing entries
		Firewalls
Service interworking	VPC communication	Layer 3 communication across VPCs and fabric networks
SDN-based automatic deployment	Multi-DC resource management	SDN controller + network virtualization orchestrator + service orchestration
	Service orchestration	
Multi-active service DR across DCs	Application-level DR	Global Server Load Balance (GSLB)
	Network-level DR (Cross-DC DR with the same IP addresses)	Large Layer 2 network across fabric networks
		Egresses deployed in active/standby or active-active mode

Table 7.1 lists the main multi-DC service requirements. Details on the four requirements are as follows:

- Cross-DC service deployment: Customers may deploy some services across DCs; for example, a customer may plan an independent VPC that crosses multiple fabric networks for a large website. For such scenarios, intra-VPC traffic needs to be transmitted across fabric networks, and routing entries and firewalls need to be isolated.

- Service interworking: Customers usually allocate different services to different VPCs that may be deployed on different fabric networks. To implement service interworking, the VPCs need to communicate with each other across fabric networks at Layer 3. Layer 3 communication is typically used between VPCs. Therefore, for VMs that need to communicate with each other at Layer 2, you are advised to allocate them to the same VPC.

- SDN-based automatic deployment: Customers want automatic deployment through SDN. Automatic deployment involves orchestration of a virtual network across DCs and instantiation in each DC. An orchestrator evenly orchestrates cross-DC services, and the SDN controller orchestrates networks in a single DC.

- Multi-active service DR across DCs: Two modes are available: active/standby and active-active. For new service systems, customers can use GSLB to perform active-active DR. Specifically, two DCs are deployed with the same services but different IP addresses. As a result, the two systems can perform DR, which places no special requirements on the network. However, some old service systems require that the IP addresses remain unchanged after the services migrate to the DR DC. To achieve this, Layer 2 communication across fabric networks is required. In addition, gateways working in active/standby and active-active modes need to be deployed so that services can access external networks.

According to the multi-DC requirements described above, Layer 2 or Layer 3 interconnection across DCs is the most important requirement for a multi-DC network. However, different layers have different interconnection requirements. The multi-DC interconnection needs to solve the following problems:

- Data synchronization and backup require storage interconnection.

- The internal heartbeat of an HA cluster or VM migration across DCs requires large Layer 2 interconnection.

- Mutual access between services requires cross-DC Layer 3 interconnection.

- The frontend network of different DCs, the external egress of a DC, is interconnected through IP technology.

Different technologies are used for interconnection:

1. Storage interconnection is implemented through DWDM devices or bare optical fibers.

   DWDM or bare optical fibers provide direct physical links. The advantage of this interconnection mode is that exclusive channels (the channels are only used for traffic interaction between data centers) can fully meet requirements for high bandwidth and low latency for traffic interaction between DCs. In addition, the channels can carry data transmission of multiple protocols and provide flexible SAN/IP service access. This mode supports traffic transmission of both the IP SAN and FC SAN, and supports both Layer 2 and Layer 3 interconnection, meeting requirements for multi-service transmission. The disadvantage is that new optical fiber resources need to be built or leased, which increases the investment cost of DCs.

2. Application clusters and VM migration across DCs require a large Layer 2 network across DCs. The large Layer 2 network technologies include the following:

   Bare optical fibers/DWDM: The cost is high. It is mainly used between intra-city sites and is difficult to expand.

   The Virtual Private LAN Service (VPLS) is a type of L2VPN technology based on MPLS and Ethernet technologies. VPLS connects multiple Ethernet networks across a public network to enable the Ethernet networks to function as a single LAN. Layer 2 VPN channels are encapsulated on existing public and private network resources to carry data exchanged between DCs. This mode applies to interconnection scenarios of cloud computing DCs. Using this interconnection mode, you do not need to create an interconnection plane. You only need to add a VPN channel over the current network channel to isolate the existing data traffic on the network. However, deployment is complex, and the MPLS network needs to be leased from a carrier or built by yourself.

VXLAN is an advanced MAC-in-IP overlay technology that can be carried on an IP network, and provides L2VPN services on the IP core network through VXLAN tunnels. It can provide Layer 2 interconnection for scattered physical sites based on existing carriers' private line networks or the Internet. VXLAN is low cost and enables long-distance connections and easy expansion. In addition, VXLAN supports split horizon to prevent loops and broadcast storms. This mode does not depend on optical fiber or MPLS network resources, but requires Layer 3 reachability. This mode is flexible, scalable, cost-effective, and easy to deploy and maintain. However, the network quality is restricted by the IP network, and the bandwidth efficiency is low because the overlay technology is used.

3. Layer 3 interconnection between DCs is classified into the following types:

In traditional Layer 3 interconnection, IGP or BGP routes are used to achieve Layer 3 interconnection between service network segments of different DCs.

MPLS L3VPN is a virtual L3VPN constructed on the MPLS network. It enables devices on service network segments in different DCs to communicate with each other at Layer 3 and is used to carry IDC data. This mode applies to traditional service DC interconnection scenarios, and has the same advantages and disadvantages as VPLS.

VXLAN tunnels are established on the IP network to provide L3VPN services.

4. External egress of a DC

Egress devices of DCs connect to various private line networks of carriers or the Internet. IP technologies such as dynamic routing or static routing are used to implement interconnection.

Based on the preceding comparison, VXLAN technology is still most widely used in the DCI solution. Therefore, VXLAN is recommended as the DCI technology in the subsequent solution design.

7.1.3 Architecture and Classification of the Multi-DC Solution

The following uses the architecture and classification of Huawei multi-DC solution as an example.

1. Architecture of the multi-DC solution

Huawei multi-DC solution focuses on services across DCNs. The solution uses virtualization and SDN technologies to implement automatic deployment of cross-DC interconnection and cross-DC DR and active-active services.

Figure 7.5 shows the architecture of the multi-DC solution.

The architecture of Huawei multi-DC solution consists of the service control, infrastructure, and forwarding implementation layers.

- At the service control layer, the SDN controller controls the network of a DC and implements interworking of the network across DCs. To implement interworking between compute and network resources and between DCs, the SDN controller connects to the service orchestrator and VMM, with the former orchestrating services across DCs and the latter managing the lifecycle of VMs.

- The infrastructure layer consists of physical and logical networks. The physical network in a DC uses a spine-leaf architecture. Multiple DCs are connected through the DCI backbone

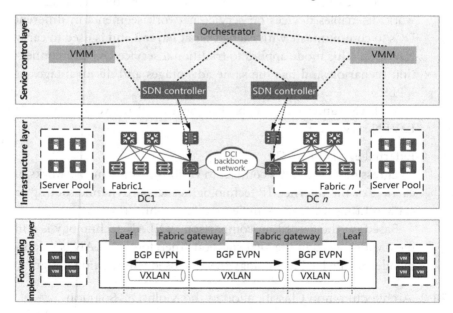

FIGURE 7.5   Architecture of the multi-DC solution.

network. A logical network is a virtual network that connects VMs based on service requirements, and uses network virtualization and VXLAN technologies.

- The forwarding implementation layer connects VMs in a DC through the VXLAN network and VMs between DCs. BGP EVPN is used as the control plane protocol of the VXLAN network.

For users, the service control layer is the focus. Users can divide a service network into multiple VPCs based on service requirements. Users use the orchestrator to orchestrate VPCs and the SDN controller to provision VPCs' logical networks in different DCs.

2. Classification of the multi-DC solution

The multi-DC solution falls into multi-site and multi-PoD solutions. A point of delivery (PoD) is a group of independent physical resources. The multi-PoD solution uses one SDN controller to manage multiple PoDs, and a VXLAN domain is formed using end-to-end VXLAN tunnels. PoDs are usually close to each other in the same city.

A site is a resource pool managed by the SDN controller consisting of one or more PoDs and is a VXLAN domain using end-to-end VXLAN tunnels. The multi-site solution uses the management domains of multiple SDN controllers, implements interworking between multiple PoDs, and forms multiple VXLAN domains. The solution can be deployed remotely regardless of the distance.

1. Multi-site solution

The multi-site solution functions as the remote multi-DC interconnection solution when two or more DCs are located in different regions or are a long distance away from each other and cannot be managed by one SDN controller.

The multi-site solution is applied to large-scale networks. In this solution, one orchestrator is used to centrally manage the networks that are managed by multiple controller clusters. All services are orchestrated by the orchestrator and then delivered to each controller, which then delivers the configurations to the corresponding physical network (Figure 7.6).

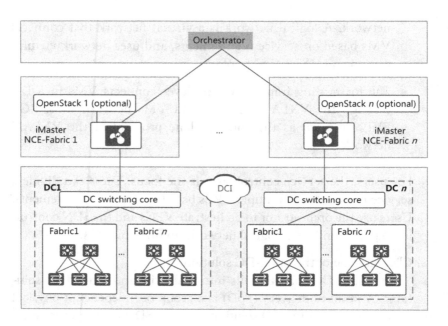

FIGURE 7.6   Networking of the multi-site solution.

2. Multi-PoD solution

In Figure 7.7, the multi-PoD solution is applicable to a scenario where DCs or resource modules are within a short distance of each other and can be managed by the same SDN controller. On a small- or medium-scale network, only one SDN controller is required to manage multiple DCs, without using an orchestrator. This is called a multi-PoD scenario, in which network configurations within and between DCs are performed on the SDN controller. The solution provides DR and active/standby egresses between multiple DCs. The multi-site and multi-PoD solutions apply to different scenarios: The multi-site solution uses management domains of multiple SDN controllers, whereas the multi-PoD solution uses a management domain of one SDN controller. Table 7.2 compares the two solutions.

3. Hierarchical multi-DC solution

The hierarchical multi-DC solution combines the multi-site and multi-PoD solutions. In the multi-site scenario, a single SDN controller cluster uses the multi-PoD solution to manage multiple PoDs. In this scenario, a single domain has multiple fabric networks and centralized egresses, and multiple domains communicate with each

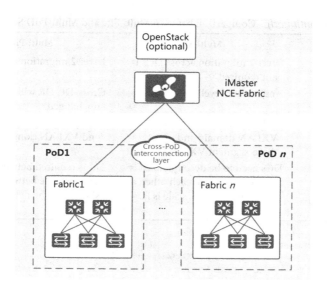

FIGURE 7.7   Networking of the multi-PoD solution.

TABLE 7.2   Comparison between Multi-Site and Multi-PoD Solutions

Item	Multi-Site	Multi-PoD
Management zone	Multiple management domains (SDN controller)	Single management domain (SDN controller)
Service orchestration	Unified orchestration performed by an orchestrator	Orchestration on the SDN controller GUI or by a single OpenStack platform
Network scale	Large-scale physical network involving multiple leaf nodes and fabric networks	Small-scale physical network (the total number of leaf nodes is subject to the specifications of the SDN controller)
Server scale	The number of servers exceeds the management capability of a cloud management platform	The number of servers is small. A single cloud management platform manages approximately 500 servers
Fault domain	Fault domain decoupling between DCs	Strong coupling of fault domains between DCs
Distance	Long distance, insensitive to the latency	Short distance and involves latency requirements
Large Layer 2 network	The large Layer 2 network is in the same VXLAN domain. In most cases, the large Layer 2 network does not cross DCs	The large Layer 2 network is in the same VXLAN domain, and the large Layer 2 network crosses DCs

(Continued)

TABLE 7.2 (*Continued*)   Comparison between Multi-Site and Multi-PoD Solutions

Item	Multi-Site	Multi-PoD
Migration	Layer 2 migration across DCs is not required	Layer 2 migration across DCs
DR	Application-level multi-active	Cross-DC DR with IP addresses unchanged
Forwarding plane	Each DC is an independent VXLAN domain and uses a three-segment VXLAN	Multiple DCs form an end-to-end VXLAN domain
Application scenario	DCs need to be decoupled, DCs are far away from each other, and the entire network scale is large	DR is required between DCs. The distance between DCs is short, and the network scale is small

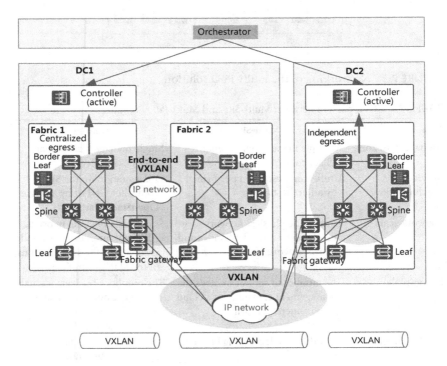

FIGURE 7.8   Networking of the hierarchical multi-DC solution.

other, as shown in Figure 7.8. The solution can be used in an ultra-large network design. For example, it can be used for geo-redundancy DC construction, in which the intra-city active-active DCs are deployed using the multi-PoD solution to implement network DR and multi-active egresses. Both the intra-city active-active DCs and remote DR DC are deployed using the multi-site solution to implement interconnection between them.

7.2 MULTI-SITE SOLUTION DESIGN

This section focuses on the detailed design of the multi-site solution, as well as the application scenario, detailed design, and recommended deployment of the multi-site solution.

7.2.1 Application Scenario of the Multi-Site Solution

In a virtualization scenario, a VPC is allocated to a service system. VPCs isolate different users or service systems to prevent them from affecting each other. With the development of services, compute resources required by service systems are increasing. When the capacity of a DC is exceeded, multiple DCs are required to deploy the service system. In this case, the VPC corresponding to the service system needs to be deployed across DCs. For example, VPCs are allocated based on service security levels, intranet, and DMZ. Services in the DMZ are placed in VPC 1, and intranet services are placed in VPC 2. In the multi-active multi-DC DR scenario, the VPCs are distributed to multiple DCs. As a result, a cross-DC large VPC is formed. This is known as the large VPC scenario.

In addition to the abovementioned scenario, different service systems of the same tenant usually need to communicate with each other. For example, traffic from the intranet and DMZ (which need to communicate with each other) is first transmitted to the DMZ and then to the intranet. The DMZ and intranet need to communicate with each other. Because the two VPCs are deployed across DCs, the VPCs across DCs also need to communicate with each other. This is known as the VPC communication scenario.

7.2.1.1 Deployment of a Large VPC

In the multi-DC scenario, service VPCs need to be deployed across DCs, as shown in Figure 7.9.

The management scope of each SDN controller is limited. Therefore, in a large-scale DC, multiple SDN controllers need to be deployed in multiple DCs. In some cases, multiple controllers are deployed for a single DC. In this instance, network services of the large VPC cannot be delivered by an SDN controller. In addition, an orchestrator is required to collaborate

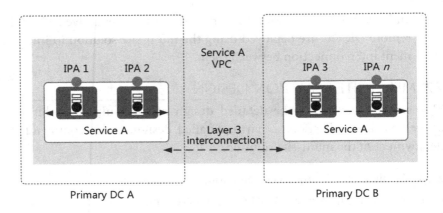

FIGURE 7.9   Networking of a large VPC.

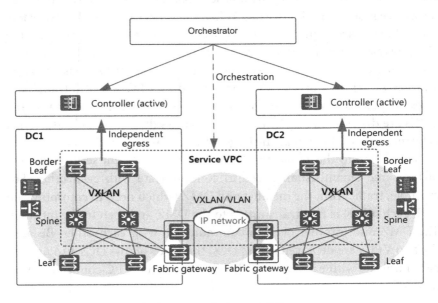

FIGURE 7.10   Cross-DC deployment of a large VPC.

with SDN controllers of multiple DCs to orchestrate VPCs across DCs. In Figure 7.10, one of the major applications of the multi-site solution is cross-DC deployment of a large VPC.

An orchestrator centrally orchestrates networks in two DCs and between DCs. After orchestration is complete, the orchestrator delivers instructions to the corresponding SDN controller to provision VPC and cross-DC interworking instances. The entire network appears as a large VPC.

In the network solution, an independent VXLAN domain is deployed in each DC and is managed by an SDN controller. DCs communicate with each other through three-segment VXLAN or the underlay network.

Two independent SDN controllers and VXLAN domains on the forwarding plane can be deployed to isolate two DCs' fault domains. This enables customers to deploy DCs in batches or in modular mode.

7.2.1.2 VPC Communication

Different services need to communicate with each other, as shown in Figure 7.11. In cross-DC scenarios, communication issues also need to be resolved.

Services are deployed by VPC, so service interworking corresponds to VPC communication, as shown in Figure 7.12. In the multi-site scenario, multiple SDN controllers need to collaborate with each other. The orchestrator can orchestrate VPC communication, collaborate with multiple SDN controllers to configure their own network devices, and enable the logical network between VPCs.

7.2.2 Multi-Site Solution Design

7.2.2.1 Service Deployment Process in the Multi-Site Scenario

Figure 7.13 shows the process of deploying services in the multi-site scenario. Before detailing the service deployment process, we introduce the following concepts.

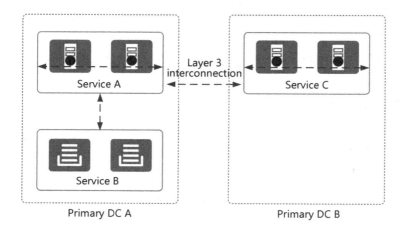

FIGURE 7.11   Networking of service interworking.

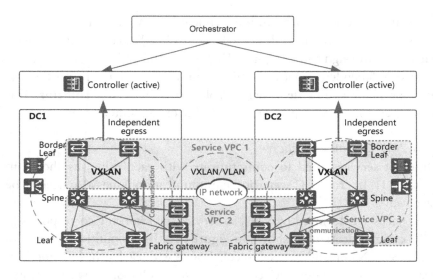

FIGURE 7.12 Networking of VPC communication.

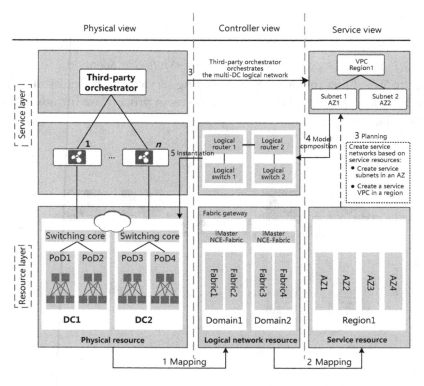

FIGURE 7.13 Service deployment process in the multi-site scenario.

- Service layer and resource layer: The service layer refers to what customers see from the perspective of services. In terms of the solution, the service layer refers to the service orchestration performed on the orchestrator. After orchestration, the corresponding configuration is transmitted to the corresponding SDN controller for service delivery. The resource layer deploys physical devices and consists of PoDs and switches. To associate physical devices with service orchestration, identifiers need to be added for the physical devices. Details of physical network, logical network, and service resources are as follows:

- Physical network resources: They are a collection of physical devices such as switches and firewalls, as well as a physical network composed of these devices.

- Logical network resources: On the SDN controller, physical networks are virtualized into domains, each of which is a collection of physical networks managed by an SDN controller, and fabric networks, several of which comprise a domain. A collection of physical networks managed by an SDN controller is called a domain, and a domain contains multiple fabric networks. The SDN controller maps the domain, fabric networks, and components on fabric networks to specific physical devices and delivers the configurations to them.

- Service resources: In the multi-site scenario, the orchestrator is used to divide customers' services logically. To ensure that a customer's service provisioning scope is controllable and that the logic of the orchestrator and fabric networks maps each other, regions and AZs are defined for the orchestrator. A region can be understood as the provisioning scope of a customer's service VPC. A service VPC is limited within a region. Essentially, an AZ can be understood as a zone that controls the provisioning scope of a subnet (logical switch) in a customer's service VPC. A subnet is limited in an AZ. When provisioning a service VPC, an administrator can provision the service VPC to a region, which contains two subnets that correspond to AZ 1 and AZ 2, respectively. In this case, regions and AZs correspond to the domains and fabric networks of the SDN controller, and the region model and AZ model correspond to the logical router and logical switch of the fabric network. The orchestrator can convert the customer's services into instructions and sends them to the SDN controller, which subsequently delivers the configurations to the physical network. The concepts and logic of regions and AZs are the same as those in the industry. This facilitates interconnection with third-party orchestrators.

After introducing these concepts, we will use an example to illustrate the working process of each component in the multi-site scenario. Figure 7.14 shows the service model in the multi-site scenario.

- A customer deploys a physical network that consists of DC1 (Fabric1) and DC2 (Fabric2), defined as Domain1 and Domain2, respectively. Each DC is deployed with one SDN controller for management.

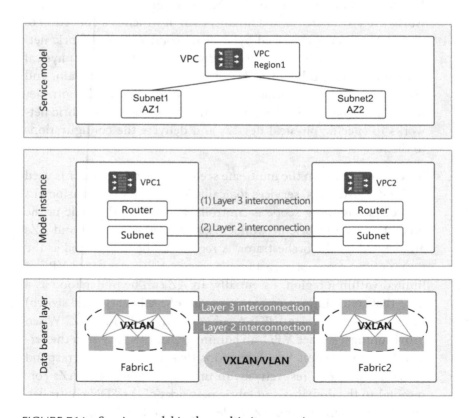

FIGURE 7.14   Service model in the multi-site scenario.

- The customer defines fabric information on the SDN controller. Assume that DC1 and DC2 both only have one fabric network, Fabric1 and Fabric2, respectively.

- The customer defines Region1 on the orchestrator. The scope of Region1 can map DC1 and DC2. In addition, the customer defines multiple AZs for Region1 on the orchestrator.

- The customer creates a service VPC on the orchestrator and limits the scope of the service VPC within Domain1. The VPC consists of two network segments that, respectively, belong to DC1 and DC2. The orchestrator maps the two network segments to two logical switches that belong to different AZs.

- Because the scope of Domain1 maps DC1 and DC2 and the two network segments belong to DC1 and DC2, respectively, the orchestrator delivers services to two SDN controllers, respectively. On each SDN controller, a VPC and corresponding logical routers and logical switches are created. Finally, the configurations are delivered to physical devices.

7.2.2.2 VMM Interconnection Design

The logical network of a DC serves VMs, and associates compute and network resources through the SDN controller. In the cloud DCN solution, the SDN controller can interconnect with multiple VMMs, including OpenStack and vCenter. In the multi-site solution, one site is a fabric network and is managed by one SDN controller. One SDN controller can connect to multiple VMMs, and each VMM manages only the VMs and connects only to the SDN controller at the local site. For details, see Figure 7.15.

7.2.2.3 Deployment Solution Design

The preceding describes the conceptual model and service model of the multi-site scenario. The following section describes how Layer 2 and Layer 3 communication between fabric networks is implemented.

In Figure 7.16, DC1 and DC2 are deployed independently, and each DC is configured with an independent network resource pool. If the two DCs need to communicate with each other at Layer 3, Layer 3 DCI can be deployed.

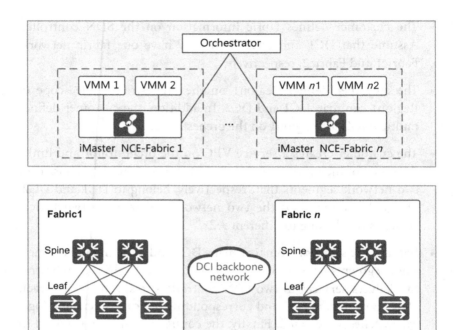

FIGURE 7.15   VMM interconnection design.

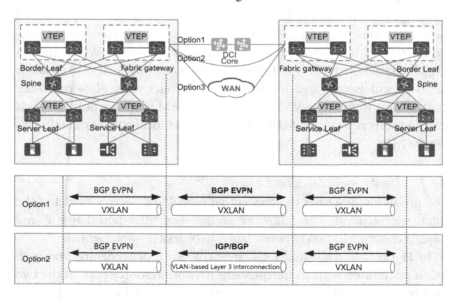

FIGURE 7.16   Layer 2 and Layer 3 communication between fabric networks.

In each DC, standard spine-leaf networking is used and a fabric gateway needs to be configured for DCI.

Three solutions are available to implement DCI physical network interconnection:

- Solution 1: For a campus with multiple sites, interconnection can be implemented through core switches.

- Solution 2: In a scenario where a small number of sites are deployed and DC fabric gateways are directly connected, direct interconnection through bare optical fibers or DWDM is applicable.

- Solution 3: In a scenario with many remote sites, WAN interconnection is applicable. It is recommended that the DCI logical network interconnection be implemented using three-segment VXLAN. Alternatively, it can be implemented through Layer 3 routing (an IGP or BGP). The VXLAN solution is not affected by customer's network planning and is applicable on a large scale.

Three-segment VXLAN is recommended for Layer 2 and Layer 3 communication between DCs. BGP EVPN is configured on a DCI gateway to create a VXLAN tunnel. VXLAN packets received from one DC are decapsulated, re-encapsulated, and then sent to the other DC. This mode ensures that end-to-end VXLAN packets are successfully transmitted across DCs and VMs in different DCs communicate with each other.

An IGP or BGP is deployed between DCI devices, which function as VTEPs on fabric networks. The IGP or BGP is used to ensure connectivity of service network segments between fabric networks. Based on IGP or BGP routes, the DCI devices directly decapsulate VXLAN packets and forward the packets. This mode, however, requires strict IP address planning. That is, routes corresponding to IP addresses in all DCs must be reachable on the Internet. This mode is applicable to interconnection between multiple DCs in an enterprise and places stringent requirements on the overall plan.

7.2.2.4 Forwarding Plane Solution Design

This section illustrates how to enable Layer 2 and Layer 3 communication based on three-segment VXLAN for logical network interconnection.

1. Inter-DC Layer 3 communication through three-segment VXLAN

The following is an example to describe the interworking process. In Figure 7.17, DC A and DC B need to communicate at Layer 3 through three-segment VXLAN.

VXLAN and BGP EVPN are deployed in each DC as the control plane protocol. Leaf1 to Leaf4 function as VTEPs in the two DCs, and Leaf2 and Leaf3 function as VTEPs between DCs.

1. Control plane

– Leaf4 learns the IP address of VMb2 in DC B and saves it to the routing table for the L3VPN instance. Leaf4 then sends a BGP EVPN route to Leaf3.

– In Figure 7.18, Leaf3 receives the BGP EVPN route and obtains the host IP route within it. Leaf3 then establishes a VXLAN tunnel to Leaf4 according to the VXLAN tunnel establishment process. Leaf3 sets the next hop of the route to the VTEP address of Leaf3, re-encapsulates the Layer 3 VNI of the L3VPN instance, and sets the source MAC address to the

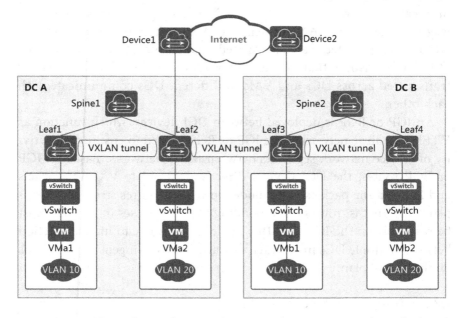

FIGURE 7.17 Networking of inter-DC Layer 3 communication through three-segment VXLAN.

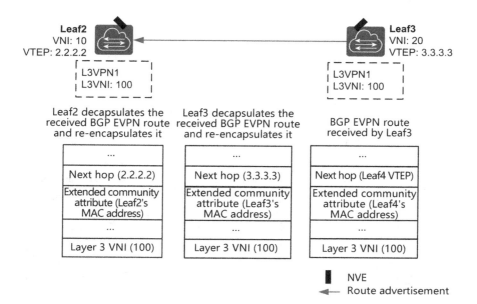

FIGURE 7.18   Networking of the control plane.

MAC address of Leaf3. Then, Leaf3 sends the re-encapsulated BGP EVPN route to Leaf2. This process is subsequently repeated from Leaf2 to Leaf1. Leaf1 then receives the BGP EVPN route and establishes a VXLAN tunnel to Leaf2.

2. Data plane

   – Leaf1 receives Layer 2 packets destined for VMb2 from VMa1 and determines that the destination MAC addresses in these packets are all gateway interface MAC addresses. Leaf1 discards the Layer 2 packets and finds the L3VPN instance corresponding to the BDIF interface through which VMa1 accesses the BD. Leaf1 then searches the routing table of the L3VPN instance for VMb2's host route, encapsulates the received packets into VXLAN packets, and sends them to Leaf2 over the VXLAN tunnel.

   – In Figure 7.19, Leaf2 receives and parses the VXLAN packets, finds the L3VPN instance corresponding to the Layer 3 VNI of the packets, and searches the routing table of the L3VPN instance for VMb2's host route. Leaf2 re-encapsulates the

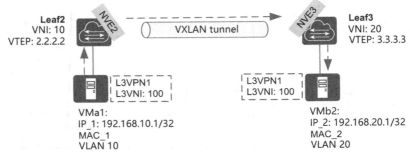

The following tables appear below the diagram:

Leaf2 receives VXLAN packets sent by Leaf1		Leaf2 decapsulates the received VXLAN packets and re-encapsulates them		Leaf3 decapsulates the received VXLAN packets and re-encapsulates them	
DMAC	NET MAC	DMAC	NET MAC	DMAC	NET MAC
SMAC	NVE1 MAC	SMAC	NVE2 MAC	SMAC	NVE3 MAC
SIP	Leaf1 IP	SIP	2.2.2.2	SIP	3.3.3.3
DIP	2.2.2.2	DIP	3.3.3.3	DIP	Leaf4 IP
UDP S_P	HASH	UDP S_P	HASH	UDP S_P	HASH
UDP D_P	4789	UDP D_P	4789	UDP D_P	4789
VNI	100	VNI	100	VNI	100
DMAC	Leaf2 MAC	DMAC	Leaf3 MAC	DMAC	Leaf4 MAC
SMAC	Leaf1 MAC	SMAC	Leaf2 MAC	SMAC	Leaf3 MAC
SIP	IP_1	SIP	IP_1	SIP	IP_1
DIP	IP_2	DIP	IP_2	DIP	IP_2
payload		payload		payload	

– – –▶ Traffic forwarding path

FIGURE 7.19   Networking of the data plane.

VXLAN packets in which the Layer 3 VNI is carried in VMb2's host route sent by Leaf3 and the outer destination MAC address is the MAC address carried in VMb2's host route sent by Leaf2. Then Leaf2 sends the packets to Leaf3. Leaf3 sends the packets to Leaf4, and then, Leaf4 sends the packets to VMb2 based on routing information, both in a similar manner.

2. Three-segment VXLAN is used to implement Layer 2 interconnection between DCs (an example is shown in Figure 7.20).

   The scenario is similar to that of Layer 3 interconnection. VXLAN tunnels are established in DC A and DC B as well as between transit leaf nodes in DCs. To enable communication between VM1 and VM2, implement Layer 2 communication between DC A and DC B.

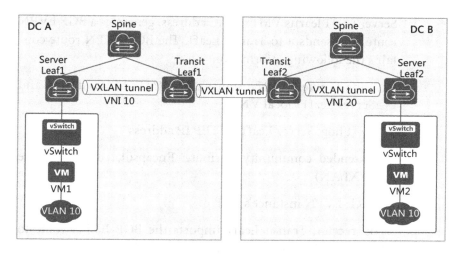

FIGURE 7.20   Three-segment VXLAN for Layer 2 interconnection between DCs.

If the VXLAN tunnels within DC A and DC B use the same VNI, the VNI can also be used to establish a VXLAN tunnel between Transit Leaf1 and Transit Leaf2. Unlike Layer 3 interconnection, in real-world situations, different DCs have their own VNI spaces. Therefore, VXLAN tunnels in DCs A and B may use different VNIs. In this case, the VNIs need to be converted to enable a VXLAN tunnel to be established between Transit Leaf1 and Transit Leaf2.

Figure 7.21 shows the control plane.

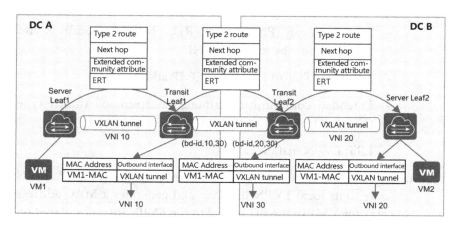

FIGURE 7.21   Control plane.

1. Server Leaf1 learns VM1's MAC address, generates a BGP EVPN route, and sends it to Transit Leaf1. The BGP EVPN route contains the following information:

   - Type 2 route: EVPN instance's RD, VM1's MAC address, and Server Leaf1's local VNI

   - Next hop: Server Leaf1's VTEP IP address

   - Extended community attribute: Encapsulated tunnel type (VXLAN)

   - ERT: EVPN instance's ERT

2. Upon receipt, Transit Leaf1 imports the BGP EVPN route to its local EVPN instance and generates a MAC address entry for VM1 in the BD bound to the EVPN instance. Based on the next hop and encapsulated tunnel type, the outbound interface in the MAC address entry recurses to the VXLAN tunnel destined for Server Leaf1. The VNI in VXLAN tunnel encapsulation information is the local VNI.
   Transit Leaf1 re-originates the BGP EVPN route.

3. When modifying the VNI, Transit Leaf1 searches the mapping table for the desired mapping VNI based on the BD ID and local VNI, and then changes the VNI in the re-originated route to the mapping VNI. The re-originated BGP EVPN route contains the following information:

   - Type 2 route: EVPN instance's RD, VM1's MAC address, and local VNI for the mapping VNI

   - Next hop: Transit Leaf1's VTEP IP address

   - Extended community attribute: Encapsulated tunnel type (VXLAN)

   - ERT: EVPN instance's ERT

4. Upon receipt, Transit Leaf2 adds the re-originated BGP EVPN route to its local EVPN instance and generates a MAC address entry for VM1 in the BD bound to the EVPN instance. Based on

the next hop and encapsulated tunnel type, the outbound interface in the MAC address entry recurses to the VXLAN tunnel destined for Transit Leaf1. The VNI in VXLAN tunnel encapsulation information is the mapping VNI.

5. Transit Leaf2 re-originates the BGP EVPN route. When modifying the VNI, Transit Leaf2 searches the mapping table for the desired local VNI based on the BD ID and mapping VNI and then changes the VNI in the re-originated route to the local VNI. The re-originated BGP EVPN route contains the following information:

   – Type 2 route: EVPN instance's RD, VM1's MAC address, and local VNI for the mapping VNI

   – Next hop: Transit Leaf2's VTEP IP address

   – Extended community attribute: Encapsulated tunnel type (VXLAN)

   – ERT: EVPN instance's ERT

6. Upon receipt, Server Leaf2 imports the re-originated BGP EVPN route to its local EVPN instance and generates a MAC address entry for VM1 in the BD bound to the EVPN instance. Based on the next hop and encapsulated tunnel type, the outbound interface in the MAC address entry recurses to the VXLAN tunnel destined for Transit Leaf2. The VNI in VXLAN tunnel encapsulation information is Server Leaf2's local VNI.

3. Forwarding plane

   Figure 7.22 shows the forwarding plane.

   • After receiving a Layer 2 packet from VM2 through a Layer 2 subinterface in a BD, Server Leaf2 searches the MAC address table in the BD based on the destination MAC address for the outbound interface of the VXLAN tunnel, and obtains information on the VXLAN tunnel encapsulation, such as local VNI, and source and destination VTEP IP addresses. Based on the obtained information, the Layer 2 packet is encapsulated through the VXLAN tunnel and then forwarded to Transit Leaf2.

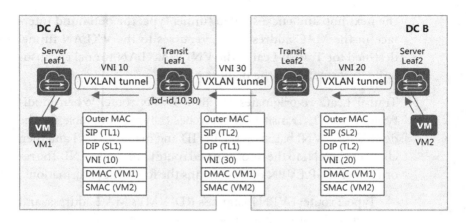

FIGURE 7.22   Forwarding plane.

- Upon receipt, Transit Leaf2 decapsulates the VXLAN packet and finds the target BD based on the VNI in the packet. The preceding process is then repeated, and the Layer 2 packet is forwarded to Transit Leaf1, where the process is repeated once again and the packet is forwarded to Server Leaf1 and then to VM1 at Layer 2.

7.2.2.5 External Network Multi-Active Model

Multiple DCs in the multi-site solution can provide services and carry the same services simultaneously. This improves the overall service capability and system resource utilization of DCs. Multiple DCs support each other. If a DC fails, services can be automatically switched to the other DC, ensuring service continuity.

The same service applications are deployed in the active-active DCs, and IP network segments of the two DCs are different. Therefore, Layer 3 interconnection is required between DCs. Service applications are accessed using domain names, so the Global Server Load Balancer (GSLB) needs to be deployed. Dynamic or static load balancing policies are used to parse different site IP addresses from access requests, as shown in Figure 7.23.

The GSLB detects the status of the application system through health status detection or by interworking with the software load balancer (SLB). If an application server in a DC is faulty, traffic is switched to other servers

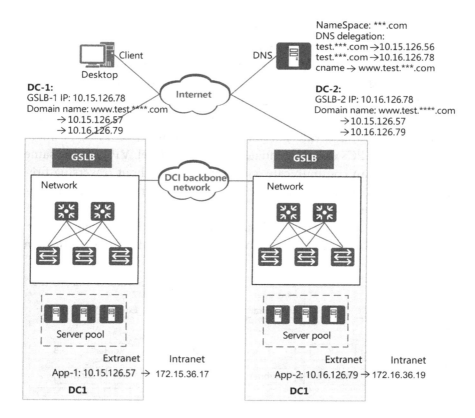

FIGURE 7.23   Multi-site solution.

in the SLB cluster of the same DC. This mitigates faults within the DC. If all the application servers in a DC are faulty, traffic is switched to the other DC.

This is called service-level active-active, which is commonly used, especially in the finance industry, Internet content providers, and CDN service providers. In these industries, technologies have to be mature.

This mode can also be extended to multiple DCs, enabling frontend nodes to provide services for end users within a short distance, reducing latency in application loading and improving the service access experience. In addition, this mode greatly improves the resource utilization of frontend service servers and network egresses in the backup DC. It uses GSLB detection and switchover policies to implement site-level fast failover, improving the availability of services.

7.2.3 Recommended Deployment Solutions

7.2.3.1 VPC Service Model by Security Level

The VPC service model by security level is mainly used in private cloud scenarios, as shown in Figure 7.24. In private cloud scenarios, networks of most enterprises in the finance industry are divided into several large VPCs (service VPCs) based on service security levels, such as the DMZ, internal network, and external network. The number of VPCs is small, while service VPCs are differentiated by security level. VPCs of the same security level can communicate with each other without traversing firewalls. VPCs of different security levels need to traverse firewalls to communicate with each other.

In most cases, private cloud networks are planned by enterprises. IP addresses are planned in a standard manner and do not overlap. In

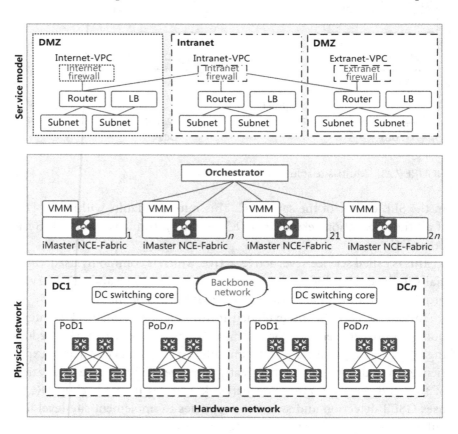

FIGURE 7.24   VPC allocation by security level.

addition, private cloud networks often need to interconnect with traditional networks, and services have to be deployed across old and new DCs. Therefore, the DCI solution can be implemented through the underlay network, and an IGP or BGP is deployed to connect DCs together. Services provided by multiple DCs can be processed in active-active mode through GSLB, as shown in Figure 7.25.

In private cloud scenarios, one VPC is deployed for each security level. Three VPCs are deployed in this example. The routing plane in each VPC is independent. Communication traffic across fabric networks in a VPC does not pass through firewalls, and a large VPC is formed across DCs. All VPCs share the same external routing plane and traffic from the external network to the internal VPC needs to pass through firewalls. BGP EVPN is deployed on each fabric network as the VXLAN control plane protocol, and each fabric network is a VXLAN domain.

IP addresses on the entire network must be strictly planned. An IGP or BGP is used to connect core devices in the DC and these devices to backbone devices. Layer 3 interconnection is implemented through VLANs.

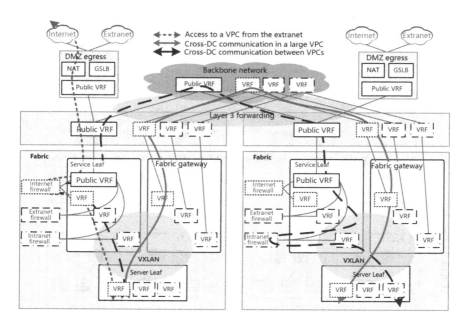

FIGURE 7.25 Forwarding plane for Layer 3 interconnection on the underlay network between DCs.

7.2.3.2 Multi-Tenant VPC Model Analysis

The multi-tenant service model is mainly applied to the carrier and Internet industries, as shown in Figure 7.26. In the carrier industry, requirements are mainly formulated by the design institute of each carrier. Because a design institute is not responsible for specific services, the requirements are extensive. The main requirements are as follows:

- One PoD has one resource pool, and services are deployed across PoDs. Each PoD is managed by one cloud platform and one controller.

- Service traffic between two PoDs needs to pass through the DC switching core. PoDs need to communicate with each other at Layer 2 and Layer 3.

- SDN PoDs need to be decoupled, and the fault and upgrade scope is controllable. Since each PoD is managed by one cloud platform and

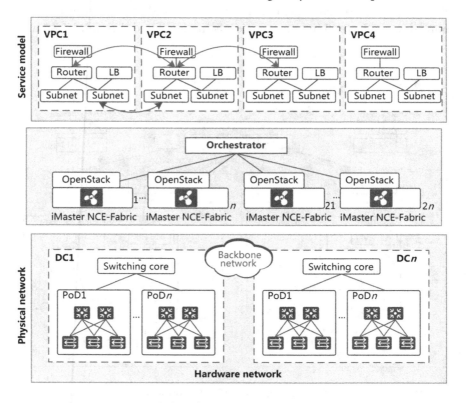

FIGURE 7.26 Recommended multi-tenant VPC solution.

one controller, each PoD is an independent VXLAN domain. This requirement can typically be met.

- Private IP addresses of different services can overlap. Therefore, PoDs need to communicate with each other through VXLAN. Routing is configured on the VXLAN underlay network and VNIs are used for isolation, without taking service IP addresses into account.

- To support multiple vendors' devices and orchestration on third-party cloud management platforms, the controller needs to develop APIs for Layer 2 and Layer 3 communication.

Figure 7.27 shows the implementation on the forwarding plane.

VPC instances on each fabric network are centrally orchestrated by the cloud management platform. The cloud management platform is also responsible for orchestrating Layer 2 and Layer 3 interconnection between PoDs to form a large VPC. BGP EVPN is deployed on each fabric network as the VXLAN control plane protocol, and each fabric network is a VXLAN domain. Fabric networks communicate with each other through

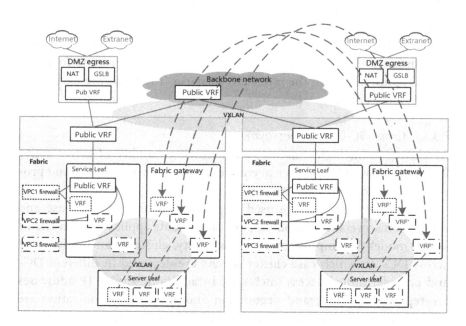

FIGURE 7.27 Forwarding plane of the inter-DC interconnection solution using three-segment VXLAN.

VXLAN. The fabric gateway of each fabric network functions as the NVE node of the interconnection VXLAN tunnel.

Similarly, all VPCs share an external routing plane. A public VRF is deployed to connect to the external network, and traffic from the external network to the VPC needs to pass through firewalls.

7.3 MULTI-POD SOLUTION DESIGN

This section covers the detailed design of the multi-PoD solution from the following aspects: application scenario, specific design, and deployment design.

7.3.1 Application Scenario of the Multi-PoD Solution

In the multi-PoD solution, when multiple DCs are within a short distance from each other, a single SDN controller remotely manages multiple DCs, and multiple DCs are in the same resource pool. This way, a large Layer 2 network across DCs can be constructed, deploying a VPC across DCs. In addition, SDN controller clusters are deployed in active/standby mode, and DR and backup are performed on the management plane. The multi-PoD solution supports cross-DC deployment of clusters and migration of VMs, as well as network-level active/standby DR.

Logically, network features and behaviors in the multi-PoD scenario are similar to those on a single-DC network. You only need to focus on special requirements in the multi-DC scenario. The following details the application scenarios.

7.3.1.1 Cross-DC Cluster Deployment

As cluster technologies mature, an increasing number of active-active DCs are deployed in a cluster to ensure cross-region availability and provide load balancing capabilities.

Physical IP addresses are used to provide services for database and application servers. This mode applies only to data applications (in client/server mode). To improve service continuity, clusters can be deployed across DCs, in which case cluster servers are deployed in different DCs and provide unified access interfaces. In addition, service IP addresses are replaced by VIPs, and negotiation and status synchronization are performed between server clusters through the cross-DC interconnection network. The cluster heartbeat link and public network usually need to access the same Layer 2 domain. Therefore, a large Layer 2 network needs

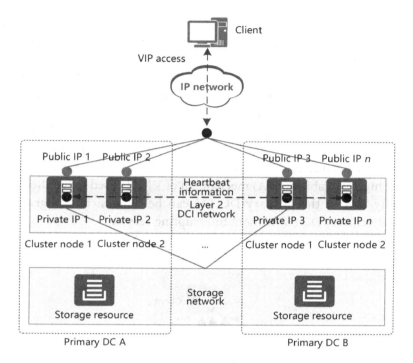

FIGURE 7.28   Cross-DC cluster deployment.

to be deployed across DCs. Layer 2 interconnection can be implemented using technologies such as bare optical fibers, Wavelength Division Multiplexing (WDM), VPLS, and EVPN VXLAN, as shown in Figure 7.28.

The server cluster solution has the following requirements on DCNs:

- Low latency: The round-trip time (RTT) and distance of deployment are limited.

- Layer 2 interconnection: Layer 2 interconnection is required, and the latency in interconnection is limited.

- High network reliability: The interconnection network is highly reliable to avoid split-brain occurring.

7.3.1.2 Cross-DC VM Migration

For data application systems carried by VMs, services are directly accessed through VMs. Therefore, the IP address of a VM is the access IP address of a service.

After servers are virtualized, they become dynamic and their resources can be reused. VMs provide services for applications, so Server Load Balancing (SLB) cannot be used to allocate resources, and VMs can be only migrated to idle PMs. Application administrators can flexibly schedule and adjust VM running locations and online or offline status based on requirements for application resources, such as CPU and memory requirements.

When a VM resource pool is expanded to two DCs, physical server resources in the backup DC can be fully utilized and virtual resources can be flexibly scheduled across DCs, greatly improving the utilization of resources.

The high availability (HA) mechanism of VMs is used to handle faults. A faulty VM in the primary DC (A) can be dynamically migrated to another primary DC (B) without interrupting services.

Figure 7.29 shows the process of migrating VMs across DCs. Network requirements for live migration of VMs across DCs are as follows:

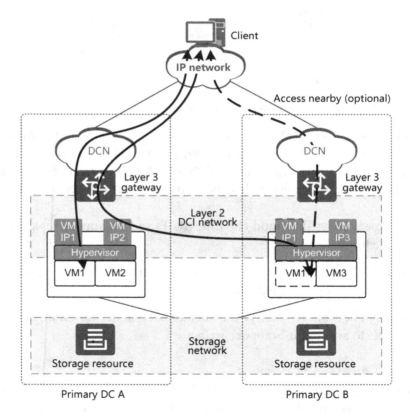

FIGURE 7.29   Cross-DC VM migration.

- Layer 2 interconnection: IP or MAC addresses and the TCP session status remain unchanged.

- Low latency: VM status synchronization requires low latency.

- High bandwidth: High bandwidth is required for VM migration to ensure rapid migration of status and storage data.

7.3.1.3 Network-Level Active/Standby DR

When the server cluster is deployed across DCs, the cluster provides a unified VIP for external access. This IP address needs to be configured on the gateway. Taking cross-DC deployment and DR into account, the service gateway needs to be created in multiple DCs and the DR relationship such as the active/standby relationship needs to be set up. Similarly, in a scenario in which a VM is migrated across DCs, a gateway of the VM also needs to be deployed across DCs, and a DR relationship needs to be established. This is because an IP address of the migrated VM remains unchanged. If north-south traffic of the service needs to be protected by a firewall, the firewall needs to be deployed across DCs, and security policies need to be synchronized.

When the DC is faulty, the active gateway is faulty, in which case traffic will be switched to the standby gateway for security protection, ensuring the north-south traffic remains uninterrupted and services can be provided externally.

Figure 7.30 shows the DR networking.

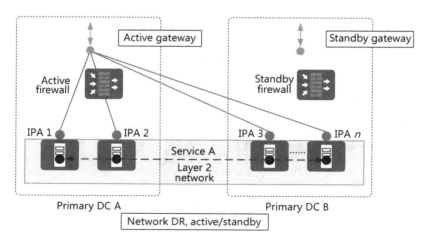

FIGURE 7.30 DR networking.

Network-level active/standby DR has the following requirements on a network:

- Active and standby egresses: Two DC egresses can work in active/standby mode. One egress is used as the external gateway, and the other is on standby. If the active egress is faulty, services are automatically switched to the standby egress.

- Active and standby firewalls: Firewall security policies are synchronized. When the active egress is faulty, services are switched to the firewall bound to the standby egress automatically.

7.3.2 Multi-PoD Solution Design

7.3.2.1 Architecture of the Multi-PoD Solution

Multi-PoD is mainly used in scenarios where DCs are deployed in the same city within a short distance, and where the network layer provides network DR.

In Figure 7.31, compute and network resources of multiple DCs form a unified resource pool and are managed by one SDN controller. VPCs and subnets can be deployed across DCs and can communicate with each other at Layer 2 or Layer 3.

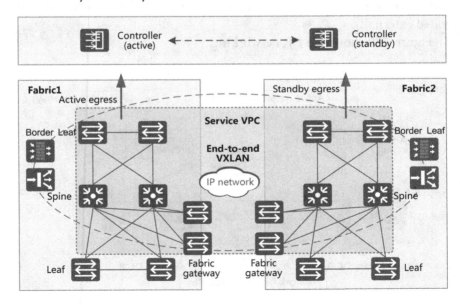

FIGURE 7.31 Architecture of the multi-PoD solution.

To improve the reliability of the management plane, SDN controllers can be deployed in the two DCs. The two SDN controller clusters work in active/standby mode. The active SDN controller cluster manages the network. If the active SDN controller cluster is faulty, an active/standby switchover is performed between the two clusters, and the original standby SDN controller cluster becomes active to take over services, performing DR on the management plane.

In this scenario, the physical networks of each DC are independent of each other in terms of architecture. Devices and channels for interconnection between DCs need to be added on the basis of a single DC. Multiple DCs are unified end-to-end VXLAN domains, and the network and computing resources are unified resource pools.

It is recommended that active and standby egress gateways be deployed; that is, all north-south traffic is diverted to the active egress gateway. If the egress gateways work in active/standby mode, firewalls are also deployed in active/standby mode to ensure high service availability.

Figure 7.32 shows main features of the multi-PoD solution.

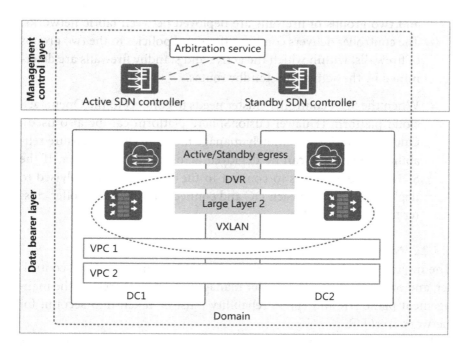

FIGURE 7.32 Networking of the multi-PoD solution.

- DR on the management plane of a single domain: Multiple DCs share a resource pool. In a VXLAN domain, only one resource pool is available to services, and VPCs can be deployed across DCs. In addition, SDN controller clusters are deployed in active/standby mode, performing DR on the management plane.

- Cross-DC VPC deployment: Similar to the DR on the management plane of a single domain, a VPC can be deployed across fabric networks because there is one resource pool.

- Cross-DC deployment of egresses in active/standby mode within a VPC: Multiple DCs can have multiple egresses. One egress can be selected as the external gateway for a VPC, while the other egresses are used as standby egresses. The active and standby egresses are subsequently determined based on route priorities. You can also select only one centralized egress through which all north-south traffic is transmitted to northbound APIs.

- Cross-DC deployment of firewalls in active/standby mode in a VPC: Firewalls are deployed in active/standby mode on the fabric network, and two groups of firewalls are deployed between fabric networks. The controller delivers configurations and policies to the two groups of firewalls, within which the active and standby firewalls are determined by the active and standby routes.

- When the SDN controller cluster needs to connect to the OpenStack cloud platform (Huawei FusionSphere platform can be also used): One OpenStack can remotely manage multiple DCs. To ensure reliability, OpenStack can be deployed across DCs in a cluster. If the SDN controller needs to connect to the VMM, you are advised to deploy one VMM in each DC and connect one SDN controller cluster to multiple VMMs.

7.3.2.2 Network-Level DR

The multi-PoD solution uses a management domain of an SDN controller, and one SDN controller cluster manages all DCs. Therefore, the management plane DR and egress reliability must be taken into account for network-level DR.

Figure 7.33 shows DR on the management plane. The standby SDN controller is used to perform DR. Although one SDN controller cluster

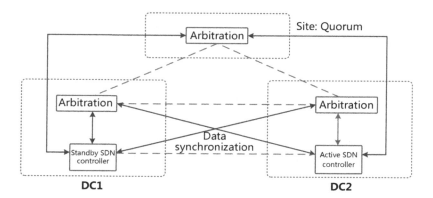

FIGURE 7.33   DR on the management plane.

manages two DCs centrally, to improve reliability on the management plane, a standby SDN controller cluster can be deployed in different DCs. The active and standby SDN controller clusters synchronize data in real time and support active/standby switchovers. The active SDN controller cluster manages devices and connects to arbitration devices. In each SDN controller cluster, nodes support each other. If a DC is faulty, for example, the power goes down, the standby SDN controller cluster becomes active to manage the network. The arbitration service is used to determine whether split-brain has occurred if the link between the two DCs goes Down and clusters split. The active cluster is determined based on the pre-configured priority to prevent two active nodes from operating.

In the egress DR design, the multi-PoD solution provides active and standby egresses. In the multi-PoD active/standby egress scenario, multiple DCs managed by one SDN controller cluster are interconnected to form a unified resource pool. One DC functions as the active egress of the external network, and the other DC functions as the standby egress of the external network, as shown in Figure 7.34.

Two DCs are connected through end-to-end VXLAN, and logical routers are deployed on distributed VXLAN gateways across DCs. In this way, VMs can exchange east-west traffic in the VPC across DCs, as well as migrate across DCs, and cloud hosts can be deployed in HA mode. The two DC egresses are configured to work in active/standby mode, in which multiple external networks can be configured. For external networks, DC1 and DC2 can be used interchangeably as the active and standby egresses.

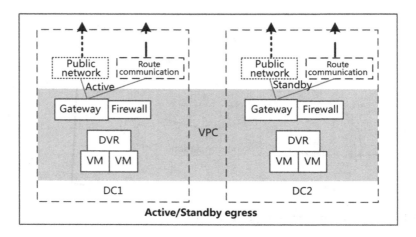

FIGURE 7.34   Active and standby egresses.

You can select different pairs of active and standby egresses for different VPCs. Either external network can be selected for VPCs. The egresses of the two DCs perform VPC-based load balancing.

Route priorities are configured based on the active/standby relationship of external networks. The active egress' route has a higher priority than the standby egress. In most cases, north-south traffic is preferentially transmitted through the active egress based on the high-priority route. If the active DC or the active egress is faulty, the active egress' route becomes invalid, and high-priority route traffic is then switched to the standby outbound interface based on the low-priority route.

7.3.2.3 Security Policy Synchronization Design

The egress of the DC needs to be protected by a firewall, and network-level DR provides active and standby egresses. After the active/standby switchover, traffic passes through the standby firewall through the egress of the standby DC. To ensure that the traffic is protected by the firewall of the standby DC and pass through the firewall, firewall policies of the active and standby DCs must be consistent.

In the multi-PoD solution, it is recommended that a group of firewalls be deployed in each DC to perform active/standby mirroring. The controller delivers identical configurations and policies to the two groups of firewalls; therefore, these firewalls work in active/standby mode through active/standby routes (Figure 7.35).

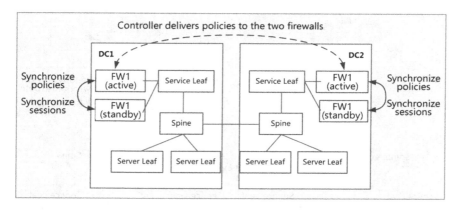

FIGURE 7.35   Firewall deployment in the multi-PoD solution.

7.3.2.4 Forwarding Plane

In a single VXLAN domain, the implementation of the multi-PoD forwarding plane is similar to that in a single DC. BGP EVPN functions as the VXLAN control plane protocol; border, server, and service leaf nodes function as VTEPs; and spine nodes function as RRs. VTEPs function as RR clients and establish BGP peer relationships with spine nodes to advertise EVPN address family routes. The BGP EVPN routes trigger automatic VXLAN tunnel establishment between VTEPs, eliminating the need for manual configuration of tunnels. BGP EVPN advertises host and MAC routes, as shown in Figure 7.36.

One end-to-end VXLAN tunnel is established between two DCs. In Figure 7.37, Leaf1 in DC A and Leaf4 in DC B run BGP EVPN to transmit MAC or host routes without changing their next hop addresses of the MAC or host routes. As a result, an end-to-end VXLAN tunnel is established between the VTEPs on Leaf1 and Leaf4 across DCs. VMb2 and VMa1 in Figure 7.37 are used as an example to illustrate the VXLAN tunnel establishment process and data packet forwarding process in a subnet.

1. Control plane

   - Leaf1 obtains information about VMa1, generates a BGP EVPN route, and sends it to Leaf2. This BGP EVPN route carries the export VPN target of the local EVPN instance, and its next hop is the VTEP address on Leaf1.

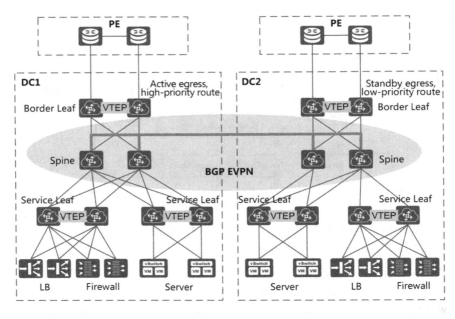

FIGURE 7.36   Forwarding plane design.

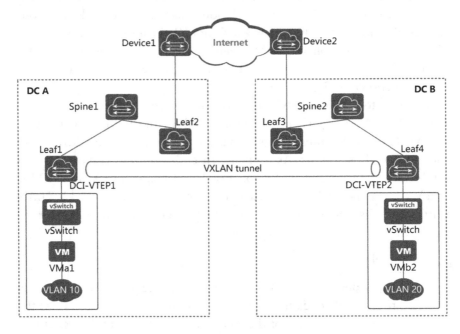

FIGURE 7.37   Data forwarding.

- Upon receipt of the BGP EVPN route, Leaf2 sends it to Leaf3 without changing the next hop of the route.

- Upon receipt of the BGP EVPN route, Leaf3 sends it to Leaf4 without changing the next hop of the route.

- Upon receipt of the BGP EVPN route, Leaf4 checks the export VPN target of the EVPN instance that it carries. If the export VPN target carried by the route is the same as the import VPN target of the local EVPN instance, Leaf4 accepts the route. If not, the route is discarded. After accepting the BGP EVPN route, Leaf4 obtains the next hop of the route, which is the VTEP address of Leaf1. Leaf4 then establishes a VXLAN tunnel to Leaf1 according to the VXLAN tunnel establishment process.

2. Data packet forwarding

End-to-end VXLAN supports inter-subnet packet forwarding as well as forwarding of known unicast packets and BUM packets on the same subnet. The data packet forwarding process in the end-to-end VXLAN scenario is the same as that in a DC configured with the distributed VXLAN gateway and is therefore not covered here.

7.3.3 Recommended Deployment Solutions

1. Basic network

Essentially, the network model of a single enterprise DC is divided into the service, management, and storage zones.

If there are other functional areas during actual deployment, you can refer to the network design of the three zones. In Figure 7.38, the service zone uses SDN and the spine-leaf architecture, and the management zone and storage zone use the star topology. This solution is also accepted by most enterprises. According to the network construction of most DCs that Huawei has participated in, networks in the management zone and storage zone are generally deployed using traditional network protocols, such as traditional Layer 3 routing or SAN; therefore, the traditional deployment and O&M mode is used.

In the multi-PoD scenario, this basic network is required to form an end-to-end VXLAN domain. Therefore, the basic networks of the two DCs need to be interconnected. The service networks of the two

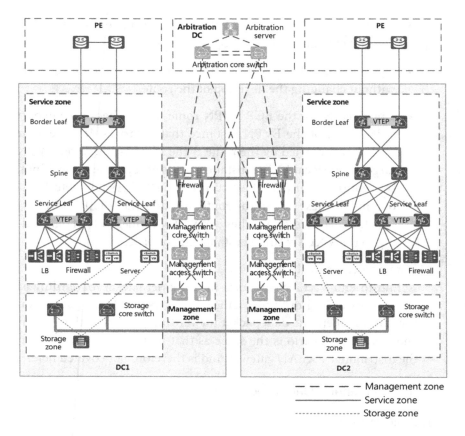

FIGURE 7.38   Basic network design.

DCs are connected through spine nodes, and the storage and management networks of the two DCs are connected based on customer habits. The network is divided into three planes: service, management, and storage network planes.

- Service network plane: The spine-leaf model is used. The service interfaces of servers are dual-homed to server leaf nodes, and firewalls and LBs are connected to service leaf nodes in bypass mode.

- Management network plane: This plane includes service control and network management. The cloud platform, SDN controller, VMM, and NMS are, respectively, dual-homed to management access switches. The arbitration server is deployed at a third-party site, an arbitration DC. The arbitration core switches are connected to the management core switches in the two DCs directly through optical fibers (WDM).

- Storage network plane: The storage network, a traditional network, connects to servers and storage devices.

For the two DCs, it is recommended that spine nodes in the service zones, management core switches, and storage core switches in the two DCs be directly connected through optical fibers (WDM).

2. Management network design

The management zone uses the two-layer architecture and deploys the cloud platform and SDN controller. The management access switches only function as Layer 2 access switches, and the management core switches function as the gateways of the management zone of each DC. The cloud platform requires Layer 2 communication between DC1 and DC2. Therefore, VRRP needs to be configured on the core switches of the two DCs to ensure Layer 2 communication between servers in the management zones of the two DCs.

Dynamic routes are deployed between the management core switches in the DCs and the firewalls on the management network to achieve connectivity in the management zone. Static routes are configured between the third-party site and the core switches in the management zones of the two DCs, as shown in Figure 7.39.

3. Routing design of the underlay network in the service zone

Because the two DCs are in the same VXLAN domain, connectivity of the underlay network between the two DCs must be achieved and it is recommended that OSPF be deployed on the underlay network. To facilitate route convergence and limit the fault domain, multiple OSPF areas are used to separate the OSPF routing domains of the two DCs. In Figure 7.40, OSPF area 1 is deployed in DC1, OSPF area 2 is deployed in DC2, and OSPF area 0 is deployed between DCs.

On the underlay network, IP addresses of loopback interfaces (VTEP IP addresses) and directly connected interfaces must be reachable.

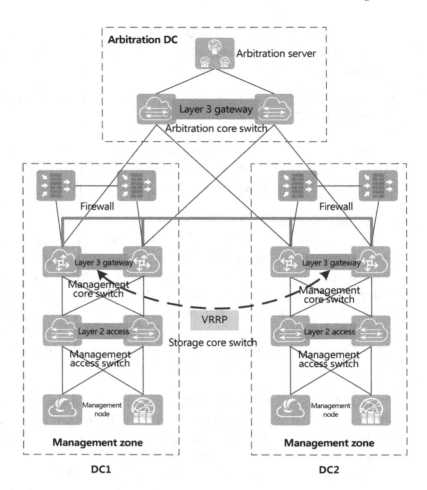

FIGURE 7.39 Management network design.

4. Routing design of the overlay network in the service zone

Two DCs are in the same VXLAN domain. Distributed VXLAN gateways learn the ARP entries of servers or F5 LBs, convert them into 32-bit host routes, and synchronize the routes on the entire network through EVPN. On service leaf nodes, default routes to firewalls are configured to access the external network, and priorities are configured for the default routes to the firewalls and imported to EVPN so that traffic is transmitted through active and standby egresses.

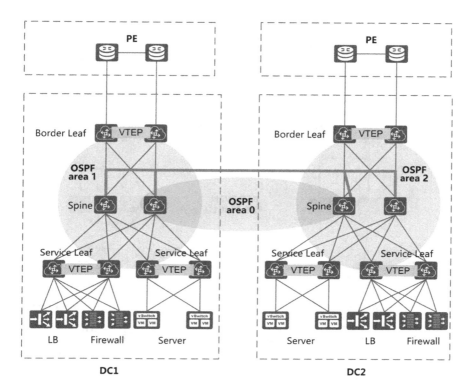

FIGURE 7.40   Routing design of the underlay network in the service zone.

The two DCs are in the same BGP EVPN domain. Similar to a single DC, all host and network segment routes can be flooded to the two DCs through BGP. The spine nodes function as RRs and the two spine nodes flood routes.

EBGP dynamic routes are deployed between border leaf nodes and external PEs. As the active and standby status of routes is pre-configured on external networks, similarly, the priorities of these routes are also manually pre-configured.

The active and standby status of routes is implemented through a specific BGP attribute according to the actual situation. Both the egress gateway and peer PE must be able to identify the specific attribute.

Figure 7.41 shows the routing design on the overlay network.

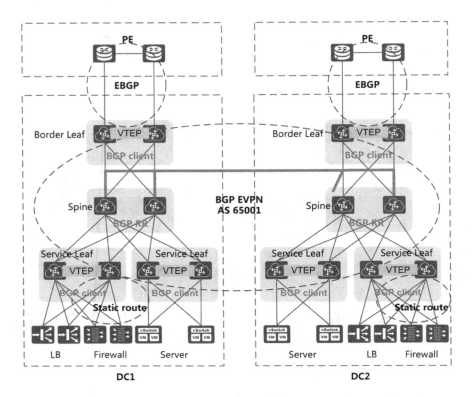

FIGURE 7.41   Routing design on the overlay network.

CHAPTER 8

Building E2E Security for Cloud DCNs

S ECURITY HAS ALWAYS BEEN at the heart of every DCN, and this is set to continue as SDN introduces even greater security challenges. Huawei provides security sub-solutions for the forwarding, control, and application layers, and integrates Huawei firewall, IPS/IDS, sandbox, security manager, and CIS series security products to further implement security across the network. This provides a closed loop from single-point defense to DCN-wide collaborative defense, enabling a rapid response to internal threats as well as effective isolation to prevent spread.

8.1 CLOUD DCN SECURITY CHALLENGES

1. Providing effective security services is challenging, particularly those capable of satisfying the new requirements of cloud and SDN-based DCs.

    As DCs evolve toward cloudification and SDN, DCNs are deployed and provisioned in a dramatically different manner, and the same is true of applications. Specifically, this new deployment will be dynamically on demand and feature elastic scaling capabilities. However, this also poses new requirements for the deployment and provisioning of DCN security services.

    a. Security services are complex and time-consuming to provision.

Security devices are usually deployed in a dispersed manner, covering risk points such as different DC zones and borders. When deploying a security service for a new application, it is necessary to plan and configure the service on multiple security devices. The entire process is both complex and time-consuming.

b. Security capabilities offer limited automation.

Security capabilities provided by OpenStack include only basic firewall capabilities such as security policies, with additional security services requiring complex and inefficient manual configuration.

c. Security capabilities are redundant.

In order to deal with traffic bursts, each security device is required to maintain extra capabilities which are often redundant and a waste of resources.

2. Blurred security borders and increasingly complex threats.

a. In cloud-based and SDN-based DCNs, the virtualization of servers and Layer 2 networks makes it impossible to define tenant network borders as before, which, in turn, leads to blurred security borders.

The physical borders of static and native networks are now replaced by dynamic and virtual logical borders. Virtual Layer 2 traffic on a virtual Layer 2 network is forwarded by vSwitches, and so traffic becomes invisible and difficult to detect. In addition, the logical network topology can change at any time based on moment-to-moment service requirements, meaning legacy security architecture based on traditional physical borders cannot provide appropriate protection.

b. Internal attack sources lead to zero trusted zones.

Cloud DCN resources are highly integrated. As tenants can no longer be isolated at the physical layer, it is difficult to protect them against security threats. If VMs in a DCN become zombie hosts (online hosts controller by an attacker), attacks may be launched from within the network. Traditional security countermeasures cannot offer defense in these cases.

c. Legacy security methods cannot detect and defend against new threats such as advanced persistent threats (APTs).

An APT is a network attack and intrusion behavior directed against enterprises, and designed to intercept core data.

APTs usually exploit zero-day vulnerabilities or utilize advanced evasion techniques (AETs). Through internal penetration and privilege escalation, APTs covertly mine data over a long period of time. Remotely controlled APTs can lead to data damage, loss, and leakage. Legacy detection based on accurate signatures cannot detect and defend against APTs.

3. Security management and O&M become increasingly complex.

   a. Security policy management is complicated and applies without awareness of applications.

   b. DCs usually operate a large number of security policies (such as in the financial industry), and managing all of these policies based on IP addresses would be a daunting task. In addition, these policies cannot be associated with the status of service applications, meaning they cannot be adjusted according to service changes.

      DC security devices are separately deployed in order to form security silos, and individual security threats are detected and processed by DC zones. As such, central prevention and processing is not possible, and a security protection system with a larger protection scope is beyond reach. As a result, there are no effective guidelines for security O&M.

   c. Investigating and handling security threats is challenging, leading to slow response times.

      A large number of security device threat logs are stored separately and can only be analyzed by professionals, which is both inefficient and ineffective. The Mean Time to Recovery (MTTR) from security threats is long, security response is slow, and the loop closure rate is low. In addition, manual investigation is required to trace and analyze threat events, leading to high OPEX.

8.2 CLOUD DCN SECURITY ARCHITECTURE

8.2.1 Overall Security Architecture

A legacy DC's security protection system is designed based on the principles of multi-layer protection, zone-based planning, and hierarchical deployment. On a cloud DCN, resources are integrated and shared, requiring a high degree of security isolation. Security must meet the virtualization requirements, where the security boundaries have become blurred. Strict control is required for access to cloud DCN resources, regardless of whether the access requests come from internal or external networks. In addition, security threat detection and countermeasures have a wider scope. On cloud DCNs, security threat detection and countermeasures are unified, information is comprehensively shared, and the security protection system is more extensive than legacy DCs to cover a larger scope. Such cloud DCN security needs to be delivered to users as a service, capable of adapting to dynamic on-demand deployment and elastic scaling of cloud DCN resources.

Under these principles, the security system of a cloud DCN differs greatly from that of a legacy DCN. Figure 8.1 shows the overall security solution for cloud DCNs.

The following describes SDN-related security technologies and solutions, including security groups offering virtualization security, network security, advanced threat detection and defense, and security management. The key to enterprise network security is to locate the security boundary (attack point). On the left side of the boundary is the attacker (script kiddie, hacker, and APT), and on the right side are the network and information assets. Enterprise network security is usually built at the security border for maximum defense; however, such security boundaries become blurred as a result of increased services, technology evolution, and continuous mode adjustment. Overcoming these challenges and maintaining enterprise network security boundaries is of vital importance in order to ensure effective security protection.

8.2.2 Architecture of Security Components

Figure 8.2 shows the architecture of security components in a cloud DCN.

1. Application layer

    At the application layer, cloud security services are provisioned on demand by the cloud OS, which orchestrates computing, storage, and security resources in a unified manner. Administrators dynamically provision security services using the cloud OS. In the southbound direction, the cloud OS interconnects with the controller to deliver security service requirements. The controller then converts the security service requirements into forwarder configurations and delivers them to network devices.

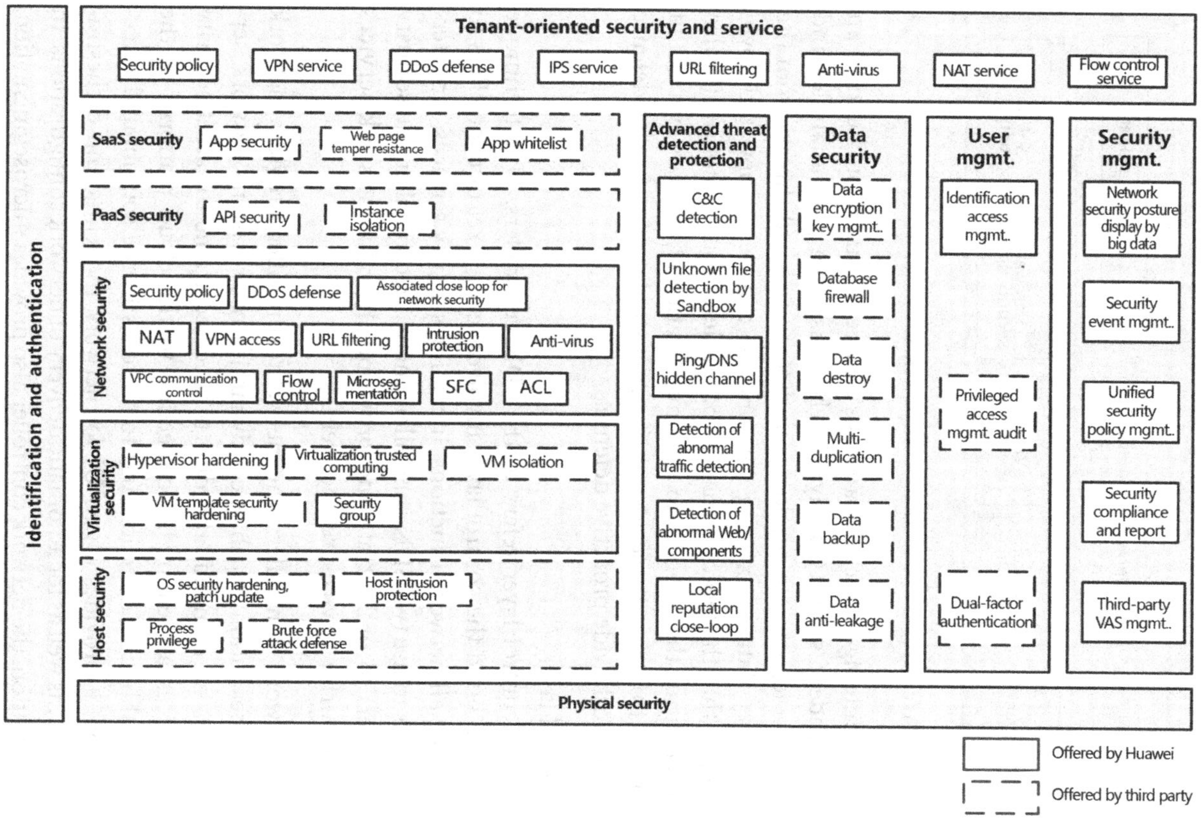

Tenant-oriented security and service

Security policy | VPN service | DDoS defense | IPS service | URL filtering | Anti-virus | NAT service | Flow control service

Security mgmt.
- Network security posture display by big data
- Security event mgmt.
- Unified security policy mgmt.
- Security compliance and report
- Third-party VAS mgmt.

User mgmt.
- Identification access mgmt.
- Privileged access mgmt. audit
- Dual-factor authentication

Data security
- Data encryption key mgmt.
- Database firewall
- Data destroy
- Multi-duplication
- Data backup
- Data anti-leakage

Advanced threat detection and protection
- C&C detection
- Unknown file detection by Sandbox
- Ping/DNS hidden channel
- Detection of abnormal traffic detection
- Detection of abnormal Web components
- Local reputation close-loop

SaaS security
- Security policy
- App security
- App whitelist
- Web page temper resistance

PaaS security
- API security
- Instance isolation

Network security
- Security policy
- DDoS defense
- Associated close loop for network security
- Anti-virus
- NAT
- VPN access
- URL filtering
- Intrusion protection
- ACL
- VPC communication control
- Flow control
- Microsegmentation
- SFC

Virtualization security
- Hypervisor hardening
- Virtualization trusted computing
- VM isolation
- VM template security hardening
- Security group

Host security
- OS security hardening, patch update
- Host intrusion protection
- Process privilege
- Brute force attack defense

Physical security

Identification and authentication

Legend:
- Offered by Huawei
- Offered by third party

FIGURE 8.1 Overall security solution for cloud DCNs.

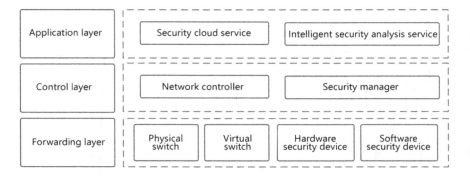

FIGURE 8.2 Architecture of security components.

The intelligent security analysis service is usually provided by an independent security analyzer, which uses big data analytics and machine learning technologies to detect advanced threats from the telemetry data of existing networks. It can detect new unknown threats that the legacy signature-based detection cannot, while also accelerating the response to, and investigation of, security events. Put simply, this security service shifts from single-point passive defense to network-wide proactive defense.

2. Control layer

The control layer refers to the SDN controllers. One technique used to build the control layer involves using a set of controllers to provide all network functions, including security services. An alternative is to use two SDN controllers (network controller and security manager) to orchestrate and provision the basic network services and security services, respectively.

- Network controller: The network controller provides network orchestration capabilities, dynamically provisions network services on demand, and diverts specified traffic to target security devices as required. Connecting to the cloud platform in the northbound direction and to switches in the southbound direction, it provides a web UI for network modeling and orchestration, as well as for automatic delivery of network configurations. In addition, the network controller also provides microsegmentation, SFC orchestration, and configuration delivery. SFC enables users to divert traffic to security devices to provision security services.

- Security manager: The security manager orchestrates security services (such as IPsec, security policies, anti-DDoS, security content detection, and address translation) and dynamically provisions security services on demand. It provides security service orchestration and unified security policy management, models and orchestrates security services, and automatically delivers security service configurations. In addition, the security manager collaborates with the network controller, security devices, and security analysis system to provide network-wide proactive security defense, which comprehensively detects, analyzes, and responds to threats.

Two sets of SDN controllers are deployed but decoupled from one another, facilitating future service upgrades and scale-out. In addition, these controllers can connect to external systems to provide differentiated functions and facilitate functional scale-out and secondary development. For example, the security manager can connect to a dedicated big data analysis platform to receive results and provide guidelines on security service deployment and adjustment, delivering higher-level security protection capabilities as a result.

This solution is recommended by Huawei, and a large number of enterprises are now decoupling network controllers from security managers when building their cloud DCNs. Further chapters in this document will elaborate on the controller decoupling solution in more detail.

3. Forwarding layer

The forwarding layer consists of physical and virtual network devices, including DC switches and security devices. They work together to implement security protection at the DC or tenant border, or within a tenant.

DC switches carry network traffic and provide security capabilities such as ACL, security group, and microsegmentation. They can also redirect specified traffic to security devices, such as firewalls, IDS, IPS, and WAFs, which then perform security detection and defense using their security capabilities.

A cloud DCN supports a large number of users, and the DC's switches and firewalls support virtual systems, such as VRF or vSys/

VD. While a set of virtual systems can be allocated to a user, virtual systems of different users do not affect each other. Inter-user traffic is blocked by default, which narrows down security zones. In addition, differentiated security policies can be formulated based on the service characteristics and requirements of different users. These security policies can then be further refined and managed, improving their overall effectiveness.

8.3 BENEFITS OF THE CLOUD DCN SECURITY SOLUTION

A comprehensive cloud DCN security solution offers the following benefits:

1. Pooled security resources and automated configurations.

   Cloud DCNs can pool hardware and virtual security resources in order to minimize the differences of underlying hardware. Based on the varied security requirements for east-west and north-south traffic, east-west and north-south firewall resource pools can be created to deploy equally varied security policies.

   Elastic scaling of the vFW or firewall cluster enables on-demand automatic scaling, improving resource utilization and reliability.

   Security services are orchestrated by tenant, and the security services of different tenants are isolated from each other. Tenants can implement automatic configuration of OpenStack's basic firewall capabilities, in addition to other threat detection capabilities such as IPS and antivirus.

2. Rich security capabilities for layer-by-layer defense.

   The cloud DCN security solution provides tenant-oriented security capabilities while also supporting tenant-based firewall security capabilities, which include security policies, IPsec VPN encryption, NAT policies (hidden addresses), IPS, antivirus, URL filtering, DDoS, and ASPF, as well as tenant-based microsegmentation and SFC. Microsegmentation refers to refined isolation and control on inter-DC mutual access traffic in order to implement security control on east-west traffic on cloud DCNs. SFC can flexibly divert specified traffic to multiple security service nodes and orchestrate traffic diversion.

   The cloud DCN security solution also provides infrastructure-oriented security capabilities for tenants. In terms of virtualization, security groups are deployed on cloud DCNs to isolate VMs. In terms of network, the following security capabilities are provided:

- Anti-DDoS devices can be deployed at the network border to implement refined traffic cleaning and defense.

- IPS/IDS series intrusion detection and defense devices can be deployed at the network border to provide a comprehensive signature database of vulnerabilities and threats, while also interworking with the sandbox to detect new network attacks (such as APT and zero-day attacks).

- A dedicated sandbox (for example, Huawei FireHunter) can be deployed to use multi-engine virtual detection and Hypervisor behavior capture to detect unknown threats.

- An advanced intelligent security analysis platform (for example, Huawei CIS) can be deployed to utilize big data analytics and AI intelligent detection algorithms to detect security threats through self-learning.

The cloud DCN security solution associates network devices with security devices to build a closed-loop security system. A network device collects security analysis and also executes security policies. The network and security analysis platform work together to detect security threats and enact appropriate security policies. This technique implements network-wide defense, instead of traditional single-point defense, which, in turn, enables rapid response to, and isolation of, internal threats.

3. Intelligent defense, visualized security, and simplified security O&M.

1. Intelligent analysis, detection, and defense

Complex systems and networks result in DC security vulnerabilities that are difficult to detect and defend against with legacy security monitoring.

The cloud DCN security solution interconnects with the CIS and sandbox (FireHunter), and based on big data technologies, comprehensively analyzes data from multiple dimensions, such as logs, files, traffic, and terminal behavior. Security threats are then identified using AI detection algorithms and self-learning, and proactive defensive measures are taken accordingly.

2. Visualized security

Legacy security O&M is performed on various security devices, making it difficult to effectively monitor the security posture of an entire DC. Intuitive monitoring is available once a cybersecurity intelligence system (CIS) is deployed.

Network-wide security status detection: The CIS collects system information, including that related to network and security devices, servers, terminals, and hosts, summarizes and correlates the information, and presents the security posture of the entire network in multiple dimensions. The collected information includes vulnerability information, asset information, access records, Internet access logs, and security events.

Visualized attack paths: The CIS uses big data analytics technology to associate threat information, graphically display attack paths, and provide attack backtracking and investigation capabilities.

3. Unified management and simplified O&M

The cloud DCN security solution centrally manages physical and virtual resources, as well as various types of security devices from multiple vendors.

The security manager provides refined security policies for administrators and hierarchical management of security policies and O&M capabilities for tenants, all of which is accessible through the unified management entrance. The security manager also supports infrastructure-based security policy optimization, as well as service application-based policy simulation and optimization, to detect security policies and simplify security policy O&M.

8.4 CLOUD DCN SECURITY SOLUTION

This section describes the technical solutions available for virtualization security, network security, advanced threat detection and defense, border security, and security management.

The following uses Huawei iMaster NCE-Fabric as the network controller and Huawei SecoManager as the security manager.

8.4.1 Virtualization Security

A large number of VMs are deployed in a DC to run various applications. Virtualization security refers to VMs that are grouped to implement security control and isolation.

1. Security group overview

   Security groups can control and isolate mutual access between VMs so that tenants can customize security isolation for their leased VMs.

   As shown in Figure 8.3, VMs can be added to or removed from a security group. By matching the five-tuple of packets, the system can filter the incoming and outgoing traffic of VM ports, and implement access control and isolation between VMs. Security group rules can also be dynamically migrated with VMs.

   Security groups have the following characteristics by service model:

   - A security group is an abstract set of VMs or BMs with the same security attributes, and which also contains a set of security policies applied to these VMs or BMs.

   - Each security group can define both ingress and egress directions, as well as the actions to be taken in each direction.

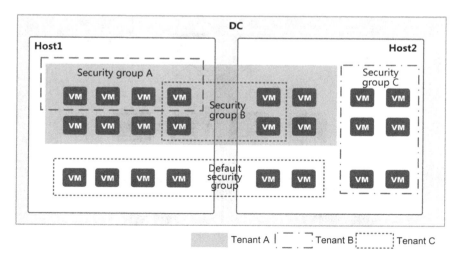

FIGURE 8.3   Security groups in a DC.

- Ingress policy rule: the security policy for incoming traffic destined for a local security group

- Egress policy rule: the security policy for outgoing traffic of a local security group

- Action: permit or deny action taken on selected traffic

- By default, the members of a security group can communicate with each other.

- If the members of a security group initiate north-south access, they are allowed to access the external network by default.

- If traffic comes from external networks to security group members, it must match the whitelist policy and the action must be set to permit for mutual access.

2. Implementation of the solution

Figure 8.4 shows implementation of security groups on the network overlay.

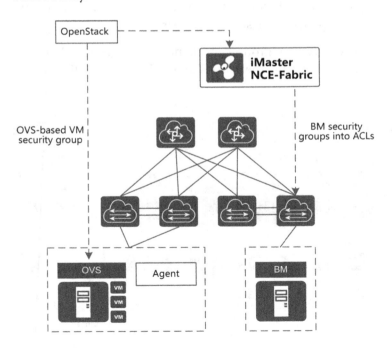

FIGURE 8.4   Security group implementation on network overlay.

- In a VM access scenario, the cloud platform orchestrates security groups and delivers stateful IP tables to the OVS.

- In a BM access scenario, the cloud platform orchestrates BM security groups, and the SDN controller converts the orchestration into an ACL policy before delivering it to TOR switches.

On a hybrid overlay, OVS is replaced by a virtual switch (for example, Huawei CloudEngine 1800V, or CE1800V for short). Figure 8.5 shows implementation of security groups.

- In a VM access scenario, the cloud platform orchestrates security groups and the SDN controller provisions security group services to the CE1800V.

- Security groups are implemented in a BM access scenario in the same way as on network overlay.

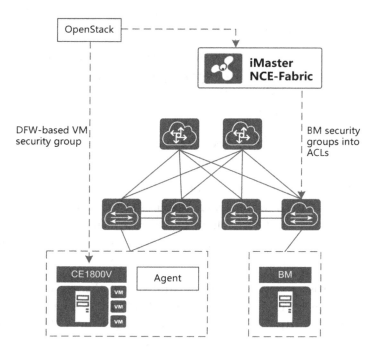

FIGURE 8.5   Security group implementation on a hybrid overlay.

8.4.2 Network Security

8.4.2.1 Network Security Overview

DCN network security has three dimensions: intra-DC intra-VPC security, intra-DC inter-VPC security, and inter-DC security.

1. Intra-DC intra-VPC security

   Traffic in a VPC can be classified by model into north-south traffic, east-west traffic between subnets, and east-west traffic within a subnet. Each type of traffic requires different defenses.

   The north-south traffic in a VPC can be protected using the following approaches:

   - A logical firewall is provisioned on the logical network and functions as the default firewall, providing security services for north-south traffic in the VPC. These services include IPsec VPN, SNAT, EIP, security policies, content security detection, anti-DDoS, and bandwidth management.

   - SFC flexibly diverts traffic to multiple security devices and adds multiple security detection and protection services, as illustrated by the solid lines in Figure 8.6.

   - Microsegmentation creates internal and external end point groups (EPGs) and inter-group policies to implement access control and isolation on the north-south traffic, as illustrated by the dotted lines in Figure 8.6.

   The east-west traffic between subnets in a VPC can be protected using the following approaches:

   - SFC, as illustrated by the solid lines in Figure 8.6, enables switches to use ACLs to implement stateless security access control and isolation (traffic unable to pass through firewalls), and flexibly divert traffic to multiple VAS devices while utilizing multiple security detection and protection services.

   - Microsegmentation, as illustrated by the dashed lines in Figure 8.6, uses stateless EPGs and policies for access control and isolation (traffic not passing through firewalls). This approach applies to scenarios that require moderate security but a high degree of forwarding efficiency.

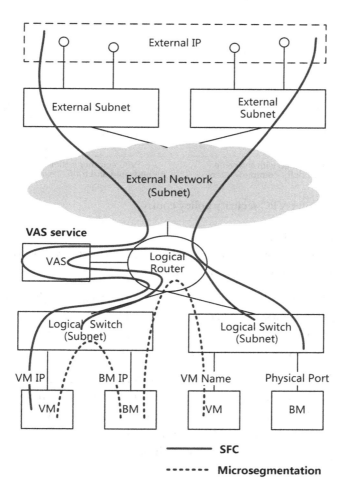

FIGURE 8.6   SFC and microsegmentation traffic models.

The east-west traffic on the same subnet in a VPC can be protected using the following approaches:

- Microsegmentation creates EPGs based on discrete IP addresses and VM attributes to implement fine-grained access control and isolation, as illustrated by the dotted lines in Figure 8.6.

- If the devices do not support microsegmentation, SFC can be used to redirect traffic on the same subnet to security devices in order to implement security detection and protection, as illustrated by the solid lines in Figure 8.6.

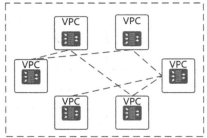

VPCs are associated with the same external network for communication

VPC communication instances are orchestrated and traffic flexibly passes firewalls

FIGURE 8.7 Inter-VPC security policy control.

2. Intra-DC intra-VPC security

There are two scenarios where security policy control applies between VPCs, as shown in Figure 8.7.

- Scenario 1: By default, the VPCs associated with the same external network are interconnected at Layer 3. The tenant administrator configures security policies to protect north-south traffic that passes through the firewall.

- Scenario 2: VPC communication instances are created and orchestrated, and traffic can flexibly pass through firewalls (none, one firewall in a VPC, or two firewalls in two VPCs). Specifically, the tenant administrator plans interconnections and security policies between VPCs in each pair.

Table 8.1 compares these two scenarios.

3. Inter-DC security

A multi-DC scenario is further divided into multi-PoD and multi-site scenarios. This section describes security protection relating to both scenarios.

In the multi-PoD scenario, as shown in Figure 8.8, the controller cluster remotely manages multiple fabric networks, and a unified VXLAN domain is configured between them.

- If the centralized egress solution is used, security protection will be the same as that for a single DC.

- If the active/standby egress solution is used, a group of active/standby firewalls must be deployed in each fabric. The controller

TABLE 8.1 Comparisons between Scenarios Where Security Policy Control Applies between VPCs

(a) Item	Scenario 1	(b) Scenario 2
(c) Strength	The security policies functioning between VPCs are decoupled to enable elastic VPC scaling	End-to-end security policy management can be implemented between VPCs
Weakness	There is no visualized view of end-to-end security policy management	During elastic VPC scaling, security policies must be configured for interconnection with other VPCs, which is a complex and labor-intensive process
Application scenario	Mutual accesses between applications are complex and changeable, and security policies cannot be completely effective for mutual accesses between VPCs	Mutual accesses between applications are constant and clear

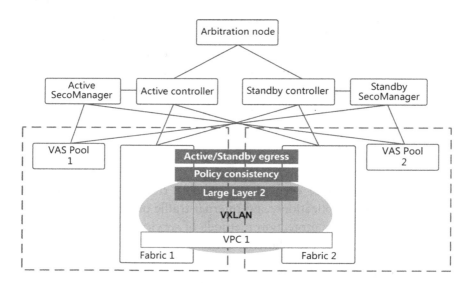

FIGURE 8.8 Multi-PoD scenario.

delivers security configurations to the firewalls in all DCs for synchronization.

- If the multi-active egress solution is used, a firewall cluster must be deployed across fabric networks. These clustered firewalls will then synchronize security configurations (such as security policies) as well as data entries (such as firewall sessions).

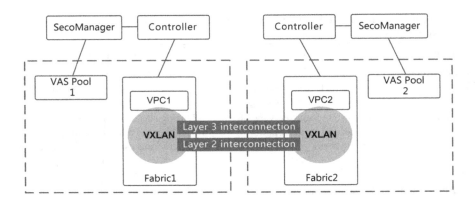

FIGURE 8.9   Multi-site solution.

In the multi-site scenario, as shown in Figure 8.9, each controller cluster independently manages its own DC, and each DC is an independent VXLAN domain, which uses three-segment VXLAN to implement Layer 2 or Layer 3 interconnections between DCs.

When DCs interconnect at Layer 3, traffic can be orchestrated to flexibly pass through the firewalls in one or more DCs, with security policies deployed for inter-DC traffic.

8.4.2.2 Microsegmentation

1. Microsegmentation overview

Enterprises face even greater security risks as the amounts of stored data, applications, and internal traffic on DCNs continue to increase. Legacy protection methods of service isolation by subnet and ACL leads to the following issues:

- A network is divided into service subnets by VLAN or VNI to isolate services. Such isolation is based on subnets, but applications within a subnet cannot be isolated.

- ACLs can be configured to isolate applications. However, as a DCN usually has a large number of servers, a massive number of complex ACL rules must be configured for isolation purposes. Network devices have limited ACL resources, and as such this type of legacy protection does not meet all requirements.

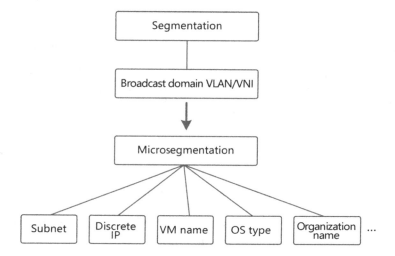

FIGURE 8.10   Legacy protection versus microsegmentation.

Microsegmentation, also known as refined group-based security isolation, addresses these issues. Servers in a DCN are grouped by application, and traffic control policies are deployed by server group. Microsegmentation simplifies O&M and security control, as shown in Figure 8.10.

In VXLAN, microsegmentation provides grouping rules featuring smaller granularities than subnet-level, including discrete IP addresses and MAC addresses. Microsegmentation also supports a wide range of grouping granularities expressed in network and IT languages such as VM attributes, which are easy to deploy and better suited to service requirements. For example, to implement access control and isolation between applications, simply group them on a VXLAN based on rules, with traffic control policies then deployed by application group.

2. Microsegmentation orchestration model

Microsegmentation typically applies to the zero-trust security model where internal protection needs to be ensured against both internal and external security risks. Whitelists are generally used for access control, with internal groups allowed to access one another as required, enabling east-west access in a DC. Figure 8.11 shows the microsegmentation orchestration model.

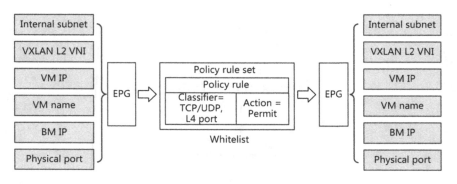

FIGURE 8.11 Microsegmentation orchestration model.

The following concepts enable a better understanding of the microsegmentation orchestration model:

- EPG: An EPG is an abstract set of ports with the same attributes. EPGs can be created based on items such as logical switches, subnets, MAC addresses, discrete IP addresses, VM attributes, and host attributes. Ports in an EPG have the same security attributes.

- Policy rule set (PRS): provides multiple security policies for inter-group access control.

- Classifier: classifies traffic by protocol and port number.

- Action: refers to a Permit or Deny action taken on selected traffic.

3. Microsegmentation process

On a cloud DCN, a deployed OpenStack-based cloud platform can provision the microsegmentation service based on the GBP model. If no cloud platform is deployed, the SDN controller can orchestrate and provision the microsegmentation service.

As shown in Figure 8.12, the microsegmentation solution of the cloud DCN implements security control and isolation based on the GBPs configured on switches. Specifically, the solution delivers GBPs to the TOR switches connected to the source and destination EPGs, and delivers GBP (permit) to the TOR switch connected to the destination EPG.

The TOR switch connected to the source EPG follows the ingress microsegmentation process below:

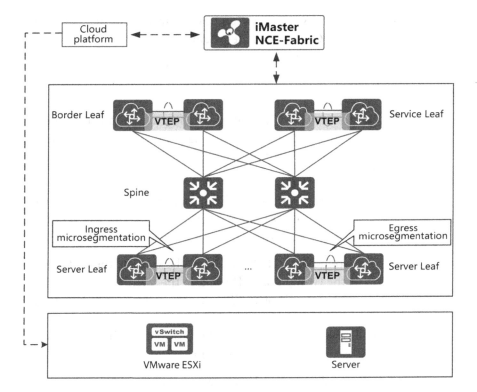

FIGURE 8.12   Microsegmentation process.

1. Searches for the VRF matching the source IP address of user traffic.

2. Searches for the source EPG using the VRF and source IP address.

3. Confirms if inter-device or local forwarding should be performed.

- If inter-device forwarding should be performed, the TOR switch encapsulates the source EPG information into the reserved field in the VXLAN packet header and forwards the traffic to the downstream device.

- If local forwarding should be performed, the TOR switch takes the following actions:

  - Searches for the destination EPG using the VRF and destination IP address of user traffic.

– Searches for GBPs between EPGs using the source and destination EPGs, and matches them with the traffic.

– Permits or denies the traffic based on the GBPs between EPGs.

Note:

The source VTEP sends microsegmentation information to the destination VTEP through the G flag bit and Group Policy ID field in the VXLAN packet header, as shown in Figure 8.13.

– G flag: Defaults to 0. If the value is 1, the Group Policy ID field in the VXLAN packet header carries the ID of the EPG to which the source server belongs.

– Group Policy ID field: When the value of the G flag bit is 1, the Group Policy ID field in the VXLAN packet header carries the ID of an EPG that the source server belongs to.

The TOR switch connected to the destination EPG follows the egress microsegmentation process below:

1. Obtains the destination IP address in the VXLAN header.

2. Searches for the VRF using the destination IP address.

3. Obtains the destination EPG using the VRF and destination IP address.

4. Parses the G flag bit in the packet and, if the value is 1, continues to parse the Group Policy ID field in the packet to obtain the source EPG ID.

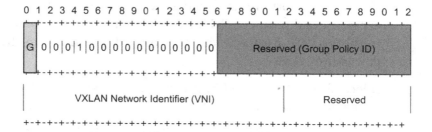

FIGURE 8.13   G flag bit and Group Policy ID field in a VXLAN packet header.

5. Searches for GBPs between EPGs using the source and destination EPGs.

6. Permits or denies the traffic based on the GBPs between EPGs.

8.4.2.3 SFC

1. SFC overview

An SFC is an ordered set of service functions that allows certain service flows to pass through specified VAS devices in a particular sequence, ensuring the service flows access one or more VASs. Figure 8.14 shows a typical SFC instance. An Internet user wants to access the web server in the DC. To ensure security and reliability, the administrator specifies that the traffic passes through the firewall, IDS, and LB in sequence before reaching the web server.

The logical model of an SFC includes the following objects:

- Service function (SF): provides VASs and SFs, including firewalls and LBs.

- Service classifier (SC): identifies the traffic to be imported to the SFC.

- Service function forwarder (SFF): connects to the SF and forwards packets based on service path information in SFCs. SFFs include TOR switches, gateways, and vSwitches.

Figure 8.15 shows a logical SFC model formed by the objects mentioned above.

In the cloud DCN solution, SFC allows administrators to orchestrate SFCs based on logical network elements (NEs), decouple various VASs from the physical topology, and scale out services on demand. In addition, the controller provides a GUI for administrators to directly deploy SFCs, simplifying deployment and operations.

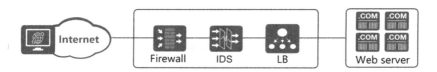

FIGURE 8.14   Example of SFC.

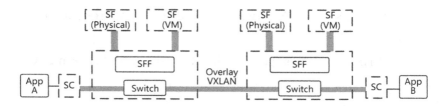

FIGURE 8.15   Logical model of SFC.

2. SFC orchestration model

Figure 8.16 shows the SFC orchestration model. The following concepts enable a better understanding of the SFC orchestration model:

- EPG: created by attributes such as subnets, logical routers, and logical switches. Ports in an EPG have the same security level.

- Policy rule set (PRS): provides multiple security policies for access control between EPGs.

- Classifier: classifies traffic and matches traffic with classes by five-tuple.

- Action: refers to the action to be taken on selected traffic. If SFC specifies that traffic does not pass through the VAS device, it will be permitted or dropped. If SFC specifies that traffic passes through the firewall, it will be redirected to a VAS device.

- Service function path (SFP): refers to the path along which SFs are deployed. The SF type can be customized, and the position of the SF in the SFC path, as well as the logical VAS that carries the SF, can be selected.

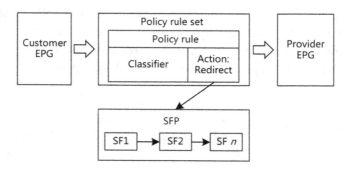

FIGURE 8.16   SFC orchestration model.

3. SFC process

In a cloud DCN virtualization scenario, the SDN controller can directly orchestrate SFCs, including EPG, SFP, and PRS. In a cloud-network integration scenario, the SDN controller restores the logical network provisioned by the cloud platform and then orchestrates SFCs for a second time. If Huawei VAS devices function as SFs, SF policies can be orchestrated on the SecoManager. However, if third-party VAS devices function as SFs, the SF policies must be orchestrated on third-party VAS management platforms. For details about SFC orchestration, see Figure 8.17.

The SFC deployment process is as follows:

- On the SDN controller, manage the Layer 2 or Layer 3 fabric established by switches, create an EPG for SFC, and configure a specific service path and traffic diversion policy. The SDN controller then delivers configurations to implement bidirectional (or unidirectional) network provisioning and traffic diversion between the Layer 2/Layer 3 fabric and VAS devices.

- The security manager manages VAS devices and orchestrates and configures SF policies, while firewalls provide functions such as ACL, EIP, SNAT, IPsec VPN, and content security detection.

In the cloud DCN solution, SFC can be implemented in two modes: policy-based routing (PBR) and network service header (NSH).

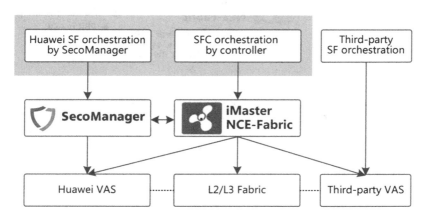

FIGURE 8.17   SFC orchestration.

4. SFC in PBR mode

PBR matches and redirects service traffic so that it is forwarded along a specified path, instead of being forwarded at Layer 2 or Layer 3. In this mode, PBR configurations must be delivered at each hop and without any change in the original packets. Figure 8.18 shows the SFC process based on PBR traffic diversion.

The process of SFC in PBR mode is as follows:

- VXLAN tunnels are established between the SC and SFFs, and between SFFs.

- The SDN controller delivers PBR policies to each SC and SFF node to divert service traffic.

- When receiving service traffic, the SC classifies it according to the matched traffic classification rule, searches for a PBR policy, and determines that the next hop of redirection is SFF1 and the outbound interface is a VXLAN tunnel interface. SC then encapsulates the service traffic into VXLAN packets and forwards them.

- After receiving the VXLAN packets, SFF1 decapsulates them, matches the service traffic with the traffic classification rules, and determines that the next hop of redirection is SF1. SFF1 then queries ARP entries based on the next hop IP address and

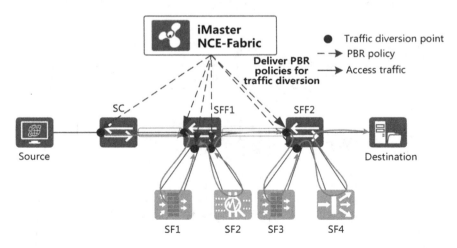

FIGURE 8.18   SFC process based on PBR traffic diversion.

forwards the service traffic accordingly. Similarly, when the service traffic is forwarded back to SFF1 by SF1 through IP forwarding, the traffic will go through the same redirection process and be forwarded to SF2. When the traffic is forwarded from SF2 by SFF1, SFF1 searches for a PBR policy based on the traffic classification rules and finds that the next hop is SFF2 and the outbound interface is a VXLAN tunnel interface. SFF1 then encapsulates the service traffic into VXLAN packets and forwards them.

- After receiving the VXLAN packets, SFF2 removes the VXLAN header, classifies it by the matched traffic classification rule, searches for a PBR policy, and redirects the traffic to SF3 and SF4 in sequence. When the traffic is forwarded to SFF2 by SF4, packet forwarding in the SFC domain is terminated. The traffic is then forwarded to the destination in IP encapsulation and forwarding.

5. SFC in NSH mode

The original packets are encapsulated using NSH to carry SFP information, ensuring they pass through SFs along the specified SFP. In this mode, the SC needs to deliver classification rules to redirect traffic along the SFP, and an NSH header must be added to the original packets.

NSH packets can be carried over multiple types of networks, such as VLAN and VXLAN.

Figure 8.19 shows the NSH encapsulation format when NSH packets are carried over a VXLAN network.

Figure 8.20 shows the NSH encapsulation format when NSH packets are carried over a VLAN.

Outer Ethernet Header	VXLAN Header	Inner Ethernet Header	802.1Q ET=0x894f	NSH	Payload Packet

FIGURE 8.19   NSH encapsulation format — VXLAN carrying NSH packets.

Outer Ethernet Header	802.1Q ET=0x894f	NSH	Payload Packet

FIGURE 8.20   NSH encapsulation format — VLAN carrying NSH packets.

To implement the SFC function, packets must be encapsulated into NSH packets when entering the SFC domain. Figure 8.21 shows the NSH packet format.

Table 8.2 describes the fields in an NSH packet.

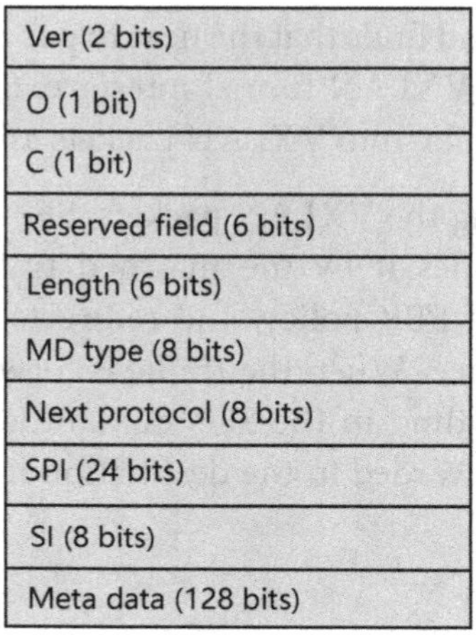

Ver (2 bits)
O (1 bit)
C (1 bit)
Reserved field (6 bits)
Length (6 bits)
MD type (8 bits)
Next protocol (8 bits)
SPI (24 bits)
SI (8 bits)
Meta data (128 bits)

FIGURE 8.21 NSH packet format.

TABLE 8.2 Fields in an NSH Packet

(d) Field	Indication
(e) Ver (2 bits)	NSH version number, which is currently 0
O (1 bit)	Packet type: 0 indicates data packets and 1 indicates OAM maintenance packets
C (1 bit)	Key metadata. When the MD type is 1, the value of this field is 0
Reserved field (6 bits)	This field is set to 0 when an NSH packet is sent, and is ignored when an NSH packet is received
Length (6 bits)	NSH packet length. When the MD type value is 1, this field must be 6, indicating that the NSH packet length is 6×4 bytes. When the MD type is 2, this field can be 2 or any other value
MD type (8 bits)	Metadata field format. 1 indicates a fixed metadata field length, which is 16 bytes, and 2 indicates a variable metadata field length
Next Protocol (8 bits)	Packet type prior to NSH encapsulation: 0x1 indicates IPv4 packets, 0x2 IPv6 packets, 0x3 Ethernet packets, 0x4 NSH packets, 0x5 MPLS packets, and 0x6 to 0xFD are undefined
SPI (24 bits)	Service path identifier
SI (8 bits)	Service index. Index to the SF through which traffic is passing
MetaData (128 bits)	Metadata field. The MD field can have a fixed or variable length, as identified by the MD type field

Figure 8.22 shows the SFC process for NSH-based traffic diversion. The SFC process for NSH-based traffic diversion is as follows:

- VXLAN tunnels are established between SC and SFF1, and between SFF1 and SFF2.

- The SDN controller delivers the NSH forwarding table to SC and SFF to establish SFC forwarding paths, and delivers classification rules to SC to redirect traffic to the NSH forwarding path.

- When receiving service traffic, the SC classifies it according to the matched traffic classification rule and then redirects this classified traffic to SFC. The SC then queries the NSH forwarding table based on SPI and SI in the packet prior to NSH encapsulation (the SI can be used to calculate the sequence number of the SF point that the traffic passes through), and determines that the next hop is SFF1 and the outbound interface is a VXLAN tunnel interface. After the packets are encapsulated into NSH and VXLAN packets, they are forwarded.

- When receiving the IP over NSH over VXLAN packets, SFF1 removes the VXLAN header, queries the NSH forwarding table based on SPI or SI, and determines that the next hop is SF1. SFF1 obtains the ARP information based on the next hop, creates a packet header accordingly, and then forwards the packets.

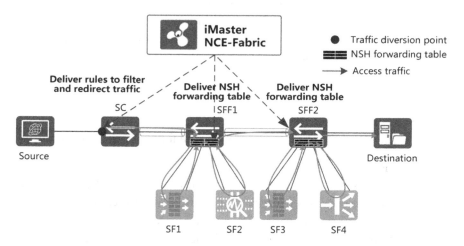

FIGURE 8.22   SFC process for NSH-based traffic diversion.

- SF1 analyzes the packets when they are received, reduces SI by 1, encapsulates the packets with an NSH header and the new packet header, and then forwards the packets to SFF1. Similarly, the packets forwarded back to SFF1 are forwarded to SF2 in the same process. SF2 reduces SI by 1 and forwards the packets back to SFF1.

- When receiving the packets, SFF1 queries the NSH forwarding table based on the SPI in the NSH and the new SI, and determines that the next hop is SFF2. SFF1 then encapsulates the packets with NSH and VXLAN headers and forwards them.

- SFF2 removes the VXLAN header when receiving the IP over NSH over VXLAN packets. The packets are then forwarded to SF3 and SF4 in sequence based on SPI and SI in the NSH header, before being forwarded back to SFF2. The SI in the NSH header is the same as that from the last hop, and packet forwarding in the SFC domain ends as a result. The device then removes the NSH packets and encapsulates them with an IP header before forwarding the packets to the destination.

  Note:

  If the VAS device is an NSH-unaware device, which does not support NSH, SFF needs to perform an NSH proxy. In this case, SFF removes the NSH header when forwarding traffic to the VAS device. When receiving traffic from the VAS device, SFF matches it with the NSH SC policy and sends it along the NSH path again. In this application scenario, metadata information in the NSH header will be lost.

8.4.2.4 Security Services

1. Security services overview

    The cloud DCN solution provides the following security services:

    1. IPsec VPN

        This is a VPN tunnel access mode in which the IPsec protocol is used to provide encryption. If secure communication is required between an enterprise DC and an enterprise private network, IPsec VPN can be used to implement secure communication and transmission of tenant data on the Internet. The solution uses north-south firewalls to provide IPsec VPN services, and the key used for IPsec encryption and authentication can be dynamically negotiated using the Internet Key Exchange (IKE) protocol. IKE can automatically generate a shared key, which improves the security of the key and simplifies IPsec management.

2. Source NAT (SNAT)

SNAT translates private IP addresses into public IP addresses and supports QoS of bidirectional traffic limiting based on public IP addresses. When an internal IP address in a DC needs to access services on the public network (for example, accessing an external website), the internal IP address is used to initiate a connection request. The firewall translates this address into a public IP address and hides the original one. In some scenarios, SNAT is also used for east-west traffic in a DC.

3. EIP

EIP provides elastic IP address services for VMs while also supporting QoS of bidirectional traffic limiting based on elastic IP addresses. EIP binds a public IP address to a tenant network's private IP address, so that various tenant network resources can provide services to the external network using a fixed public IP address and devices on the tenant network can access the external network. In some scenarios, EIP is also used for east-west traffic in the DC.

4. Firewall-based bandwidth management

Bandwidth management refers to firewall management and control of traffic based on the inbound interface/source zone, outbound interface/destination zone, service, and interface number. Bandwidth utilization is improved, while the bandwidth exhaustion caused by attacks is avoided. The cloud DCN solution supports a bandwidth management service that restricts the number of firewall sessions based on five-tuple.

5. Anti-DDoS

DDoS attacks use zombie hosts to send a large number of malicious attack packets to a target. These attack packets may congest network links and exhaust system resources, resulting in the target failing to provide services for authorized users. Zombie hosts are online hosts controlled by an attacker, and a network of zombie hosts is known as a botnet. The cloud DCN solution provides firewall-based anti-DDoS to defend against common types of DDoS attacks.

6. Security policy

The cloud DCN solution provides integrated firewall policies for traffic filtering and security inspection, as well as supporting security policies in a VPC, between VPCs, and between a VPC and an external network. In addition to the 5-tuple, the solution supports security policies based on EPGs, address pools, and a combination of attributes such as the service type and time range.

7. Content security detection

The cloud DCN solution also provides a firewall-based intelligent awareness engine to implement integrated inspection and processing of traffic contents, involving such security functions as antivirus, IPS, and URL filtering.

As a security mechanism, the IPS analyzes network traffic and detects intrusions (including buffer overflow attacks, Trojan horses, and worms). It can automatically discard intrusion packets or block attack sources to terminate intrusions in real time. The IPS identifies packet application layer information, as well as specific characteristics based on the packet content, and compares this information with IPS signatures. If they match, the IPS blocks the packets or generates an alarm based on the configured action.

Antivirus software identifies and processes virus files to ensure network security. Viruses are malicious programs that can be attached to applications or files, and are generally transmitted through emails or file sharing protocols. Antivirus software utilizes a professional intelligent awareness engine and a continuously updated virus signature database to detect and process virus files, preventing data damage, permission changes, and system crashes. When a file containing a virus is detected, the system either blocks that file or generates an alarm based on the action specified in the antivirus configuration file. If file reputation detection is enabled, the FireHunter sandbox can be used to further inspect related files.

URL filtering controls the URLs accessible to users, prohibits them from accessing certain web page resources, and standardizes Internet access behaviors to prevent unauthorized or malicious website access which may result in disclosure of confidential information and even threats such as viruses, Trojan horses, and worms. URL filtering can identify URL access requests, and may block the request or generate an alarm if a certain rule is matched.

2. Security service architecture

Figure 8.23 shows the security service architecture in the cloud-network integration scenario of the cloud DCN solution.

The security service architecture has the following components:

- Cloud management platform (for example, Huawei ManageOne): supports the orchestration and provisioning of standard security services (such as NAT, FWaaS, and IPsec VPN) defined by OpenStack, as well as extended security services (such as IPS and antivirus).

- OpenStack: orchestrates and provisions standard security services defined by OpenStack, including NAT, FWaaS, and IPsec VPN.

- Network controller (such as iMaster NCE-Fabric): manages DC switches, connects to OpenStack L2 to L7 plug-ins, orchestrates and provisions basic networks, and distributes standard NAT, FWaaS, and IPsec VPN service requests orchestrated and provisioned by the cloud platform to the security manager.

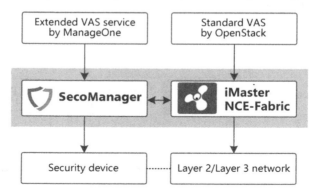

FIGURE 8.23   Security service architecture in the cloud-network integration scenario of the cloud DCN solution.

- Security manager (such as SecoManager): manages security devices and connects to the network controller to implement standard NAT, FWaaS, or IPsec VPN service modeling and automatic configuration defined by OpenStack. In some scenarios, the security manager can also provide an SDN controller to directly connect to the cloud management platform in order to implement modeling and automatic configuration of extended security services.

- Security device: A firewall provides security services such as NAT, FWaaS, IPsec VPN, and content security detection.

- Layer 2 and Layer 3 network: A network resource pool is abstracted from the spine-leaf network consisting of DC switches.

Figure 8.24 shows the security service architecture in the network virtualization scenario of the cloud DCN solution.

The security service architecture has the following components:

- Network controller (such as iMaster NCE-Fabric): manages DC switches and third-party security devices, orchestrates bidirectional or unidirectional networks between the network and security devices, and delivers configurations.

- Security manager (such as SecoManager): manages security devices, orchestrates and delivers security services, and provides VAS functions such as NAT, FWaaS, IPsec VPN, bandwidth management, anti-DDoS, and content security detection.

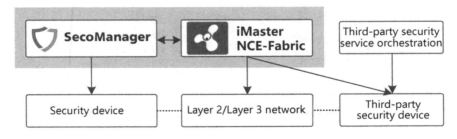

FIGURE 8.24 Security service architecture in the network virtualization scenario of the cloud DCN solution.

- Security device: (Huawei) Firewalls provide security service functions such as NAT, FWaaS, IPsec VPN, bandwidth management, anti-DDoS, and content security detection.

- Third-party security device: The scope of third-party security devices depends on the security capabilities of the devices.

- Third-party security service orchestrator: orchestrates and delivers third-party security services. The scope of orchestration depends on the capabilities of the third-party security management platform.

  Note:

  There are three common management solutions available for third-party VAS devices (including security devices) within the industry.

  - Service Manager Mode: The network controller manages the network and orchestrates the Layer 2 and Layer 3 networks between the network and third-party VAS devices, while the third-party management platform orchestrates and delivers third-party VAS policies.

  - Service Policy Mode: The network controller manages networks and VASs, orchestrates policies, and delivers configurations for all third-party VASs.

  - Network Policy Mode: The network controller does not manage third-party VAS devices, and is only responsible for unidirectional interconnection and traffic diversion orchestration from the network to third-party VAS devices.

3. Security policy service of the SDN security manager

   The security manager supports fine-grained legacy security policy management for firewalls, tenant-oriented OpenStack standard FWaaS service based on logical firewalls, and tenant-oriented security policy automation service by service scenarios. The security policy automation service synchronizes logical network information, EPG information, and SFP information from the network controller to detect the network topology and implement automatic orchestration and distribution of security policies. The network and related security are highly collaborative, and logical VASs do not require manual selection. This, in turn, simplifies the security policy configuration. In addition, security policies can be dynamically updated and adjusted based on network changes. The following describes the process of security policy automation based on service scenarios provided by the security manager.

1. The security manager detects logical network information through the network controller, including the associated mapping information of the logical router, logical switch, logical firewall, external network, and subnet. In this way, the security manager can obtain the protected network segment data of the logical firewall (the associated mapping between the subnet and the logical firewall).

2. The security manager uses the network controller to automatically learn the mapping between subnets and EPGs, SFC instance information, and VPC communication instance information, and in order to detect the logical firewall selected during the orchestration of SFC instances and VPC communication instances.

3. When orchestrating security policies based on EPGs or IP addresses, the security manager automatically selects a logical firewall based on the detected logical network, SFC instance, and VPC communication instance information and delivers security policy configurations based on the following principles, as shown in Figure 8.25.

   – IP-based security policy (for mutual access within a VPC): The security manager matches traffic with the protected network segment, which is obtained during logical network information synchronization, and automatically identifies the protection type (east-west or north-south). For east-west traffic, the security manager delivers configurations to the east-west logical firewalls on the corresponding logical network in order to implement security policies. For north-south traffic, the security manager selects the north-south logical firewalls.

   – IP-based security policy (for cross-VPC access traffic): The security manager automatically selects the logical firewall

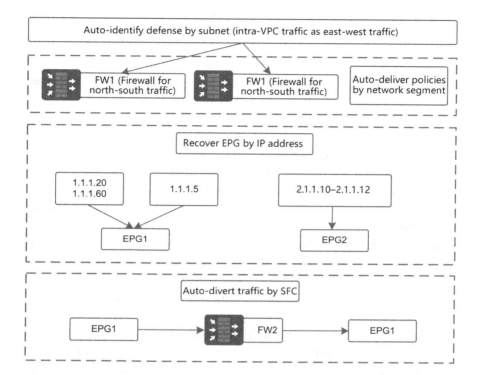

FIGURE 8.25 Security policy configuration.

that is selected during VPC communication instance orchestration, delivering configurations and implementing security policies based on the synchronized VPC communication instance information.

– Application group EPG-based security policy: The security manager automatically selects the logical firewall that is selected during SFC instance orchestration based on the SFC instance information synchronized in order to deliver configurations and implement security policies.

– If an IP-based security policy matches an EPG but does not match any logical firewall, the security manager uses an algorithm to select a logical firewall to carry the security policy. The SDN controller of the network controller is then instructed to automatically create an SFC instance and divert traffic to that logical firewall.

8.4.3 Advanced Threat Detection and Defense

Security threats have changed greatly in recent years. Hacker attacks have evolved from mere pranks and exhibitions of technical prowess, into the profit-oriented, commercial, and organized criminal behavior we see today. Such APTs are growing at an alarming rate.

Unlike traditional attacks, APTs mainly originate from organized crime groups and are targeted at high-value information assets, such as business secrets, intellectual property, and even political or military secrets. Utilizing formidable organization and resources, APT attackers combine various types of intelligence and hacking technologies, as well as advanced social engineering, to launch complicated attacks on valuable information assets and systems. Administrators, being in charge of valuable information systems, are also targeted by APTs, usually through the aforementioned social engineering. Out-dated security systems are unable to effectively deal with such advanced threats, as the range of technical capabilities employed by APTs, such as zero-day and (AETs), cannot be identified by traditional signature-based detection methods. In addition, APTs do not exhibit typical attack patterns or characteristics, with intrusion methods often appearing legal or displaying a low security threat. Today, more than 80% of modern enterprises have experienced some form of APT attack without realizing they were under attack.

1. Unknown threat detection

    A professional APT detection system is recommended to defend against APTs and protect core information assets.

    Huawei FireHunter is a next-generation, high-performance APT detection system, with support for independent traffic restoration and the capable to identifying mainstream protocols and over 50 file types. Using a signature database and a heuristic detection engine, the system executes suspicious files in a virtual environment (VE) designed to simulate common OSs and applications and captures hypervisor behaviors. Leveraging these capabilities, the system detects malicious files and behaviors across multiple layers, and identifies potential attacks. Figure 8.26 shows how Huawei FireHunter operates.

    Huawei FireHunter operates as follows:

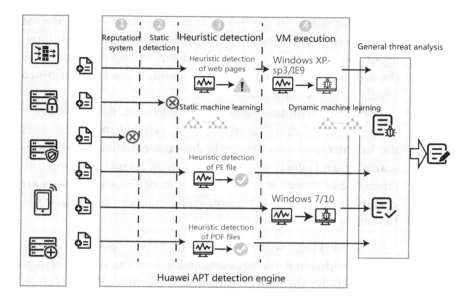

FIGURE 8.26   Huawei FireHunter operation process.

1. An antivirus detection engine is used to scan incoming files.

2. Following detection, files of various types are sent to either the static engine or heuristic sandbox for further identification.

   - Office, image, and flash files are sent to the static sandbox.

   - Portable executable (PE) files are sent to the PE sandbox.

   - Web files are sent to the web sandbox.

   - PDF files are sent to the PDF sandbox.

   - Files of other types are moved to the virtual execution environment.

3. Following static or heuristic sandbox detection, files may be sent to the virtual execution environment for further detection, depending on configuration. This environment is also known as the heavy-weight sandbox and is where malicious file behavior is collected.

   The threat analysis engine summarizes the behavioral data generated by the sandbox, and matches these behaviors and rules with those stored in the behavior pattern library. This process leverages hypervisor behavior capture technology and an intelligent behavior deep learning algorithm in order to maximize the accuracy of each match. At this point, it can be determined whether a file is malicious.

Huawei FireHunter can be deployed in a DC to prevent malicious files and web traffic attacks from external sources, such as the Internet, and also to rapidly detect latent attacks, malicious scanning, and penetration on the intranet. As such, the horizontal spread of threats is prevented, and core DC server assets are effectively safeguarded. Huawei FireHunter can be deployed in the following modes:

- Interworking with a firewall: When network traffic passes through the firewall, files are extracted and sent to FireHunter through an interworking protocol for detection. Once FireHunter has received and detected the files, the firewall queries the results through an associated interface, generates security policies based on the results, and either blocks or permits the network traffic containing the original files.

- Independent deployment of traffic restoration: FireHunter receives mirrored network traffic from the switch, restores the traffic, performs command and control (C&C) detection, and extracts files for purpose of detection.

- Interworking with the CIS, such as Huawei CIS: The network security intelligence system receives mirrored network traffic using a flow probe, restores the traffic, and extracts and sends the files to FireHunter using an interworking protocol. FireHunter then inspects these files, sends the inspection logs to the collector for summarization, and replays the attack paths on the GUI so that the security status of the entire network can be viewed.

2. Intelligent threat analysis based on big data

Legacy real-time detection featuring "single device, single means, and single threat" cannot adequately defend against advanced threats such as APTs. Instead, an in-depth defense system capable of analyzing the entire attack chain is required. Such a system utilizes threat detection and investigation analysis based on big data technologies.

Huawei CIS leverages big data analytics and AI detection algorithms to analyze and detect threats, and to accurately identify and defend against APT attacks. Such measures effectively prevent the loss of core information assets that result from APT attacks.

The CIS also detects and defends against repeated attacks. Maintaining a collection component deployed close to key departments and assets, the CIS detects and analyzes real-time traffic information and offline information to effectively detect advanced threats. Figure 8.27 shows how the CIS operates.

The following should be adhered to when deploying a CIS in a DC:

- Deploy the CIS in the management zone so that it detects unknown threats and displays the security posture of the entire network.

- Deploy flow probes at the Internet access border, WAN remote access border, DC service area border, and within other important

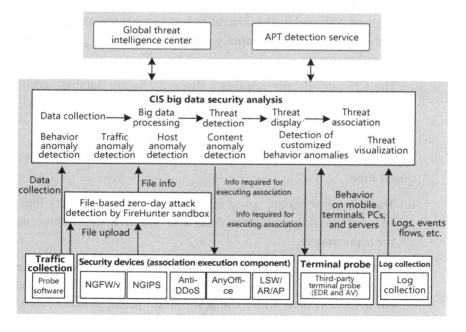

FIGURE 8.27   CIS operation process.

service areas, enabling them to collect traffic and logs to be sent to the CIS for detection.

- You are advised to deploy the CIS with security devices such as sandboxes and firewalls.

3. Associated network-and-security closed-loop

In the associated network-and-security closed-loop solution, the analyzer (CIS), controller (network controller and security manager), and execution device (firewall+switch) are deployed to automatically block and isolate APT threats.

- As a security analysis and detection device, the CIS identifies unknown threats and interworks with the security manager to deliver appropriate policies.

- After receiving security instructions from the analyzer, the security manager and network controller convert them into executable policies and deliver them to firewalls and switches.

- Active firewalls and switches provide security analysis data (original traffic and security logs) for the analyzer. Specific instructions are then delivered by the controller, security services are deployed, and internal threats are rapidly isolated. This system prevents the spread of threats and enables processing within a closed loop.

- The CIS periodically releases a threat signature database compiled by the analyzer, which firewalls can quickly share and utilize when building a unified security defense line.

Figure 8.28 shows the process of associated network-and-security closed-loop.

1. The CIS collects data using flow probes, firewall security logs, and sandbox file reputation data, and uses this information as the input source.

2. The CIS utilizes big data and AI detection algorithms to analyze security threats based on input source. Once a security threat is confirmed, it determines the appropriate threat isolation technique based on severity. In all cases, the threat can be isolated automatically or manually by the administrator.

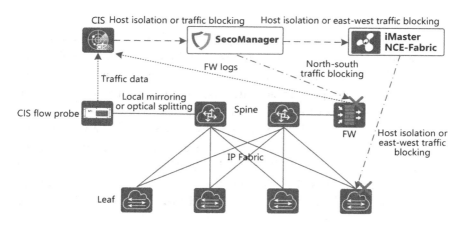

FIGURE 8.28   Process of associated network-and-security closed-loop.

- Isolation mode 1: host isolation. In this mode, the CIS notifies the security manager (SecoManager in Figure 8.28) of the target host information, and then, the security manager notifies the network controller (iMaster NCE-Fabric in Figure 8.28). The network controller locates the leaf nodes connected to the host and delivers host isolation policies to them.

- Isolation mode 2: traffic blocking. In this mode, the CIS notifies the security manager of the target traffic information. In cases of east-west traffic that does not pass through a firewall, the security manager notifies the network controller of the traffic information, which, in turn, delivers a blocking policy on the corresponding TOR to block the traffic. In cases of north-south traffic that passes through a firewall, the security manager delivers a security policy directly to the firewall in order to block the traffic.

8.4.4 Border Security

Border security refers to defense against external security threats, such as intrusions, viruses, and DDoS attacks. Currently, products such as firewalls, IDS, IPS, WAFs, and anti-DDoS devices are deployed to defend against DDoS attacks, and operate by blocking external attacks at network ingresses and egresses.

This section describes two border security technologies: anti-DDoS and intrusion detection.

1. Anti-DDoS defense

Modern DDoS attacks have split into two forms. The first involves large-packet flood attacks, such as SYN flood, UDP flood, HTTP Get flood, and ACK flood, which occupy network bandwidth. The second form incorporates slow attacks, which accurately target service systems such as Internet finance and games.

DCs are affected most severely by DDoS attacks, and they face bidirectional threats. Inbound DDoS attacks directly affect downstream bandwidth and the availability of the DC infrastructure and online services, while outbound DDoS attacks threaten the uplink bandwidth of the DC access layer and the overall reputation of the DC. To counter these threats, dedicated anti-DDoS devices and systems are required.

Huawei's anti-DDoS system consists of dedicated anti-DDoS devices and an Abnormal Traffic Inspection & Control System (ATIC). The dedicated anti-DDoS device is composed of a detection center and a cleaning center, which can be deployed separately or within the same chassis. Figure 8.29 shows the relationship between the detection center, cleaning center, and ATIC.

The detection center analyzes traffic for possible attacks and reports attack events to the ATIC, which then delivers traffic diversion policies to enable the cleaning center to clean the traffic. The detection center uses NetFlow-based traffic detection technology as well as application-based in-depth packet detection technology.

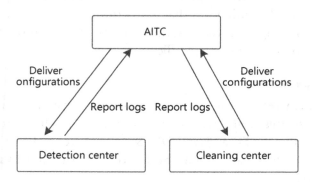

FIGURE 8.29 Relationship between the detecting center, cleaning center, and ATIC.

The cleaning center diverts and cleans traffic based on the policies delivered by the ATIC and then re-injects the cleaned traffic. During this process, the cleaning center also logs the events and reports them to the ATIC.

The ATIC manages both the detecting and cleaning centers. As the management center of the anti-DDoS system, the ATIC manages devices, policies, performance events, alarms, and reports.

The anti-DDoS system is usually deployed at the border of a DC and connected to a router in bypass mode to detect downstream traffic. Figure 8.30 shows how an anti-DDoS system operates.

An anti-DDoS system works as follows:

- In mirroring or optical splitting mode, the DC's egress router either diverts downstream traffic to the anti-DDoS detection center for per-packet detection or sends the NetFlow sampling data on the live network to the anti-DDoS detection center for per-flow detection.

- Upon detection of any attack, the detection center sends attack traffic logs to the ATIC.

- The ATIC sends traffic diversion commands to the cleaning center.

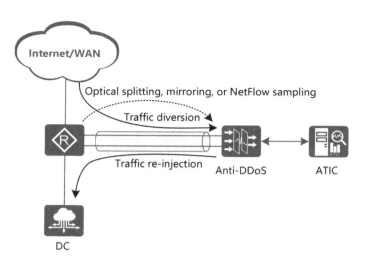

FIGURE 8.30   Anti-DDoS system operation process.

- The cleaning center delivers the host route for traffic diversion.

- The cleaning center re-injects the cleaned traffic into the network through PBR.

- The cleaning center reports relevant information such as cleaning results and logs to the ATIC, which then generates report statistics.

2. Professional intrusion detection and defense

Huawei provides professional intrusion detection and defense products, including NIP IDS and NIP IPS. These products receive regular signature database updates from the Huawei security center, which include intrusion prevention, application identification, and virus databases, enabling rapid signature updates and vulnerability detection. The NIP IDS and NIP IPS are plug-and-play products capable of flexible deployment. By default, they operate in transparent mode when transmitting normal packets. The NIP IDS and NIP IPS can also interwork with the FireHunter sandbox to effectively detect APT location threats.

- The NIP IDS is usually deployed at the DC border in offline mode. It detects but does not defend against attacks, and it is recommended that bidirectional traffic be copied to the NIP IDS for detection. The NIP IDS also supports unidirectional traffic detection, as shown in Figure 8.31.

- The NIP IPS must be deployed in inline mode in order to support detection and defense, as illustrated by No. 1 in Figure 8.32. If multiple links require protection, multiple interfaces can be

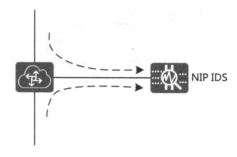

FIGURE 8.31   Working diagram of NIP IDS.

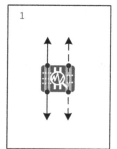 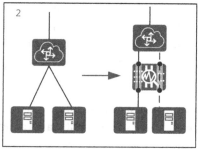 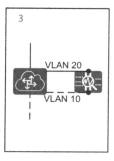

FIGURE 8.32　Working process of the NIP IPS.

used for access, as illustrated by No. 2 in Figure 8.32. It is recommended that the NIP IDS be cascaded on the line to be protected or be physically connected to the network in bypass mode, as illustrated by No. 3 in Figure 8.32.

8.4.5 Security Management

An APT is a set of lasting network attacks, utilizing a variety of advanced techniques, targeting a specific entity. While recent years have seen the spread of APTs into various industries, legacy security measures remain unable to detect or defend against them, resulting in increasing losses.

Huawei has launched the CIS, a big data-based APT defense product, to address these concerns. Leveraging big data analytics, the CIS accurately identifies and defends against APT attacks, effectively protecting core information assets. The CIS can also dynamically detect the security posture of the entire network, investigate attack events, and provide intelligent security policy management.

1. Network-wide security posture awareness

   The CIS collects vulnerability, asset, access record, Internet access log, and security event information relating to systems such as network devices, security devices, servers, and terminal hosts, and summarizes and associates this information. The security posture of the entire network is then displayed in multiple dimensions, and the CIS offers predictions and warnings regarding current security trends which enable administrators to take effective preventive security measures.

   • The CIS dynamically displays attacks and threat hotspots on the threat map, including attack events originating from external sources and occurring between internal zones. It also displays the latest threat event list in real time and maps network security threat events to the topology map. In this way, current threats and recently detected threat events are displayed intuitively on the map.

- The CIS comprehensively displays asset threat information, including threat events, log events, and asset threat degree evaluation. Threats to DC assets are displayed intuitively, and by grouping assets and classifying events, it is possible to quickly identify high-risk assets and key threats.

2. Forensic investigation

Details and attack paths related to threat events are intuitively displayed on the network. Based on this information, administrators can analyze threats, collect evidence, and initiate appropriate countermeasures.

The CIS displays attack information by link, attack behaviors by node, the stage of each APT attack, as well as the threat event and traffic list related to each attack path. In addition, attack paths can be dynamically restored.

All nodes are displayed along attack paths, and the CIS can mine node threat events, node anomaly events, node traffic metadata, and node log data. This approach delivers data source tracing along the complete path.

3. Intelligent security policy management

When maintaining security policies, the administrator needs only to focus on service policy requirements and not on the policies configured on firewalls. Instead, the policy redundancy analysis function, which is device-based, can identify redundant security policies configured on the firewall. The administrator can then clear redundant policies as required. The policy optimization function, which is application-based, helps users to automatically identify newly discovered applications, offline applications, and changed applications. Finally, the policy simulation function helps to evaluate the impact of changed policies before they are deployed to devices, enabling the administrator to quickly optimize policies.

Device-based policy redundancy analysis operates as follows:

• The administrator selects a firewall and initiates static policy optimization analysis.

• The security manager (for example, SecoManager in Figure 8.33) initiates policy analysis based on user requirements. Policy matching information is obtained from the firewall and analyzed to determine whether there are redundant policies based on the policy configuration information and policy matching principles. This policy redundancy analysis is static.

In application-based policy simulation and optimization, host probes are deployed on the hosts to be protected and flow probes on key network nodes. The CIS and security manager collaborate to implement application-based policy simulation and optimization, as shown in Figure 8.33.

Note:

NGFW refers to the next-generation firewall.

The policy simulation process is as follows:

• The administrator creates a simulation task on the security manager, and the security manager sends instructions to the CIS, which then groups applications and learns mutual access relationships.

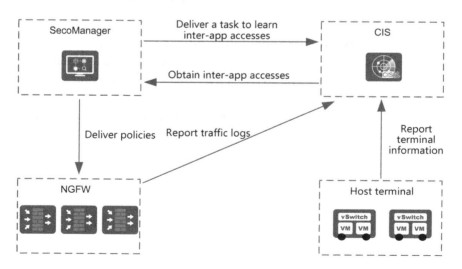

FIGURE 8.33   Policy simulation and optimization.

- The CIS collects traffic logs and terminal information, and generates application groups and mutual access relationships based on the learning algorithm. During this process, an agent collects terminal process information, including the process name, IP addresses, and ports for sending and receiving packets.

- After receiving the application groups and mutual access relationships, the security manager matches them with the policies to be deployed and displays the impact of these policies on the network.

- The administrator can identify policies that do not match simulation expectations, and change these policies accordingly.
  The policy optimization process is as follows:

- The administrator creates an optimization task on the security manager, and the security manager sends instructions to the CIS, which then groups applications and determines mutual access relationships.

- The CIS collects traffic logs and terminal information, and generates application groups and mutual access relationships based on the learning algorithm. During this process, an agent collects terminal process information, including the process name, IP addresses, and ports for sending and receiving packets.

- After receiving the application groups and mutual access relationships, the security manager matches them with the policies to be deployed and displays the impact of these policies on the network.
  The administrator can disable or delete policies based on optimization suggestions.

CHAPTER 9

Best Practices of Cloud DCN Deployment

THIS CHAPTER DESCRIBES HOW best to deploy the cloud DCN solution based on Huawei's extensive deployment experience. To begin, design partitions for the DCN and carefully plan the physical network, SDN controller, and cloud platform server for each partition. It is then possible to proceed with deployment, which will include basic network pre-configuration, controller installation, controller interconnection commissioning, and service provisioning.

9.1 DEPLOYMENT PLAN

9.1.1 Overall Plan

The deployment plan for the new cloud DCN differs from that of the legacy DCN in the following aspects:

- The SDN controller deployment plan is required for cloud DCN deployment. As the SDN controller is essentially software running on the OS, it is necessary to plan server parameters and the working network plane of the OS and SDN controller. Different network planes require different network connections. For example, the southbound network plane of the controller must connect to network devices such as switches and firewalls, for which dedicated IP addresses and routes should be planned. The northbound network plane of the controller is used for administrator web logins and interconnection with systems such as the cloud platform and VMM, for which dedicated IP addresses and routes also need to be planned. Legacy DCNs involve neither the SDN controller, nor interconnection between the SDN controller and other systems.

- A large DC is usually divided into several zones, and a cloud DC may be deployed in one or more zones. In this scenario, interconnections must be planned between multiple cloud DCs, and between cloud DCs and legacy DCs, none of which are required in a legacy DC plan.

9.1.1.1 Common User Requirements

Understanding user requirements for DCNs is critical when deploying a cloud DC based on SDN. Table 9.1 lists the common user requirements for deploying an SDN DC.

9.1.1.2 Network Zone Design

A typical large-scale DCN can be partitioned into multiple zones, each with specified functions. Each zone can then be further designed as per their functions. The following is a typical partition example.

As shown in Figure 9.1, the DC is partitioned into three zones: resource zone (production and non-production intranet zones), non-resource zone (production and non-production Internet access zones, production and non-production extranet access zones, and other network access zones), and O&M management zone.

In the resource zone, the production intranet zone carries core enterprise services, which are the most critical enterprise assets. The non-production intranet zone carries common office systems, non-critical services, and temporary services.

In the non-resource zone, the production extranet access zone is the egress zone that connects to the remote subnets of an enterprise. Devices such as firewalls, IPS/IDP, and egress routers are usually deployed in this zone for remote networking and security control.

In the non-resource zone, the production Internet access zone is the egress zone that connects to the Internet. Similar to the production extranet access zone, devices such as firewalls, IPS/IDP, and egress routers are generally deployed in this zone for remote networking and security control.

TABLE 9.1   Common User Requirements for Functions

Requirement	Description
High volumes of service data traffic	• Traffic bursts may occur during data writing and result in the replication of upper-layer applications. Consequently, a large cache is required • The network is required to provide an appropriate overload ratio, which is typically 3:1 for servers and TOR switches • The network must provide sufficient access bandwidth. For example, 10GE access and 40GE uplink • Network latency must be less than 1 ms
Network service deployment	• The controller can interconnect with a mainstream cloud platform through a standard API, enabling the cloud platform to orchestrate and auto-provision services • The controller can automatically discover network devices and add them to the management zone • The underlay network can be deployed automatically • A fabric can be divided into logical networks, which are isolated by default. Isolation zones in a fabric network cannot interwork • When the VM status changes, network service resources are adjusted on demand based on scheduling by the cloud platform • The controller provides GUIs for visualized network service orchestration and provisioning, and needs to provide network service orchestration in different dimensions, such as overlay, tenant, and application • The controller provides GUIs to control the interconnection between the fabric and external network • If a fabric network has been divided into logical networks, the controller needs to collaborate with Layer 4 to Layer 7 devices such as firewalls, LBs, and encryptors by diverting traffic. In this way, the controller manages security for the east-and-west traffic within or between tenants • A fabric can be partitioned into logical networks, and their configurations are managed by multiple homogeneous or heterogeneous cloud platforms
Basic network functions	• EVPN-based VXLAN • Distributed gateway • IPv6
Security	• Application systems under different levels of protection must communicate with each other through firewalls • The production network is isolated from the non-production network through different fabrics • The intranet and DMZ are isolated by fabric • The controller supports hierarchical rights-based user management • The controller, spine nodes, and leaf nodes support authentication, authorization, and accounting (AAA)

(Continued)

TABLE 9.1 (*Continued*)   Common User Requirements for Functions

Requirement	Description
High availability (HA)	• The network support system offers a high degree of availability and prevents Layer 2 loops (such as broadcast storms and unicast flooding) • Broadcast storms are suppressed • Load is balanced among controller cluster members. The system automatically detects a faulty controller, but the upper-layer service system will be unaware of the fault. The controller cluster can be deployed across Layer 3 • When all controllers are faulty, the forwarding tables of spine and leaf nodes are not affected, and traffic is forwarded as normal
High scalability	• The horizontal scalability of the network is improved, and the reuse efficiency of compute and storage resources is maximized in physical and virtual environments • The controller can manage at least 200 leaf nodes • The controller supports access and management of at least 50,000 VMs
Maintainability	• The controller automatically deploys software and policies onto all switches without manual intervention • The controller logs user operations • Various methods are available for fault locating and troubleshooting, and they enable quick locating of faulty devices, components, interfaces, and service communication paths in a fabric network

In the O&M management zone, a dedicated O&M management system is deployed to manage and monitor the entire DCN and IT applications. The network management platform, SDN controller, cloud platform controller node, and VMM monitoring node are typically deployed in this zone.

9.1.1.3 Physical Architecture Design

The physical architecture of each zone is designed based on the characteristics of carried services. Figure 9.2 shows a simple design example, where some zones of the same type are combined for easier demonstration. In the figure, 1 indicates the production/non-production extranet access zone, 2 is the production/non-production Internet access zone, 3 is other network access zone, 4 is O&M management zone, 5 is the production/non-production intranet zone, and 6 is the core switching zone.

In the production/non-production intranet zone, various application networks and servers are deployed. In the case of complex network

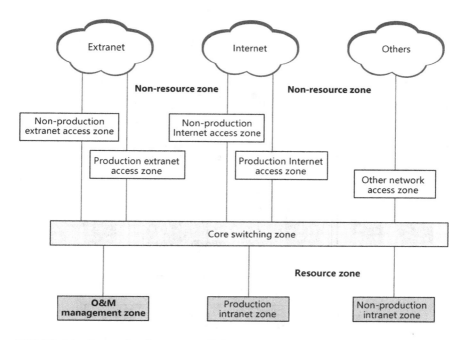

FIGURE 9.1   Example of zones in a DC.

configurations, new services are provisioned and current services are scaled out frequently. This leads to high demand for network automation, and DCs from this zone begin to evolve into SDN. An SDN controller is deployed in this zone to manage switches and firewalls, and it automatically delivers overlay network configurations based on the VXLAN service model in order to implement automatic network deployment.

9.1.1.4 SDN Design

1. Deploy SDN controllers and cloud platforms based on the fabric plan.

    To ensure high security and reliability, an SDN controller and a cloud platform can be independently deployed for each fabric, and a unified service platform matching user services can be developed based on the open interfaces of the cloud platforms. The service platform interconnects with the compute, storage, and network resources of multiple OpenStack systems to implement agile service provisioning and resource pool sharing. Figure 9.3 shows the overall design.

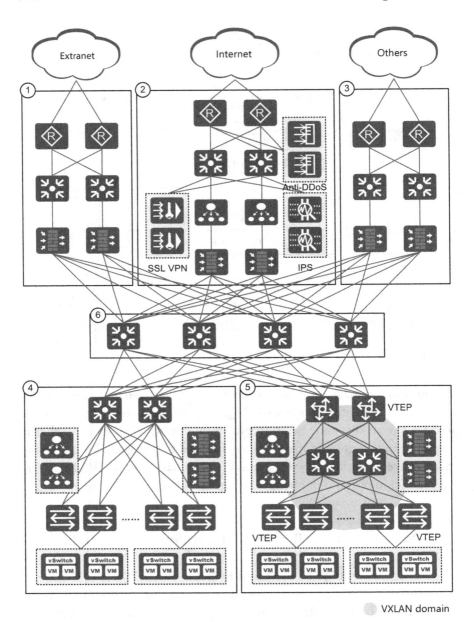

FIGURE 9.2 Overall physical architecture of the DCN zones.

In this mode, multiple controllers are deployed independently, and faults on one controller do not affect another. If a single controller fails, only new services in the fabric where that controller resides are impacted, while the current services in this fabric and services in

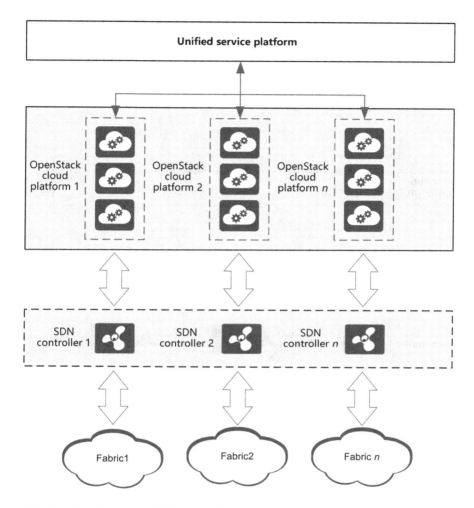

FIGURE 9.3   Deploying SDN controllers and cloud platforms based on the fabric plan.

other fabrics remain unaffected. A gray upgrade can be performed in a fabric network prior to a formal version upgrade, and fabrics and SDN controllers can be enabled as required.

2. Deploy SDN controllers and cloud platforms based on service types.
   SDN controllers and cloud platforms can be deployed based on the fabric plan, or different types of services. Independent SDN controllers are deployed in the intranet service zone, extranet zone, and development and test zone, and are connected to respective OpenStack systems. A unified service platform is developed based

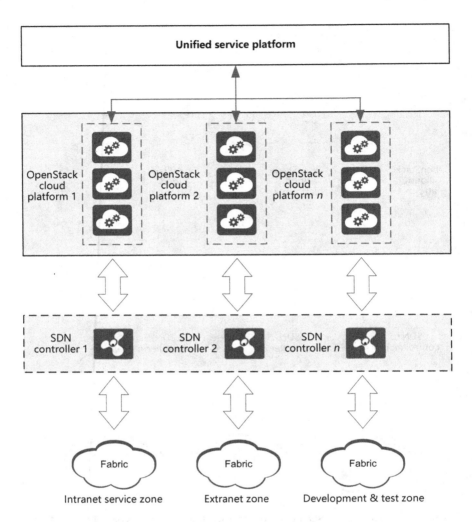

FIGURE 9.4 Deploying SDN controllers and cloud platforms based on service types.

on the open interfaces of the cloud platforms, and this platform interconnects with the compute, storage, and network resources of multiple OpenStack systems to implement agile service provisioning and resource pool sharing. Figure 9.4 shows the overall design.

In this mode, one SDN controller can manage multiple fabrics of the same type, enabling the resources of multiple fabrics to be scheduled in a unified manner and allowing these fabrics to flexibly access one another. For example, it is possible to orchestrate policies for mutual access of services of the same type, for mutual access of different types of services through firewalls, and for mutual access of services across fabrics. Services of the same type can be deployed in different fabrics in order to implement fabric-level DR, improving service deployment flexibility and reliability.

9.1.2 Recommended Service Network Plan

As mentioned previously, various application networks and servers are deployed in the production or service zones of most DCs, where the network configurations are complex, new services are provisioned, and current services are frequently scaled out. This leads to a high demand for network automation. As such, the following describes the SDN plan for DC zones.

When deploying an SDN DCN, the service network and management network should be planned separately. A service network carries service traffic and consists of nodes in four roles: server leaf, border leaf, service leaf, and spine.

9.1.2.1 Basic Principles for Designing a Physical Network

You are advised to use Huawei CloudEngine switches to build a spine-leaf structure on a DCN. The number of spine and leaf nodes can be flexibly configured based on the network scale. For details, see Figure 9.5.

1. Design of spine nodes

   In the spine-leaf structure, the number of spine nodes depends on the oversubscription of leaf nodes, which varies depending on industries and customers.

   Spine and leaf nodes are interconnected through Ethernet interfaces, which are configured to operate in Layer 3 routing mode in order to build an all-IP network.

2. Design of leaf nodes

   It is recommended that two leaf nodes form a group, with each group using M-LAG active-active networking. If one leaf node is faulty, the other leaf node in the group is not affected, and service traffic is forwarded normally. In addition, an upgrade of devices in one group does not affect other groups.

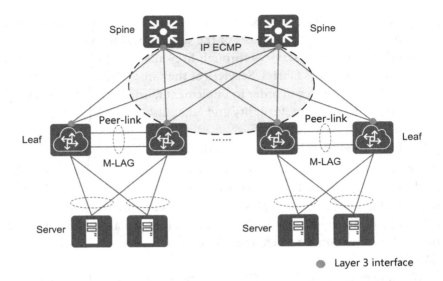

FIGURE 9.5 Spine-leaf structure in a fabric network.

3. Design of forwarding

   OSPF or BGP can be used for routing on the underlay network. Between spine and leaf nodes, IP ECMP paths can be created for load balancing. Through this approach, high bandwidth is available, traffic is forwarded without blocking, and faults are quickly converged. If any link fails, traffic is forwarded over other normal links, resulting in improved link reliability.

9.1.2.2 Recommended Service Network Architecture

Distributed VXLAN architecture is recommended for a service network, where the server leaf, border leaf, service leaf, and spine nodes are deployed separately. As shown in Figure 9.6, the server leaf, border leaf, and service leaf nodes function as NVE nodes on a distributed VXLAN.

The following describes the components on the recommended network architecture.

- 1 refers to the egress network, with border leaf indicating the egress device of the local fabric. In a border leaf group, two switches are deployed as an M-LAG or in active-active mode and connect to PE devices (external interconnection devices of the fabric) in dual-homed

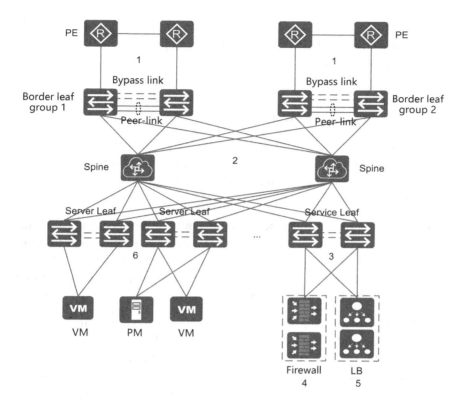

FIGURE 9.6   Service network architecture with separate deployment of roles.

or square-looped mode. Layer 3 bypass links are configured between the two border leaf nodes, enabling traffic to be forwarded over these links in cases of service uplink or PE failure, and ensuring network reliability. In addition, multiple groups of border leaf nodes can be deployed as required (two groups of border leaf nodes are shown in the figure).

- 2 refers to spine nodes, which function as NVE nodes on the VXLAN and implement Layer 3 interconnections between the underlay network and other nodes, over which packets are forwarded at high speeds. At least two spine nodes are deployed, and each is deployed independently. If required by the network scale and specific over-subscription requirements, multiple spine nodes can be horizontally deployed. It is recommended that BGP EVPN RR be deployed on two spine nodes.

- 3 refers to service leaf nodes, which connect to VAS devices (such as firewalls and LBs). Two switches are deployed as an M-LAG and are not connected to servers. In the case of large service scale, multiple groups of service leaf nodes can be deployed (only one group is shown in the figure) to meet scale-out requirements when VAS resources are insufficient.

- 4 refers to firewalls. You are advised to deploy Huawei firewalls and the SecoManager, Huawei's security manager. Huawei SecoManager can interwork with Huawei CIS to greatly improve network security protection and automation.

- 5 refers to LBs. In the cloud-network integration scenario, the cloud platform manages LBs to implement automatic service provisioning. In network virtualization without a cloud platform, the SDN controller directs service traffic to an LB, where a load balancing service is manually configured by the administrator.

- 6 refers to server leaf nodes, which connect to servers and take the highest number in a DC. Two switches build a group. Their downlink ports are connected to servers in M-LAG mode and uplink ports to all spine nodes. IP ECMP paths are created on the underlay network to provide sufficient bandwidth and reliability. When server NICs are connected to server leaf nodes and operate in active/standby mode, the peer-link bandwidth between server leaf node groups must be greater than or equal to the uplink bandwidth of a single server leaf node.

9.1.2.3 Routing Plan

You are advised to deploy a dynamic routing protocol on the underlay network, or specifically, OSPF for small- and medium-sized networks, and EBGP for large-sized networks. It is also recommended to deploy BGP EVPN as the control plane protocol on the overlay network.

When devices are connected through M-LAG, it is recommended that the DFS group and EVPN peer use the same loopback address. Note that this cannot repeat the VTEP IP address.

1. OSPF routing plan

Underlay network: Figure 9.7 shows the routing plan when OSPF is used as the routing protocol on the underlay network. The DCN

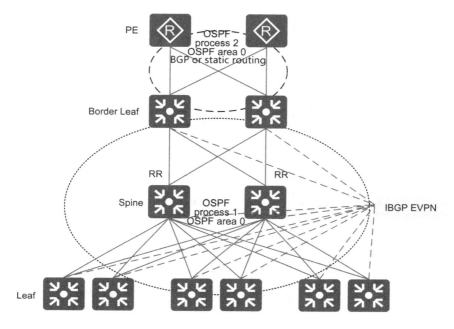

FIGURE 9.7 Routing plan when OSPF is used as the routing protocol on the underlay network.

with spine and leaf nodes is planned as the OSPF backbone area, and OSPF neighbor relationships are established between the leaf and spine nodes. Static routes, BGP and OSPF processes are configured between border leaf nodes and PEs (external routers).

Overlay network: Spine nodes function as RRs of IBGP EVPN, and both spine and leaf nodes use independent loopback addresses to establish EVPN peer relationships. If there are four or more spine nodes, only two RRs need to be deployed to ensure reliability. A BGP route advertisement delay can be set on all devices to extend the service convergence time in the case of a device fault.

2. EBGP routing plan

Underlay network: Figure 9.8 shows the routing plan when EBGP is used as the routing protocol on the underlay network. In a DCN with spine and leaf nodes, each leaf node group and all spine nodes are planned in an AS. Static routes, BGP routes, or OSPF routes are configured between border leaf nodes and PEs (external routers).

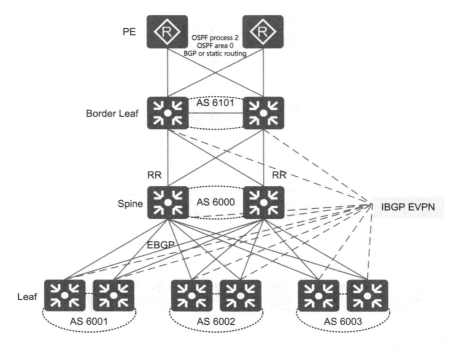

FIGURE 9.8 Routing plan when EBGP is used as the routing protocol on the underlay network.

Overlay network: Spine nodes function as RRs of IBGP EVPN. Spine nodes and leaf nodes use independent loopback addresses to establish EVPN peer relationships. If there are more than two spine nodes, only two RRs need to be deployed to ensure reliability. Delayed BGP route advertisement can be configured on all devices to extend the service convergence time in the case of a device fault.

9.1.2.4 Egress Network Plan

1. Egress network routing plan

When firewalls are connected to border leaf nodes in bypass mode, routes on the egress network can be classified into Layer 3 dynamic routes, Layer 3 static routes, and routes to external networks through firewalls. You are advised to deploy dynamic routes on the Layer 3 egress network. For public external networks, it is advised to specify egress VRFs which are isolated from the underlay network. Public VRFs and root firewalls are not recommended.

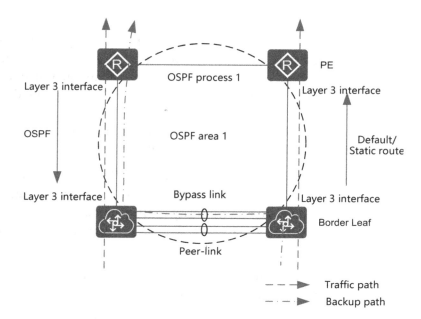

FIGURE 9.9   Deploying static Layer 3 routes on the egress network.

(1) Deploying dynamic Layer 3 routes on the egress network (recommended)

As shown in Figure 9.9, a bypass link is deployed between two border leaf nodes, and they can each be taken as a stand-alone node. Border leaf nodes and PEs are connected in square-looped or dual-homed networking (square-looped networking is used as an example).

In this networking mode, the routes for northbound and southbound traffic are described as follows:

- Northbound traffic: In the egress VRF, manually config-ure a default/static route to the PE, and configure a default/static route with a low priority to the peer border leaf node. An outbound interface should not be configured on the SDN controller.

- Southbound traffic: Deploy OSPF between PEs and between PEs and border leaf nodes. Static routes are imported to OSPF on border leaf nodes and advertised to PEs.

(2) Deploying static Layer 3 routes on the egress network

As shown in Figure 9.10, a bypass link is deployed between two border leaf nodes, and they can each be taken as a standalone node. Border leaf nodes and PEs are connected in square-looped or dual-homed networking (square-looped networking is used as an example).

In this networking mode, the routes for northbound and southbound traffic are described as follows:

– Northbound traffic: In the egress VRF, manually config-ure a default/static route to the PE, and configure a default/ static route with a low priority to the peer border leaf node. An outbound interface should not be configured on the SDN controller.

– Southbound traffic: On the PE, manually configure a sum-marized static route to the border leaf node.

2. Optimizing egress network convergence performance

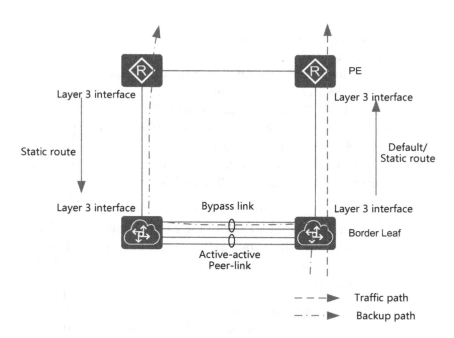

FIGURE 9.10 Deploying static Layer 3 routes on the egress network.

The recommended deployment varies depending on the device, board, and interface model of the border leaf node. The following are common deployment scenarios:

(1) Border leaf nodes are fixed or modular switches with only one LPU.

When the border leaf node has only one LPU (shared by the uplink and downlink traffic), or the border leaf node is a fixed switch, the recommended egress connection mode is shown in Figure 9.11. The egress connection is the same in both square-looped and dual-homed networking.

- In this example, a modular switch is equipped with one 10GE LPU, and firewalls or LBs and PEs are connected to the 10GE LPUs of border leaf nodes.

- A delay is set for uplink interfaces of border leaf nodes connected to PEs to go up, and also for route advertisement.

- The two member links of the peer-link in the M-LAG that is configured on the border leaf nodes are not used as bypass links (the peer-link is not shown in the figure).

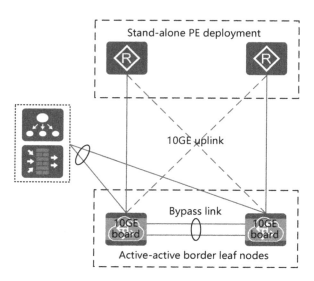

FIGURE 9.11   Egress connection when a border leaf node is a fixed switch or a modular switch with only one LPU.

(2) Border leaf nodes are equipped with multiple LPUs of the same type.

When multiple LPUs of the same type are configured on a border leaf node, the recommended egress connection mode is shown in Figure 9.12. The egress connection is the same in both square-looped and dual-homed networking.

- Firewalls or LBs and PEs are connected to the 10GE LPUs of border leaf nodes, and links are created across the chassis and cards.

- A delay is set for uplink interfaces of border leaf nodes connected to PEs to go up and also for route advertisement.

- Two border leaf nodes in a group are connected to different cards of each spine node.

- On the border leaf node, two peer-link member links in M-LAG and two bypass links are bundled across cards (the peer-links are not shown in the figure). If a 40GE card and a 10GE card are used, the 40GE port can be split into four 10GE ports and then bundled with the 10GE card.

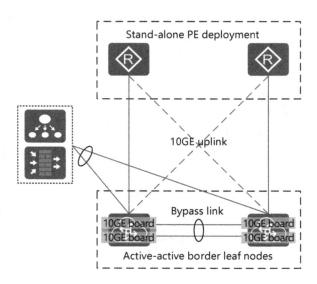

FIGURE 9.12 Egress connection when a border leaf node is equipped with multiple LPUs of the same type.

9.1.2.5 Firewall Deployment Plan

Huawei hardware firewalls can be connected to border leaf nodes or service leaf nodes in bypass mode, or to border leaf nodes and PEs in inline mode. It is recommended that firewalls be connected to service leaf nodes in bypass mode, as shown in Figure 9.13.

The controller automatically delivers configurations such as vSYS, security policies, and SNAT/EIP for the interconnection between tenants' VPCs and firewalls.

9.1.2.6 LB Deployment Plan

The following LB deployment plan is recommended:

- LBs are deployed in active/standby mode.

- LBs are connected to service leaf nodes in bypass mode.

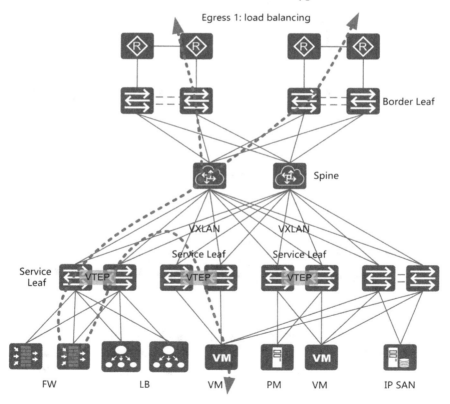

FIGURE 9.13 Load balancing on an egress network when firewalls connect to service leaf nodes in bypass mode.

- LBs are connected to the VXLAN network at Layer 2. The Self-IP address, service VIP address, and server IP address are on the same subnet.

- An LB uses Eth-Trunk to connect to the service leaf node in M-LAG mode.

Figure 9.14 shows the traffic model when LBs are used. The server forwards packets to the LBs at Layer 2 over the VXLAN tunnel, and the LBs route the packets by default to service leaf nodes, which search for a route and forward the packets to the firewall.

In cloud-network integration, the cloud platform can connect to LBs provided by some vendors and automatically deliver load balancing services.

In network virtualization, the SDN controller can manage LBs provided by some vendors and automatically deliver interconnection addresses and

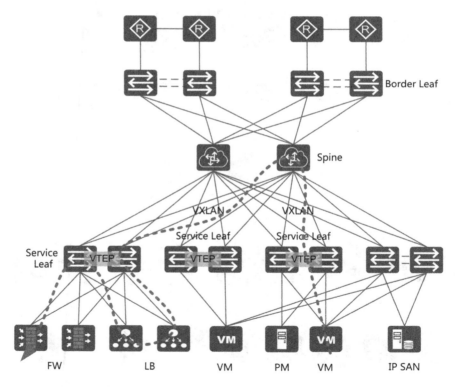

FIGURE 9.14   Traffic model when LBs connect to service leaf nodes in bypass mode.

bidirectional routes between LBs and switches. Load balancing services have to be manually configured on LBs.

9.1.2.7 Server Access Deployment Plan

Figure 9.15 shows the common access modes of x86 servers.

- (Recommended) Servers are connected to M-LAG switches working in active/standby mode, as shown in Figure 9.15a.

- (Recommended) The server is connected to M-LAG switches working in load balancing mode, as shown in Figure 9.15b.

- The server is connected to leaf nodes working in active/standby mode, as shown in Figure 9.15c.

Table 9.2 compares the three access modes of x86 servers.
Figure 9.16 shows the key configurations related to server access.

- When servers are connected in active/standby mode, physical interfaces on leaf nodes are used as interconnected interfaces. In this case, Eth-Trunks do not need to be configured. When servers are connected in load balancing mode, M-LAG is configured on leaf nodes.

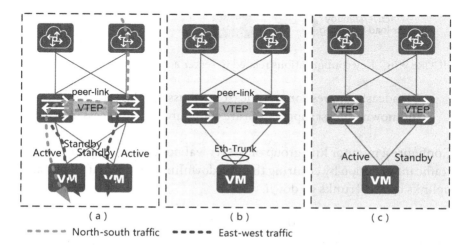

----- North-south traffic ■■■■■ East-west traffic

FIGURE 9.15   Access modes of x86 servers.

TABLE 9.2   Comparison of x86 Server Access Modes

Access Mode	Applicable Scenario	Precautions
Active/standby connections to M-LAG switches (recommended)	VMs or BMs do not proactively send notifications to the network during active/standby switchover of NICs	The peer-link bandwidth should be equal to the total uplink bandwidth, and the uplink and downlink oversubscription should be applied
Load balancing connections to M-LAG switches (recommended)	Servers are connected to the network in load balancing mode	The peer-link bandwidth should be high enough to meet demand if a downlink of the leaf node is faulty
Active/standby connections to single leaf nodes	There is no coupling among leaf nodes	During the active/standby switchover of NICs, VMs and BMs must detect the fault and proactively send notifications to the network. Otherwise, the network convergence is slow

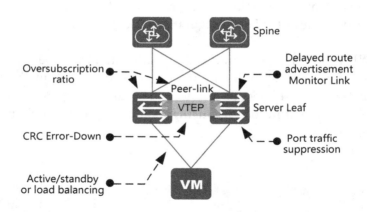

FIGURE 9.16   Key configurations related to server access.

- Broadcast suppression, multicast suppression, ARP rate limit, and unknown unicast suppression are configured on all access interfaces.

Configure a monitor link group on server leaf nodes. This helps to prevent traffic interruption by ensuring that the downlinks go down if all physical uplinks or Eth-Trunks go down.

9.1.3 Management Network Plan (Recommended)

When deploying an SDN DCN, plan the service network and management network separately.

A management network is used by DC administrators to remotely manage servers and network devices. This dedicated network may provide in-band or out-of-band management. For in-band management, which applies to small-scale networks, management traffic and service traffic share forwarding links. For out-of-band management, which applies to large-scale and complex networks, dedicated devices carry management traffic that is separated from the service network. The following uses out-of-band management as an example to describe how to plan a management network.

9.1.3.1 Management Network Deployment Plan

As shown in Figure 9.17, the SDN controller server cluster, SecoManager server cluster, cloud platform controller nodes, and VMM controller nodes are deployed on a management network, separate from the service network.

- The NICs of the SDN controller server cluster, SecoManager server cluster, cloud platform controller nodes, and VMM controller nodes are connected to the management access switches in active/standby or load balancing mode.

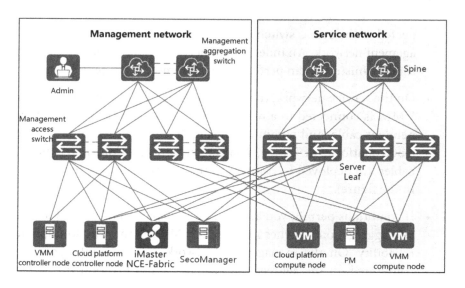

FIGURE 9.17   Independent deployment of the management network.

- If the controller node of the cloud platform must provide the DHCP function, independent cables need to be deployed to connect this node to the server leaf nodes on the service network. The administrator needs to perform only basic configurations, and the SDN controller will automatically deliver other configurations.

- The SDN controller and SecoManager are connected to the server leaf nodes of the service network with independent cables. To implement Layer 3 interconnections for the managed IP addresses of devices, the southbound network segment of the controller is deployed and advertised in the routing protocol on the underlay network.

- The SDN controller northbound gateway, SecoManager northbound gateway, FabricInsight (not shown in the figure) external access plane gateway, cloud platform, and VMM-related network planes are manually deployed on the management network using VLANIF interfaces. In this case, a dedicated management VPN is planned. The northbound network segment of the SDN controller, northbound network segment of the SecoManager, network plane used by the cloud platform, and VMM-related network plane are reachable at Layer 3, facilitating system interconnections.

- On the service network, network devices' management ports and servers' Baseboard Management Controller (BMC) ports are connected to the access switches (not shown in the figure) on the management network. An independent network plane is planned so that the administrator can perform unified remote management.

- On the service network, the compute node of the cloud platform or VMM is connected to a management access switch with an independent cable, and communicates with the controller node of the cloud platform or VMM, respectively. In addition, independent cables are deployed to connect to the storage network (not shown in the figure).

- If conditions permit, you are advised to deploy management access and aggregation switches in a stack or M-LAG in order to improve reliability. On the management network, VLANIF interfaces are configured on the aggregation switches and VLAN on the access switches.

9.1.3.2 SDN Controller Deployment Plan (Recommended)

An SDN controller cluster is typically deployed on three physical servers and, based on the services that are carried, can be divided into the following three network planes:

- Network plane for the internal communication zone: used for internal communication in the cluster, such as communication between different nodes and communication with the database.

- Network plane for northbound management: used for northbound communication and Linux management, including interconnections with the cloud platform, web access, and Linux login.

- Network plane for the southbound service zone: used to communicate with network devices through protocols such as NETCONF, SNMP, and OpenFlow.

The first two planes can be combined for deployment, whereas two independent network planes should be planned for the southbound service zone. For each network plane, allocate two physical network ports.

Figure 9.18 shows the cable connections when SDN controller is deployed.

Table 9.3 describes the dual-plane plan of server ports.

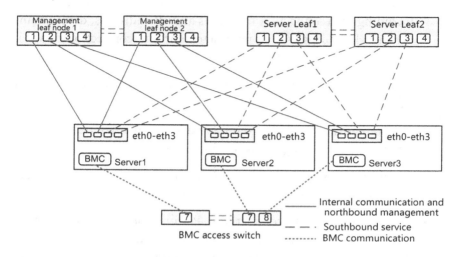

FIGURE 9.18   Connection between the SDN controller and SecoManager servers.

TABLE 9.3   Example of Dual-Plane Plan of Server Ports

Server	Network Port	IP Address (Example)	Description
Server1	bond0 (eth0 & eth1)	192.168.12.1/24	Internal communication and northbound management
	bond1 (eth2 & eth3)	192.168.2.1/24	Southbound service
Server2	bond0 (eth0 & eth1)	192.168.12.2/24	Internal communication and northbound management
	bond1 (eth2 & eth3)	192.168.2.2/24	Southbound service
Server3	bond0 (eth0 & eth1)	192.168.12.3/24	Internal communication and northbound management
	bond1 (eth2 & eth3)	192.168.2.3/24	Southbound service

9.2 Deployment Process

Figure 9.19 shows an overview of the DC deployment process. The cabling, hardware installation, and device power-on processes are similar to those of a legacy DC and are not described here.

9.2.1 Overview

The following operations are involved in the deployment process:

- Planning deployment: Plan underlay network configurations, including interconnection interfaces and IP addresses, device management IP addresses, and parameters for interconnection between devices and the controller.

- Installing hardware: Install network devices and servers, connect cables, and power on the devices.

- Pre-configuring network: Manually deploy underlay network configurations and configure the management network.

- Installing software: Install Huawei SDN controller and SecoManager on the designated servers and verify their operation.

- Commissioning interconnections: Perform pre-configurations on Huawei SDN controller and SecoManager, discover and manage devices on the service network, create resource pools, and establish interconnections with the cloud platform.

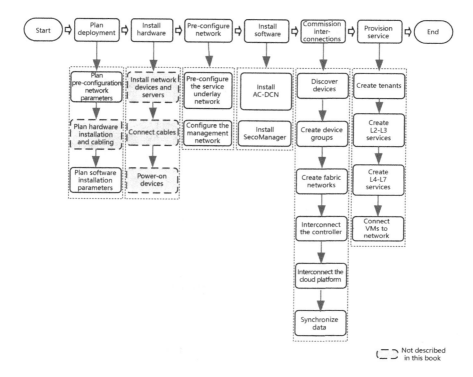

FIGURE 9.19   DC deployment process.

- Provisioning services: Create a logical network VPC for tenants and set service parameters. The SDN controller can then automatically deliver configurations to devices.

9.2.2 Basic Network Pre-Configurations

9.2.2.1 Networking

This section describes the key steps involved in manually deploying and configuring an underlay network. Zero-Touch Provisioning (ZTP) can also be used to deploy the underlay network, but this is not described here. After the underlay network is manually configured, the SDN controller, which is unaware of the underlay network configurations, can automatically provision services.

The DCN shown in Figure 9.20 is used as an example. On the DCN, a group of border leaf nodes, a group of service leaf nodes, and multiple groups of server leaf nodes are connected through two spine nodes. The gray dots in the figure refer to Layer 3 interconnection interfaces.

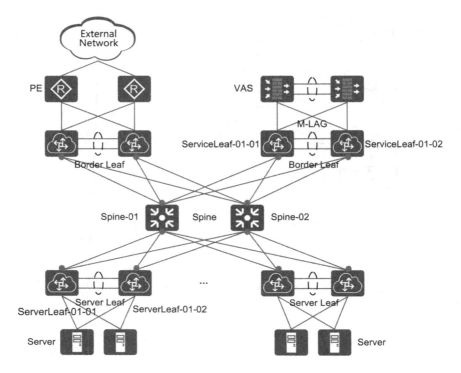

FIGURE 9.20   Example of an underlay network.

9.2.2.2 Deployment Parameter Plan

Before deploying an underlay network, plan the following:

- Physical connection table

- IP network segment

- IP addresses of devices

- IP addresses for device interconnections

- Parameters for interconnection between network devices and the controller

The following is an example of a parameter plan.

1. Plan physical connection tables.

   Table 9.4 provides an example of a physical connection table, based on the physical connections of ServerLeaf-01-01 and ServerLeaf-01-02.

TABLE 9.4 Example of a Physical Connection Table

Local Device Name	Local Interface Name	Peer Device Name	Peer Interface Name	Description
ServerLeaf-01-01	Eth-Trunk 0 • 40GE1/0/1 • 40GE2/0	ServerLeaf-01-02	Eth-Trunk 0 • 40GE1/0/1 • 40GE2/0	Used for peer-link links. The following settings are recommended • Set the number of member interfaces to the nth power of 2 to ensure optimal load balancing (hash) • For modular switches, peer-link interfaces on different cards must be selected to ensure reliability
	40GE 1/0/3	Spine-01	40GE 1/0/1	Layer 3 interface connected to a spine node
	40GE 1/0/4	Spine-02	40GE 1/0/1	Layer 3 interface connected to a spine node
	25GE 1/0/3	Computer001	Port1	Used for server access
	25GE 1/0/4	Computer002	Port1	Used for server access
	25GE 1/0/5	Computer003	Port1	Used for server access
	MEth0/0/0	Management Ethernet switch (MEth/BMC)	GE 1/0/6	Used for interconnection with the remote management device
ServerLeaf-01-02	Eth-Trunk0 • 40GE 1/0/1 • 40GE 1/0	Server Leaf-01-01	Eth-Trunk0 • 40GE 1/0/1 • 40GE 1/0	Used for peer-links. The following settings are recommended • Set the number of member interfaces to the nth power of 2 to ensure optimal load balancing • For modular switches, peer-link interfaces on different cards must be selected to ensure reliability
	40GE 1/0/3	Spine-01	40GE 1/0/2	Layer 3 interface connected to a spine node
	40GE 1/0/4	Spine-02	40GE 1/0/2	Layer 3 interface connected to a spine node
	25GE 1/0/3	Computer001	Port2	Used for server access
	25GE 1/0/4	Computer002	Port2	Used for server access
	25GE 1/0/5	Computer003	Port2	Used for server access
	MEth0/0/0	Management Ethernet switch (MEth/BMC)	GE 1/0/7	Used for interconnection with the remote management device

TABLE 9.5   Example of IP Address Plan

IP Address Plan	Value	Description
IP address of the out-of-band management network segment	192.168.100.0/24	Usage: Out-of-band management addresses of devices are used for remote login and management. The management interfaces include the server's BMC interface, the switch's MEth interface, and the firewall's GE t0/0/0 interface Tips: Plan the IP network segment based on the number of devices on the live network
Loopback0 address	192.168.0.0/32– 192.168.0.254/32	Usage: This address resource is required when the underlay network is manually deployed. It includes the VTEP address, router ID, address managed by the SDN controller, and DFS group address Tips: Set two loopback addresses for each switch
Loopback1 address	192.168.1.0/32– 192.168.1.254/32	Usage: This address resource is required in M-LAG networking Tips: Plan the following loopback addresses for each CE switch • Use the Loopback0 address as the router ID, address managed by the SDN controller, and DFS group address • Use the Loopback1 address as the VTEP address, which is the same on the two member devices in the M-LAG
Device interconnection address	10.254.0.0– 10.254.0.254	Usage: This address resource is required when the underlay network is manually deployed. It is used for Layer 3 interconnection between leaf and spine nodes Tips: Calculate the number of occupied address segments and the specific range based on the number of devices

2. Plan IP address segments.

Plan IP address segments based on networking requirements to avoid address conflicts. Table 9.5 provides an example of IP address plan.

3. Plan IP addresses of devices.

Plan an IP address for each device based on the address segment types and address ranges listed in Table 9.5. Table 9.6 provides an example of an IP address plan, based on ServerLeaf-01-01 and ServerLeaf-01-02.

TABLE 9.6   Example of Device IP Address Plan

Address Type	ServerLeaf-01-01	ServerLeaf-01-02
Management IP address	192.168.100.1/24	192.168.100.2/24
Loopback0	192.168.0.1	192.168.0.2
Loopback1	192.168.1.1	192.168.1.2
Router-ID	192.168.0.1	192.168.0.2
VTEP IP	192.168.0.1	192.168.0.2

TABLE 9.7   Example of IP Address Plan for Device Interconnections

Local Device	Local Interface	Local IP Address	Peer Device	Peer Interface	Peer IP Address
ServerLeaf-01-01	40GE 1/0/3	10.254.0.1/30	Spine-01	40GE 1/0/1	10.254.0.2/30
	40GE 1/0/4	10.254.0.5/30	Spine-02	40GE 1/0/1	10.254.0.6/30
	VLANIF9	192.168.3.2/24 Virtual-IP: 192.168.3.1	SDN controller server cluster	—	SDN controller southbound gateway
	VLANIF19	192.168.5.2/24 Virtual-IP: 192.168.5.1	SecoManager server cluster	—	SecoManager southbound gateway
ServerLeaf-01-02	40GE 1/0/3	10.254.0.9/30	Spine-01	40GE 1/0/2	10.254.0.10/30
	40GE 1/0/4	10.254.0.13/30	Spine-02	40GE 1/0/2	10.254.0.14/30
	VLANIF9	192.168.3.3/24 Virtual-IP: 192.168.3.1	SDN controller server cluster	—	SDN controller southbound gateway
	VLANIF19	192.168.5.3/24 Virtual-IP: 192.168.5.1	SecoManager server cluster	—	SecoManager southbound gateway

4. Plan IP addresses for device interconnections.

Table 9.7 provides an example of an IP address plan for device interconnections, based on ServerLeaf-01-01 and ServerLeaf-01-02.

5. Plan parameter settings for interconnections between network devices and the controller.

To connect a network device to the controller, set the following parameters:

- SNMPv3: used by the SDN controller to discover, add, and manage devices.

TABLE 9.8 Example of Parameter Setting Plan for Interconnections

Category	Parameter	Value	Description
SNMPv3	Version	SNMP v3	Huawei SDN controller uses SNMPv3 to establish an SNMP connection with a device
	snmp-agent group	dc-admin	SNMPv3 user group name
	snmp-agent usm-user	admin	SNMPv3 user
	snmp-agent usm-user authentication-mode	SHA	Authentication mode
	authentication password	Huawei@123	Authentication password
	privacy-mode	AES256	Encryption mode
	privacy password	Huawei12#$	Encryption password
	snmp-agent trap source	Loopback0	Interface used by the SDN controller to manage switches
NETCONF	manager-user	client@huawei.com	The SSH user name is *client*, and the domain name is *huawei.com*
	password	Huawei@123	SSH user authentication password
	level	15	SSH user level

- NETCONF: used by the controller to deliver command configurations to network devices. Table 9.8 provides an example of a parameter setting plan for interconnections.

9.2.2.3 Key Configuration Steps
Table 9.9 lists the key steps for basic network pre-configurations.

9.2.3 Installing the Controller
The following procedure describes how to install the SDN controller:

- Install the OS on the server and configure Redundant Array of Independent Drives (RAID).
- Download and configure the installation tool.
- Load the controller installation program to the installation tool, and complete the installation.
- Apply for and activate a license after the installation is completed.

For installation examples, visit the Huawei technical support website and search for related product documentation.

9.2.4 Commissioning Interconnections

After the SDN controller is installed, pre-configure the controller to prepare for service provisioning. To commission interconnections, perform the following operations:

TABLE 9.9   Key Steps of Basic Network Pre-configurations

No	Step	Configuration
1	Configuring basic device information	• Create a device name, which must be unique on the entire network • Configure system time and NTP synchronization information on the device • Configure a remote login management mode, user name, and password • Configure an IP address for the device management interface and a route on the management plane
2	Configuring device interconnection interfaces	• Connect spine nodes and leaf nodes, including server leaf nodes, service leaf nodes, and border leaf nodes, through Layer 3 main interfaces • Configure IP addresses for interconnections in preparation of configuring routing protocols on the underlay network
3	Configuring loopback and VTEP addresses	Create two virtual interfaces, Loopback0 and Loopback1, on each leaf node (NVE node) • Loopback0 address is used as the router ID, DFS group address of the M-LAG, and source interface for sending BGP packets during BGP EVPN peer establishment • Loopback1 IP address is used as the VTEP IP address. Members in an active-active device group must use the same VTEP IP address
4	Configuring routing protocols on the underlay network	A routing protocol on the underlay network is used to configure underlay routes to establish VXLAN tunnels on the upper-layer overlay network. This ensures that VTEP IP addresses are mutually reachable at Layer 3. OSPF is recommended for small- and medium-sized networks, and EBGP for large-sized networks
5	Configuring an M-LAG	Configure a DFS group and peer-links to establish an active-active M-LAG It is recommended that server leaf nodes, border leaf nodes, and service leaf nodes use two switches to form an M-LAG, which improves reliability and ensures service continuity even during a device upgrade

(Continued)

TABLE 9.9 (*Continued*) Key Steps of Basic Network Pre-configurations

No	Step	Configuration
6	Configuring server access	In most cases, servers are connected to the server leaf work group using two uplinks. That is, servers are connected through two network interfaces to two server leaf nodes in the same M-LAG For servers connected in load balancing mode, configure an M-LAG on leaf nodes. For servers connected in active/standby mode, do not configure Eth-Trunk on leaf nodes If a server leaf node connects to a common service server, configure only M-LAG access. The SDN controller will automatically deliver other access configurations When a server leaf node is connected to the SDN controller or SecoManager server, the southbound gateway of the controller uses VLANIF and VRRP to connect to the server leaf node. In this case, you do not need to configure VPN for the southbound gateway. Instead, the routing protocol on the underlay network imports direct routes to implement Layer 3 interconnections between the SDN controller and its managed devices
7	Configuring BGP EVPN	On a DC VXLAN network • Use BGP EVPN as the VXLAN control plane protocol • Configure leaf nodes as BGP RR clients and spine nodes as BGP RRs • Create EVPN IBGP peer relationships between the leaf nodes and spine nodes
8	Configuring firewalls	Two firewalls work in active/standby mirroring mode, where the two firewalls synchronize configurations through an interconnected heartbeat link. If the active firewall fails, the standby firewall takes over. The active and standby firewalls have the same configurations, thereby ensuring service continuity • Configure a service interface for connecting the firewall to a service leaf node • Configure a management interface for connecting the firewall to the SecoManager • Configure the active/standby mirroring mode for firewalls
9	Setting parameters for interconnections between network devices and the controller	Configure the following protocols on switches and firewalls so that the SDN controller can manage the firewalls • SNMPv3: used by the SDN controller to discover, add, and manage devices • NETCONF: used by the SDN controller to deliver configurations to network devices • Link Layer Discovery Protocol (LLDP): It automatically discovers neighbor relationships and reports them to the controller
10	Configuring the external management network	The configurations of out-of-band management network are similar to those for a legacy DCN and are not described here

- Discover switches on the SDN controller so that it can manage them.

- Discover links between devices and between devices and servers on the SDN controller, to prepare for service provisioning and topology management.

- Create a device group for the two switches that already have M-LAG deployed on the SDN controller.

- Create a fabric network on the SDN controller, and add it to switches. The fabric is similar to a network resource pool.

- Discover firewalls on the SecoManager, and create firewall device groups and firewall resource pools.

- Connect the SDN controller to the SecoManager and synchronize data between them.

- Set the firewall link type and associate the firewall resource pool with the fabric, so that the services on the fabric can use resources in the firewall resource pool on demand.

- Interconnect the SDN controller with the cloud platform or computing virtualization platform (VMM) and set interconnection parameters to prepare for inter-system association and automatic service provisioning.

9.2.5 Provisioning Services

This section describes the key process of provisioning services on an SDN DCN. In network virtualization, the SDN controller connects to the computing virtualization platform (VMM); whereas, in cloud-network integration, the SDN controller connects to the cloud platform. In the two scenarios, services are provisioned in different processes.

9.2.5.1 Service Provisioning Process

In network virtualization, the SDN controller interworks with the VMM (vCenter is used as an example). The SDN controller and SecoManager provide GUIs for provisioning network services, and the vCenter provides GUIs for provisioning compute resources. Figure 9.21 shows the process of provisioning computing services.

In cloud-network integration, the SDN controller interconnects with the cloud platform, and the administrator delivers services on the cloud

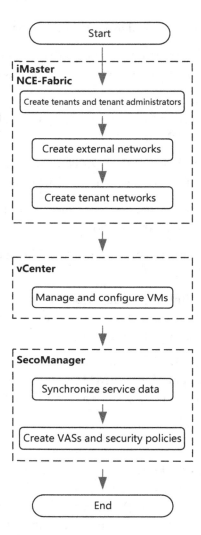

FIGURE 9.21 Process of provisioning computing services.

platform's GUI. Figure 9.22 shows the process of provisioning services in cloud-network integration.

9.2.5.2 Service Provisioning Example

In cloud-network integration or network virtualization, a service provisioning GUI is provided by the cloud platform or SDN controller. This section uses network virtualization as an example to describe the service provisioning process.

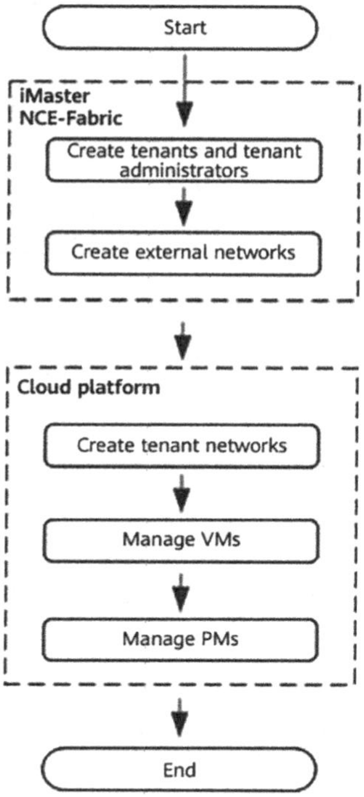

FIGURE 9.22   Process of provisioning services in cloud-network integration.

1. Scenarios

In this example, intranet VMs provide Internet-accessible services and therefore require EIPs to be used. Figure 9.23 shows the physical model and service logic model applied in this service scenario.

As shown in the physical model, the server's private IP address is 192.168.1.10/24, and the bundled EIP is 2.2.2.2. The EIP function is provided by the firewall.

As shown in the service logic model, the logical router in the VPC forwards the access traffic it receives from an external network to the vFW. After the traffic passes through the EIP, the vFW returns the traffic to the logical router, which then forwards it to the logical switch and finally to the server.

2. Data plan

Table 9.10 lists the data plan for this service scenario.

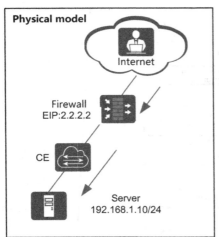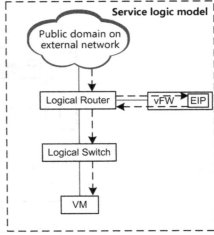

FIGURE 9.23 Physical model and service logic model of the EIP server.

TABLE 9.10 Data Plan

Item		Data
VPC plan		
Item		vpc1
Fabric		Fabric1
Internet interconnection plan		
External gateway	NAT address pool	192.168.1.10/24
	Public IP address mapping to the NAT address pool	2.2.2.2
ACL plan		
Destination address		192.168.1.10
Action		Permit
Subnet plan		
Subnet 1	Address segment	192.168.1.10/24
	Gateway IP address	192.168.1.1
	Logical router	Router1
	Logical switch	Switch1
	VNI	Automatically allocated by the SDN controller
	BD	Automatically allocated by the SDN controller

3. Service provisioning process

1. Create a tenant account for the user and authorize the tenant to use the fabric. The tenant administrator can log in to the SDN controller to create services or perform O&M operations.

2. Create an external gateway on the SDN controller, specify a network segment as the EIP address pool, and configure the interconnection interfaces and routes between the border leaf nodes and PEs. An external gateway is used for access between internal and external sites.

3. Create a Layer 2 or Layer 3 network on the SDN controller.

   - Create a VPC for a specified tenant.

   - Create a logical router in the VPC, associate the logical router with the fabric, and set the subnet and gateway associated with the logical router. The internal server is located in the subnet.

   - Create a logical switch in the VPC (equivalent to creating a Layer 2 network for connecting to servers), and associate the logical switch with a subnet set for the logical router.

   - Deploy the network to prepare for creating an EIP service.

4. Create a VM on the computing virtualization platform (such as vCenter) to provide application services to the external network.

   - For the VPC of the SDN controller, edit VMM mapping parameters (including the logical switch, VMM, and VDS).

   - Create a VM on vCenter. When configuring a network label for the VM, select a network label pushed by the SDN controller to the VMM.

   - Save the VM, and then, start it. After the VM has loaded the resources and connected to the network, you can view the VM's access point in the VPC of the SDN controller, as shown in Figure 9.24. In the figure, the Public cloud refers to an external network associated with the VPC, Router1 is the logical router, and Switch2 is a logical switch connected

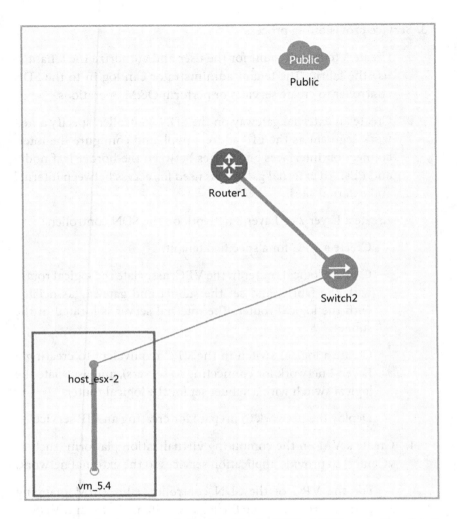

FIGURE 9.24 Logical network displayed in the VPC of SDN controller after the VM goes online.

to the logical router. The solid dot (host_esx-2) indicates the logical port (port on the server leaf node connected to the server), and the hollow dot (vm_5.4) indicates the end port (VM).

5. Create firewall-related services in the VPC of the SDN controller so that traffic passes through the firewall.

- Create a logical firewall in the VPC, and associate it with the logical router.

- Create internal and external interconnection links for the firewall.

6. Configure firewall services on the SecoManager.

- Log in to the SecoManager, and synchronize data with the SDN controller. The synchronized data includes tenants, network topology information, VPCs, logical routers, logical switches, and subnets. This enables the SecoManager to learn the associations between logical routers and logical firewalls in the VPC and subsequently set the network segment corresponding to the subnet in the VPC as the protected network segment of the firewall.

- Create an EIP policy on the SecoManager, associate the policy with the created external network (the system automatically parses the EIP address), and set the internal private network address for EIP translation.

- Create a security policy on the SecoManager to allow external users to access the internal server.

After the preceding operations are complete, the SDN controller automatically delivers configurations to the switch, and the SecoManager automatically delivers configurations to the firewall. This enables external users of the DC to access the VM through the EIP.

CHAPTER 10

Openness of DCN

A s DCs have to be built by multiple organizations and vendors, the forwarding and control layers must apply unified standards (protocols) and interfaces for system interconnections. This chapter describes the southbound and northbound openness of the SDN controller in a DC, applicable to interconnections with multiple service platforms and network devices. It also describes the northbound and interconnection openness of forwarders (network devices), applicable to interconnections with multiple control systems as well as devices from other vendors.

10.1 DCN ECOSYSTEM

As the DCN industry thrives, an increasing number of organizations and vendors are cooperating and competing with each other to promote continuous technical innovations. While providing continually improving services, the DCN industry and ecosystem are both prospering.

DCN solutions involve the cloud platform/application layer, controller layer, and forwarder layer. A currently existing issue is the cooperation between layers, and between different devices at the same layer. To ensure that all layers and devices can collaborate with each other to complete DCN tasks, unified standards (protocols) and interfaces must apply for interconnections, and full openness from the controller layer to forwarder layer is required for an open DCN ecosystem.

1. Openness of the controller layer

    The SDN controller can interconnect with the container platform, cloud management platform, virtualization platform, and various third-party applications or platforms through standard northbound APIs, and it can manage forwarders as well as third-party VAS devices through southbound APIs. Table 10.1 lists the openness capabilities of Huawei iMaster NCE-Fabric.

2. Openness of the forwarder layer

    DC switches can interconnect with the network controller, network orchestrator, cloud management platform, and GitHub through standard northbound APIs; and they can interconnect with third-party network devices through standard southbound network protocols. Table 10.2 lists the openness capabilities of Huawei DC switches.

The following sections describe the openness of the controller and forwarder.

TABLE 10.1   Openness Capabilities of Huawei iMaster NCE-Fabric

Supported Northbound Container Platform	Supported Northbound Cloud Management Platform	Supported Northbound Virtualization Platform	Supported Southbound Third-Party Device
Open-source Kubernetes mainline version and Red Hat OpenShift platform	Huawei FusionSphere, Red Hat OpenStack, community OpenStack, China Mobile OpenStack, and Mirantis OpenStack	VMware v Sphere, Microsoft System Center	F5 LB, Check Point firewall, Fortinet firewall, Palo Alto firewall, and Radware LB

TABLE 10.2   Openness Capabilities of Huawei DC Switches

Supported Northbound Network Controller	Supported Northbound Network Orchestrator	Supported Northbound Cloud Management Platform	Supported Northbound GitHub Project	Supported Southbound Standard Authentication
VMware NSX, open-source OpenFlow controller	VMware vRNI	Microsoft Azure	Ansible and Puppet	CTTL certification IPv6 Ready certification Tolly certification

10.2 OPENNESS OF THE CONTROLLER

10.2.1 Northbound Openness of the Controller

The controller's northbound openness refers to openness of interconnection with:

- A container platform

- A cloud management platform

- A virtualization platform

- Third-party applications or platforms

1. Openness of interconnection with a container platform

    The SDN controller interconnects with the container platform through SDN API Server Watcher, and containers access the virtualized overlay network in Layer 2 bridging mode (VLAN) or routing mode (BGP). This enables the container network to have automatic, visualized, and intelligent O&M. In addition, the security manager manages next-generation firewalls (NGFWs) to provide load balancing for north-south traffic and security services for containers.

    The SDN API Server Watcher should be capable of interconnecting with open-source Kubernetes, Red Hat OpenShift, and FusionStage 2.0 to provision basic network services such as networks, gateways, and subnets, as well as the container IP address management (IPAM) function. The SDN API Server Watcher monitors service objects, such as the servers and Pods of the Kubernetes API Server, and calls the SDN controller's northbound API when automatically configuring the SDN network.

    The vSwitchCNI plug-in should enable the containers to access the overlay network in bridging or routing mode. It creates and binds virtual NICs of Pods, mounts vSwitches, as well as configures BGP routes on the server side.

    Figure 10.1 shows the architecture with the SDN controller and components.

2. Openness of interconnection with a cloud management platform

    As a typical open-source cloud management platform, OpenStack occupies a large share of the private cloud DCN market. At the same time, various third-party cloud management platforms are

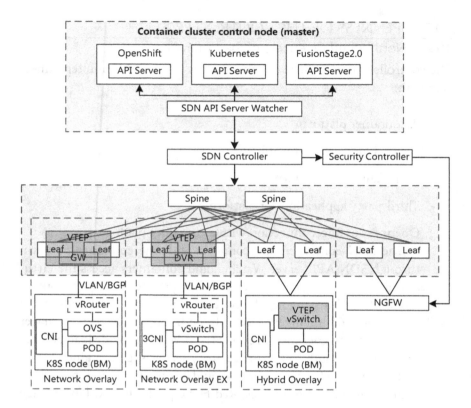

FIGURE 10.1 Architecture with the SDN controller and components such as Kubernetes, OpenShift, and FusionStage.

developed based on the native OpenStack. Therefore, the SDN controller should support OpenStack interconnections to fabricate an end-to-end SDN solution.

OpenStack's Neutron component uses the plug-in architecture, and it provides standard APIs to enable open-source communities or third-party services to manage virtual networks. Furthermore, it allows vendors to add new plug-ins to provide more advanced network functions.

In DCs with SDN and OpenStack integrated, hardware switches are used as VXLAN Tunnel Endpoints (VTEPs) to prevent CPU resource consumption (which occurs if virtual switches are used as VXLAN VTEPs) and improve CPU resource usage. The DC solution with SDN-OpenStack integration can be implemented in network overlay or hybrid overlay mode.

- Network overlay mode: All VXLAN VTEPs are deployed on physical switches. Both VMs and PMs are hierarchically bundled as defined by the OpenStack community, and they connect to VXLAN through hardware switches.

- Hybrid overlay mode: VXLAN VTEPs are deployed on hardware switches and virtual switches. Physical terminals connect to VXLAN through hardware switches, whereas VMs connect through virtual switches.

Figure 10.2 shows the architecture of the SDN-OpenStack-integrated DC solution.

The SDN controller uses standard APIs as defined by the OpenStack community, and it interconnects with the OpenStack cloud management platform by adding plug-ins to OpenStack's Neutron component. Through this process, the SDN controller

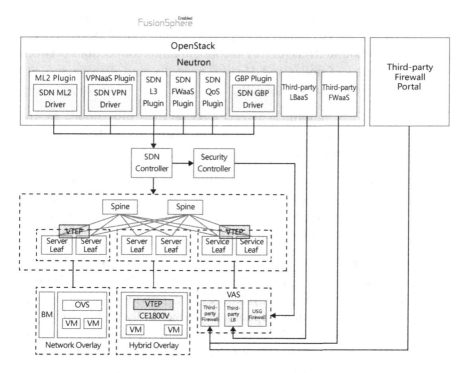

FIGURE 10.2 Architecture of the SDN-OpenStack-integrated DC solution.

implements network device configurations and visualized monitoring. The SDN controller adds the following mainstream plug-ins:

- SDN ML2 driver: It detects Neutron port events and calls the SDN controller's northbound RESTful API to complete network configurations on the access side of PMs and VMs.

- SDN VPN driver: It detects VPN services provisioned by the cloud management platform, calls the SDN controller's northbound RESTful API to configure interconnections between service leaf nodes and Huawei firewalls, and delivers VPN services to Huawei firewalls.

- SDN L3 plug-in: It detects Neutron vRouter, network, EIP, and SNAT events, as well as calls the SDN controller's northbound RESTful API to configure the routing gateway of forwarders, NAT of Huawei firewalls, and interconnections between service leaf nodes and third-party LBs or firewalls.

- SDN FWaaS plug-in: It detects firewall services provisioned by the cloud management platform, calls the SDN controller's northbound RESTful API to configure interconnections between service leaf nodes and Huawei firewalls, and delivers security policies to Huawei firewalls.

- SDN QoS plug-in: It detects QoS services provisioned by the cloud management platform and calls the SDN controller's northbound RESTful API to configure QoS for forwarders.

- SDN GBP driver: It detects GBP-based services provisioned by the cloud management platform to map these services and calls the SDN controller's northbound RESTful API to configure the GBP model on forwarders.

3. Openness of interconnection with a virtualization platform

As the mainstream VM management platforms, VMware vSphere and Microsoft System Center provide computing and storage virtualization capabilities. Specifically, they provide a vSwitch-based network virtualization solution. However, this solution cannot meet network interconnection and unified management requirements between multiple compute resources (such as

PMs, heterogeneous virtualization platforms, and container platforms) in a DC. Furthermore, the solution cannot manage existing networks.

Figure 10.3 shows the joint solution with Huawei SDN controller, VMware vSphere, and Microsoft System Center. This solution provides access to not only existing virtual servers, but also physical servers and legacy networks. In this way, this solution implements unified management of virtual resource pools, physical resource pools, and legacy networks in addition to organically interconnecting network devices and compute resources.

1. Interconnecting with VMware vSphere

   The VMware vSphere virtualization solution consists of the vCenter server and ESXi virtualization platform, as shown in

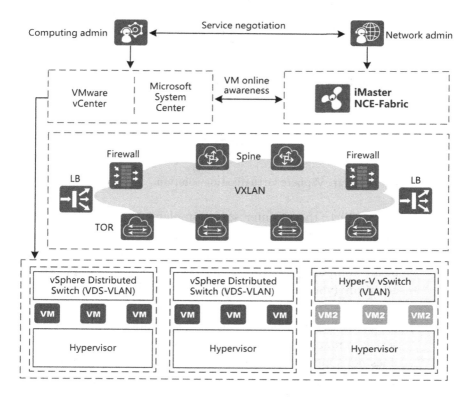

FIGURE 10.3   Joint solution with Huawei SDN controller, VMware vSphere, and Microsoft System Center.

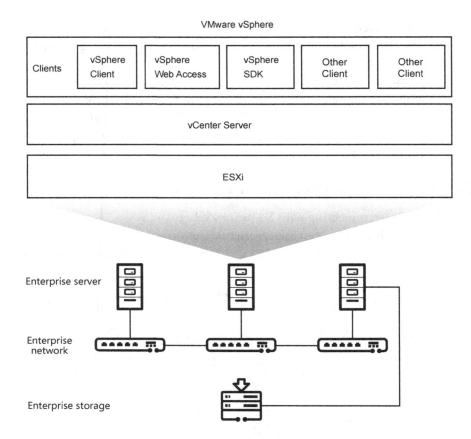

FIGURE 10.4   VMware vSphere virtualization solution.

Figure 10.4. The vCenter server implements unified management of virtual clusters and unified provisioning of network services, whereas the ESXi virtualization platform provides basic computing virtualization (vCompute), storage virtualization (vStorage), and network virtualization (vNetwork) capabilities.

The VMware vSphere virtualization solution uses the vCenter server to manage and control VMs; however, the vCenter server does not support management and control of physical network devices. Therefore, this solution cannot implement unified management and O&M of VM resources and physical network devices, let alone automatic provisioning of network services,

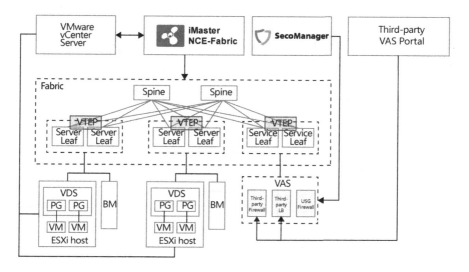

FIGURE 10.5   Architecture of components in the joint solution with Huawei SDN controller and VMware vSphere.

whereas the joint solution with Huawei SDN controller and VMware vSphere can effectively address these issues.

Figure 10.5 shows the architecture of components in the joint solution with Huawei SDN controller and VMware vSphere. The SDN controller interconnects with the vCenter server on the VMware vSphere VM management platform through open APIs. Based on VM-related events detected by the vCenter server, the SDN controller automatically delivers service configurations to network devices on the fabric, and it implements unified management of virtual resources and network devices. In addition, the SDN controller uses the SFC function to divert traffic from VMs to VAS devices.

2. Interconnecting with Microsoft System Center

In addition to supporting end-to-end service backup and IT service deployment, the Microsoft System Center DC solution configures and monitors the basic architecture and automatically provisions services. Through the Hyper-V switch in the server, the network controller automatically configures vNICs of VMs, and it deploys and configures vFWs, virtual SLBs, as

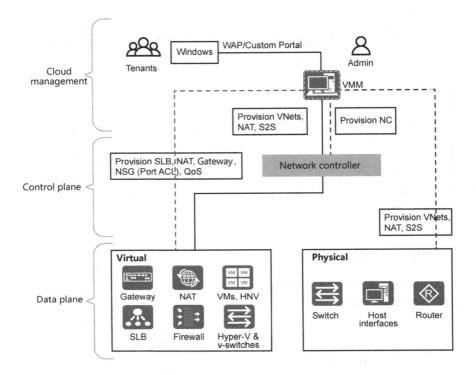

FIGURE 10.6   Microsoft System Center DC solution.

well as virtual gateway services. In addition, it manages the physical NICs (provision NC) of a server and delivers NIC team configurations.

As shown in Figure 10.6, the Microsoft System Center DC solution uses the network controller to manage and control VMs; however, the network controller does not support management and control of physical network devices. Therefore, it cannot implement unified management and O&M of VM resources and physical network devices, let alone automatic provisioning of network services, whereas the joint solution of Huawei SDN controller and Microsoft System Center can effectively address these issues.

Figure 10.7 shows the architecture of components in the joint solution of Huawei SDN controller and Microsoft System Center. The SDN controller interconnects with the network controller on

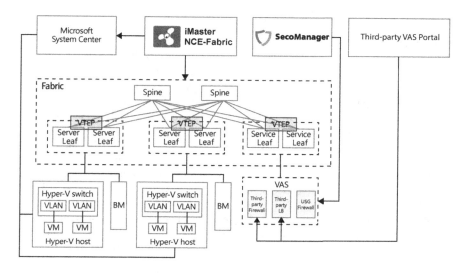

FIGURE 10.7 Architecture of the joint solution with Huawei SDN controller and Microsoft System Center.

the Microsoft System Center VM management platform through open APIs. Based on VM-related events detected by the network controller, the SDN controller automatically delivers service configurations to network devices on the fabric, and it implements unified management of virtual resources and network devices. In addition, the SDN controller uses the SFC function to divert traffic from VMs to VAS devices.

4. Openness of interconnection with third-party applications and platforms

The SDN controller also provides the standard northbound interfaces (NBIs) listed in Table 10.3 to interconnect with third-party applications and platforms (such as the cloud management platform, other SDN controllers, and gateway platforms).

10.2.2 Southbound Openness of the Controller

In the southbound direction, the SDN controller uses the standard OpenFlow, OVSDB, NETCONF, and BGP EVPN protocols to manage Huawei routers, firewalls, and switches; uses the RESTful API to manage third-party VAS devices (such as Fortinet firewalls, Check Point firewalls,

TABLE 10.3   Standard NBIs Provided by the SDN Controller

Supported NBI	Description	Applicable Scenario
OpenStack Neutron service model interface	It creates, deletes, queries, and updates networks, subnets, ports, routers, floating IP addresses, security groups, and QoS services	It connects to the cloud management platform's standard network model to implement interconnection between VMs and VAS devices
SDN controller VPC service model interface	It configures and manages logical network resources (logical ports, switches, routers, and VASs, as well as external gateways)	It interconnects with the cloud management platform's non-standard network model to provision services in scenarios such as rack leasing and multi-egress DCs
O&M API	It queries the physical and logical topologies, collects statistics on network service traffic, and performs ERSPAN mirroring for the forwarded packets	It connects to the O&M monitoring platform to manage and monitor the network topology and traffic
Network security interface	It provisions SFC services and networks, and deploys L4-L7 services	It connects to the VAS platform to provision services such as NAT, firewall, VPN, and load balancing

F5 LBs, Palo Alto firewalls, and Radware LBs); and uses the PBR traffic diversion policy to divert traffic from network devices to third-party VAS devices.

Figure 10.8 shows the southbound openness of the controller.

Figure 10.9 shows PBR traffic distribution by the controller.

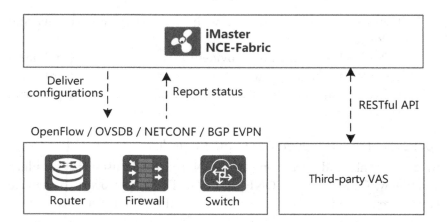

FIGURE 10.8   Southbound openness of the controller.

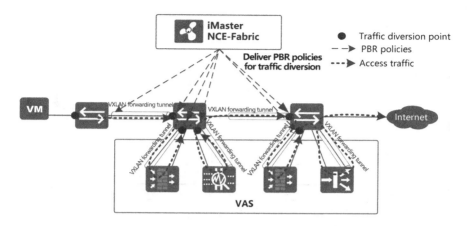

FIGURE 10.9   PBR traffic diversion by the controller.

10.3 OPENNESS OF THE FORWARDER

10.3.1 Northbound Openness of the Forwarder

The forwarder's northbound openness refers to openness of interconnection with

- A network controller

- A network orchestrator

- A cloud management platform

- A GitHub project

1. Openness of interconnection with a network controller

    As the core SDN component, the controller manages physical and virtual network devices and automatically delivers network service configurations to forwarders. The following example uses the interconnection with the VMware NSX controller to describe the interconnection between the forwarder and controller.

    As the NSX control plane, NSX Controller is deployed on three servers, and it manages the VXLAN configurations of NSX-vSwitches, border routers, and hardware switches. NSX Controller includes the API module, switch management module, logical network management module, and directory service module.

    DC switches are interconnected with the VMware NSX controller using the standard OVSDB protocol, and they are managed by

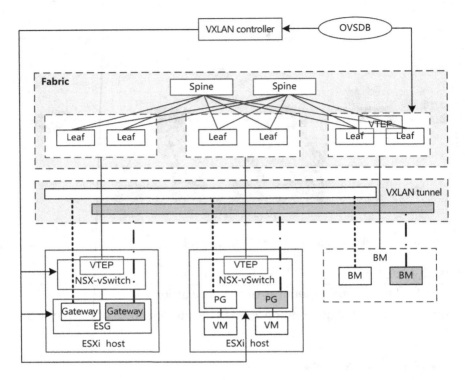

FIGURE 10.10 Architecture of the joint solution with Huawei SDN controller and VMware NSX controller.

this controller. Figure 10.10 shows the architecture of components in the joint solution with Huawei SDN controller and VMware NSX controller.

- NSX Controller: It is a control plane that manages the NSX-vSwitches, edge service gateways (ESGs), and CloudEngine switches in a unified manner.

- DC switch: A DC switch functions as a Layer 2 VXLAN gateway, and BM servers connect to VXLAN tunnels through DC switches.

- NSX-vSwitch: VMs connect to VXLAN tunnels through NSX-vSwitches, which function as Layer 2 VXLAN gateways.

- ESG: It refers to a router that implements Layer 3 communication between VMs and PMs.

2. Openness of interconnection with a network orchestrator

As an SDN component, the network orchestrator implements visualized O&M on virtual and physical networks. The following example uses the interconnection to VMware vRNI to describe the interconnection between the forwarder and controller platform.

VMware vRNI implements unified management, analysis, and fault locating for network services on the private cloud, branch offices, and public cloud. It can manage and scale out VMware NSX Controller clusters, as well as design, deploy, and provision VDS flow analysis, NSX firewalls, and microsegmentation services. In addition, it analyzes traffic paths between VMs, and analyzes as well as locates faults in AWS VPCs, security groups, and firewall policies.

Figure 10.11 shows the architecture of components in the joint solution with Huawei SDN controller and VMware vRNI. DC switches upload service information to VMware vRNI through SSH and SNMP interfaces.

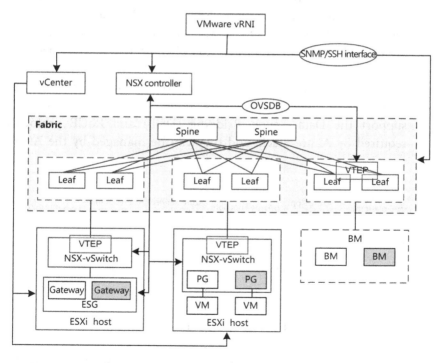

FIGURE 10.11   Architecture of the joint solution with Huawei SDN controller and VMware vRNI.

- SSH interface: used to upload the current configuration, routing table, MAC address table, FIB table, and QoS policy information of the device.

- SNMP interface: used to upload statistics on sent and received data, incoming and outgoing packet loss statistics, system error information, and status information of the interface.

3. Openness of interconnection with a cloud management platform

As an SDN component, the cloud management platform deploys, runs, and manages virtual and physical network applications on the cloud. The following example uses interconnection to Azure Stack to describe interconnection between forwarders and the cloud management platform.

As an extension of Microsoft Azure, Azure Stack integrates the hybrid cloud solution at the infrastructure and platform layers. In Azure Stack, the infrastructure layer provides computing virtualization, network virtualization, storage virtualization management, service deployment, and O&M, whereas the platform layer supports seamless deployment and migration of enterprise services between private and public clouds.

Figure 10.12 shows the architecture of components in the joint solution with Huawei SDN controller and Azure Stack. DC switches support the Data Center Bridging (DCB) and RoCE protocols required by Azure Stack, and they can be managed by the Azure

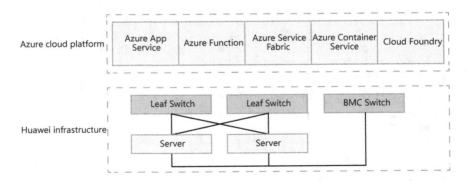

FIGURE 10.12   Architecture of the joint solution with Huawei SDN controller and Azure Stack.

Stack solution. In addition, DC switches can complete the underlay network configurations required for Azure Stack cluster deployment and running.

4. Openness of interconnection with a GitHub project

GitHub is a hosting platform for open-source and private software projects. Currently, multiple GitHub open-source projects for CloudEngine switches have been completed, providing open and standardized interfaces for integration with third-party applications.

a. GitHub project: HuaweiSwitch/CloudEngine-Ansible

Based on the Ansible framework, CloudEngine switches open the source code of the basic Playbook library for switch management and configuration, including the following modules: AAA authentication, ACL, BGP, DHCP, EVPN, NetStream, VRF, and VXLAN.

b. GitHub project: HuaweiSwitch/Puppet

Based on the Puppet framework, CloudEngine switches provide the following open-source modules: Layer 2/Layer 3 interface, VLAN, CAR and Diffserv, QoS, SSH, and Telnet.

c. GitHub project: HuaweiSwitch/OVSDB

Based on the Open vSwitch 2.5.0 framework, CloudEngine switches implement the following five sub-projects: ovsdb-server, ovsdb-client, ovs-pki, vtep-ctl, and huaweiswitch-key. These switches can be managed by a third-party controller using OVSDB. Among the preceding projects, ovsdb-server, as a standard implementation of OVSDB, stores configurations sent by the controller over the OVSDB channel; ovsdb-client provides a channel between ovsdb-server and the controller; ovs-pki, vtep-ctl, and huaweiswitch-key are auxiliary tools that provide key management and OVSDB query capabilities for interaction between the controller and CloudEngine switches.

10.3.2 Openness of Forwarder Interconnection

Forwarders must use standard protocols to communicate with third-party devices. Currently, CloudEngine switches have passed CTTL, IPv6 Ready, and Tolly Lab certifications, thereby verifying that these switches comply with IETF standards and support more than 2000 standard protocols, including BGP, Ethernet, IPv4&IPv6, VXLAN, MPLS, and FCoE.

1. CTTL certification

CloudEngine switches, as certified by CTTL, support DC features such as VXLAN, EVPN VXLAN, VXLAN over IPv6, and NetStream, and they can connect to Cisco devices using OSPF and BGP.

2. IPv6 Ready certification

CloudEngine switches are IPv6 Ready-certified and support IPv6 RFC 1981, RFC 2460, RFC 2474, RFC 3168, RFC 4191, RFC 4291, RFC 4443, RFC 4861, RFC 4862, and RFC 5095, as well as smooth DCN evolution to IPv6.

3. Tolly Lab certification

As certified by Tolly Lab, CloudEngine switches support the standard OpenFlow1.3 protocol and can interconnect with third-party controllers using the OpenFlow protocol (Ryu is used in the test). The switches support VXLAN and NVGRE hardware gateways as well as features such as TRILL, EVN, FCoE, FCF, DCB, VS, and ISSU.

CHAPTER 11

Cutting-Edge Technologies

THIS CHAPTER GIVES A glimpse into several new technologies in the DC field, including the container, hybrid cloud, and AI Fabric technologies. Upon finishing this chapter, you will have an understanding of these technologies' basic concepts, existing technical solutions in the industry, application scenarios, and development trends.

11.1 CONTAINER

11.1.1 Overview

Container technology is an OS virtualization technology that provides an isolated OS environment based on the shared OS kernel of a host. Refer to Figure 11.1 for an overview of the container technology architecture. Here, you can see that each container contains the following components: independent file system namespace (MNT NS), host and domain namespace (UTS NS), inter-process communication namespace (IPC NS), process ID namespace (PID NS), network protocol stack namespace (NET NS), and user namespace (user NS). In addition, a container is mounted to a vSwitch through a vNIC, and the vSwitch accesses a network through a pNIC.

Container technology is developed in two phases: standalone container and container cluster. In the former, a variety of OS isolation technologies have been developed based on the UNIX or Linux OS, thereby laying the foundation for the virtualization technology of a single container.

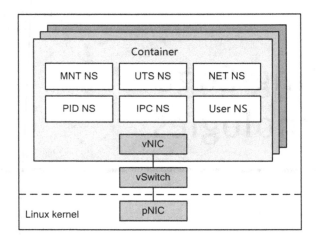

FIGURE 11.1   Container technology architecture.

Leveraging these isolation technologies, the UNIX or Linux OS is able to provide an isolated OS environment. The following describes major OS isolation technologies:

- chroot: also called chroot jail, was first developed on UNIX version 7 in 1979. chroot provides the independent file system space for each process by changing the root directory of a process and its child processes to a specific file system directory, as well as restricting the access of the process and its child processes to only the specific directory.

- FreeBSD jail: was developed by R&D Associates, a hosting provider, in 2000 to provide an independent and clean service running environment for customers on the shared FreeBSD system. FreeBSD jail can partition a FreeBSD-derived compute system into several independent mini-systems, which are isolated from each other, as well as having an independent file system and independent common users, administrators, and network space.

- Solaris Containers: also called Solaris Zones, is an independent virtual server on the Solaris system and supports the x86 and SPARC architectures. Each Solaris Container has its own host name, vNIC or pNIC, and independent storage space.

- Open Virtuozzo (OpenVZ): an open-source container platform released by Virtuozzo in 2005. OpenVZ is also called VPS or virtual environment. Each OpenVZ contains an independent file system, users and user groups, process trees, network devices, and IPC objects.

- Linux containers (LXC): an open-source container engine based on CGroup and Linux NameSpace on an open-source Linux system. In contrast to FreeBSD jail, Solaris Containers, and OpenVZ, which are based on UNIX, LXC is developed based on the Linux kernel and can be easily promoted in the industry.

- Docker: was originally an internal project initiated by Solomon Hykes, dotCloud's founder, and was released as open source in 2013. In Docker 0.9 and versions earlier than Docker 0.9, LXC is used as the container execution engine. In versions later than Docker 0.9, Libcontainer is used as the execution engine. Based on Linux CGroup and namespace technologies, Docker provides container image packaging and management, as well as container lifecycle management.

- Kubernetes: an open-source container orchestration system developed by Google. It provides container cluster management, container cluster scaling, and inter-host communication for containers through CNIs. Currently, Kubernetes is managed by the Linux Foundation.

- Docker Swarm: a container orchestration system released by Docker in 2014. It provides the following functions: Docker cluster management, container cluster scaling, service discovery, load balancing, and inter-host container network communication. In contrast to Kubernetes, Docker Swarm cannot flexibly support user-defined interfaces and has limited openness.

- Mesos: was first released in 2009 as a distributed cluster management platform. Then, with the development of container technologies, Mesos started to support container clusters in 2014. Mesos is a larger platform than Kubernetes and Docker Swarm, making it suitable for a super-large cluster system (with more than 100,000 containers).

Due to that fact that containers are more flexible and efficient than VMs, software applications are gradually being migrated from VMs to containers. Container cluster technologies provide numerous advantages, such as solving the single point of failures of standalone containers, providing unified orchestration capabilities of compute, storage, and network resources for large-scale container applications, and accelerating the development of container technologies. Major container platforms include Google's Kubernetes, Docker's Docker Swarm, and Mesos at the University of California, Berkeley, CA, USA. Of these, Kubernetes has, over years of development, become the de facto standard for container platforms.

11.1.2 Industry's Mainstream Container Network Solutions

Figure 11.2 shows the typical three-layer architecture of container-based web, apps, and DBs. A client accesses the frontend LB (ingress) cluster through a URL. Then, an LB Pod cluster performs Layer 7 load balancing based on service types and forwards requests to the service cluster of the web frontend. The Pod of the web cluster then accesses the Kubernetes service of the backend app cluster based on the service logic. After NAT and load balancing have been performed on the Kubernetes services, the actual service traffic is converted into the east-west traffic to be directly exchanged between Pods.

Similarly, when the Pod of the app cluster accesses the DB cluster service, the apps access the DB service. Then, using the Kubernetes service, NAT

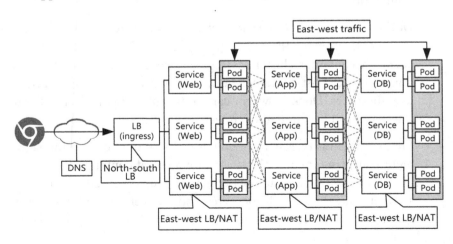

FIGURE 11.2   Typical three-layer architecture of web, apps, and DBs.

and LB service traffic is converted into east-west traffic to be exchanged between Pods.

According to the preceding service scenarios, the Kubernetes network has the following typical requirements:

- The network that carries east-west traffic between Pods must be reachable, and east-west load balancing or NAT must be supported.
- Mutual access between services and Pods, north-south load balancing, and clients' access to container clusters must be supported.

To meet these requirements, the Kubernetes community defines three principles for the Kubernetes network:

- All containers can communicate with all other containers without NAT.
- All nodes can communicate with all containers (and vice versa) without NAT.
- The IP that a container sees itself as is the same IP that others see it as.

The following example refers to Figure 11.3 to describe the three principles:

- Pod1 can directly access Pod2, Pod3, and Pod4 without NAT.
- Pod1 can directly access node1 and node2 without NAT.
- node1 can directly access Pod1, Pod2, Pod3, and Pod4 without NAT.

The Kubernetes community defines basic principles for direct IP access between Pods in a cluster and between Pods and hosts. It clearly defines east-west network requirements, without defining specific network solutions. Currently, common open-source container network solutions include Contiv, Calico, Flannel, and Huawei CNI-Genie, which can be classified into three types: host overlay solution, network overlay Layer 2 bridging solution, and network overlay Layer 3 routing solution.

1. Host overlay solution

    Figure 11.4 shows an example of the host overlay solution architecture in which PMs are connected to the fabric overlay network. In

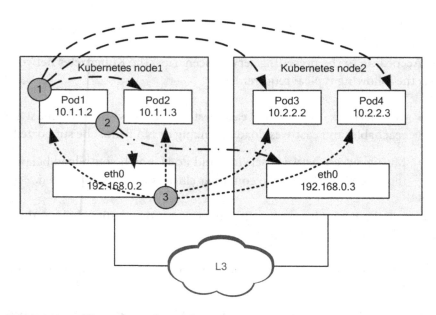

FIGURE 11.3 Three principles of the Kubernetes network.

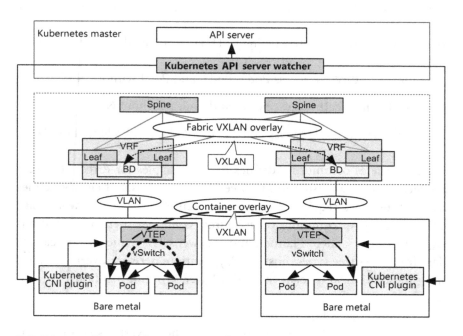

FIGURE 11.4 Host overlay solution architecture.

this architecture, a vSwitch on a physical server functions as a Pod gateway and VTEP to connect to the overlay or underlay network. Inter-Pod traffic on the same physical server is directly forwarded by a vSwitch, whereas Inter-Pod traffic on different physical servers is forwarded through the VXLAN tunnel between vSwitches. Open-source or container platform vendors automate container overlay network provisioning by providing Kubernetes API server watcher and Kubernetes CNI plugins.

Flannel, a typical host overlay solution in the industry, does not provide IP address allocation, which instead needs to be implemented by a third-party component (ETCD), as shown in Figure 11.5. On the control plane, Flannel uses ETCD to configure a routing table on each node. On the data plane, it uses the tunnel mode (such as VXLAN and GRE) for traffic forwarding. Flannel runs a daemon process called Flanneld on each host. When a host is added to a

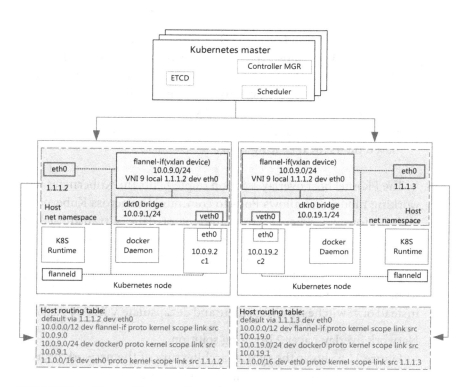

FIGURE 11.5 Flannel host overlay solution.

cluster, Flanneld applies for an IP subnet from ETCD and obtains IP subnet information of other hosts.

Because each host uses a different IP subnet, the container's IP address will be changed if a container is migrated to another host. Flannel overcomes this by using the Pod as the basic unit for network management. That is, IP addresses are allocated per Pod, while subnets are allocated per node (host or VM). Specifically, all Pods on a node must use the same subnet, and a Pod cannot be deployed across nodes.

The following describes the implementation of the Flannel host overlay solution:

- Flannel of a Kubernetes node registers its information (subnets 10.0.9.0/24 and 10.0.19.0/24) with ETCD.

- The Kubernetes node periodically queries network information from ETCD. If a subnet is added, the node updates the subnet and IP address information in the local routing table.

- When a Kubernetes container needs to access the subnet 10.0.19.0/24, it sends a packet destined for 10.0.19.2 and routes the packet to the local VXLAN tunnel for encapsulation and forwarding based on its local routing table. After the destination Kubernetes node receives the VXLAN-encapsulated packet, it decapsulates the packet and forwards it to the destination Pod based on its local routing table.

The Flannel host overlay solution implements the Kubernetes networking model and allows Pods to communicate across Kubernetes nodes. However, when a large amount of data is transmitted in highly concurrent cluster scenarios, vSwitch forwarding capabilities can hold back the overall performance. Such a bottleneck can be eliminated by deploying a network overlay solution based on the Kubernetes open-source architecture, as it uses hardware switches instead of vSwitches to encapsulate and decapsulate VXLAN packets.

2. Network overlay Layer 2 bridging solution

Figure 11.6 shows the typical architecture of the network overlay Layer 2 bridging solution. In this architecture, hardware switches on a fabric network function as Pod gateways and VTEPs, while Pods

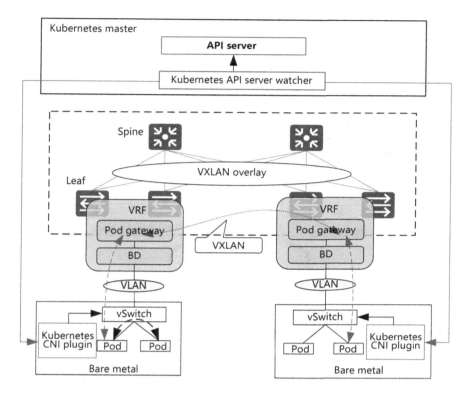

FIGURE 11.6 Typical architecture of the network overlay Layer 2 bridging solution.

connect to the overlay network through VLANs. Traffic between Pods on the same physical server is exchanged through a vSwitch, whereas that between Pods on different physical servers is forwarded through VXLAN at Layer 3. To complete server-side network provisioning of containers, open-source or container platform vendors provide Kubernetes API server watcher and Kubernetes CNI plugins.

When it comes to Layer 2 bridging, Contiv is a typical solution in the industry. It includes the Netmaster (Kubernetes API server watcher in Figure 11.6) deployed on the Kubernetes master node, as well as Netplugins (Kubernetes CNI plugins) deployed on Kubernetes nodes, as shown in Figure 11.7. Of these, Netmaster monitors the Kubernetes API server, allocates VLANs and IP addresses to Pods, and stores the information in ETCD. Each Netplugin implements

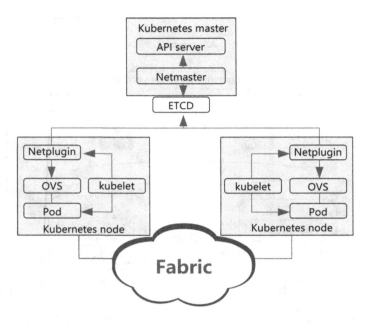

FIGURE 11.7   Contiv Layer 2 bridging solution.

the CNI networking model used by Kubernetes. Contiv uses JSON-RPC to distribute endpoints from Netplugin to Netmaster, while Netplugin handles Up/Down events from Pods and configures vSwitches.

3. Network overlay Layer 3 routing solution

Figure 11.8 shows the typical architecture of the network overlay Layer 3 routing solution. In this architecture, a Pod gateway is deployed on a vRouter of a physical server, and a hardware switch functions as a VTEP to connect the physical server to the overlay network. EBGP is used by the physical server to advertise Pod routes to the overlay VRF. Traffic between Pods on the same physical server is exchanged through vRouters, whereas that between Pods on different physical servers is forwarded through VXLAN at Layer 3. In this process, a hardware switch provides a large number of VXLAN integrated routing and bridging (IRB) routes, whereas open-source or PaaS vendors provide Kubernetes API server watcher and Kubernetes CNI plugins to implement Pod network provisioning and routing configuration on container servers.

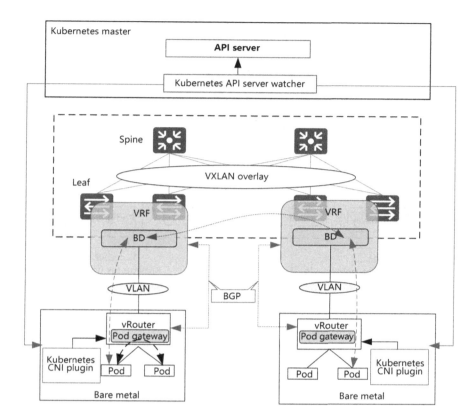

FIGURE 11.8   Typical architecture of the network overlay Layer 3 routing solution.

Calico is a typical solution in the industry for implementing open-source Layer 3 routing. It includes the Calico controller, Felix, and BGP client, which can be seen in Figure 11.9. Calico controller (Kubernetes API server watcher) monitors the API server; detects the events of services, Pods, and ingresses; allocates IP addresses; and stores the IP addresses in ETCD. A BGP client, implemented using the bird Internet routing daemon (BIRD) in an open-source BGP project, implements BGP, advertises routes of the local Pod through an RR, and learns Pod routes on other Kubernetes nodes. A vRouter is implemented based on the routing function of the Linux OS. Felix, a Kubernetes CNI plugin, is deployed on each Kubernetes node to listen to ETCD, detect the events of Pods and services, create Pod vNICs, configure IP addresses for Pods, and update the Linux kernel routing table and security policies.

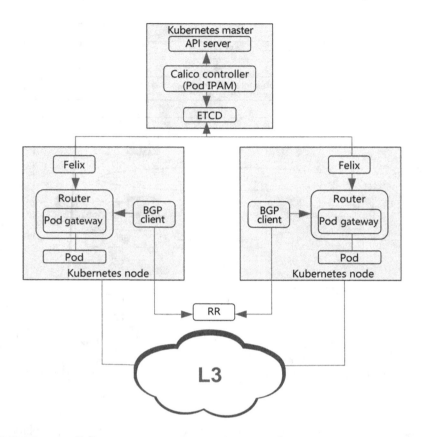

FIGURE 11.9   Calico open-source Layer 3 routing solution.

Table 11.1 provides a comparison between the three container network solutions: host overlay solution, network overlay Layer 2 bridging solution, and network overlay Layer 3 routing solution. In the host overlay solution, the container network is completely decoupled from the infrastructure layer, and the forwarding performance is poor. Such characteristics make this solution suitable for scenarios where the PaaS platform is independently deployed. The network overlay Layer 2 bridging solution is easy to deploy and provides high forwarding performance but supports only limited deployment scale, making it applicable to scenarios with a low container virtualization ratio (for example, database scenarios). The network overlay Layer 3 routing solution provides high forwarding performance and

TABLE 11.1   Comparisons between the Three Container Network Solutions

Item	Host Overlay Solution	Network Overlay Layer 2 Bridging Solution	Network Overlay Layer 3 Routing Solution
Network performance	Poor forwarding performance due to vSwitches functioning as VTEPs	High forwarding performance due to hardware switches functioning as VTEPs	High forwarding performance due to hardware switches functioning as VTEPs
Network scale	Large-scale networking, with vRouter clusters supporting horizontal expansion	Small-scale networking, limited by broadcast domains	Large-scale networking
Networking requirements	Layer 3 interconnection	Layer 2 interconnection	Layer 3 interconnection
Device dependency	None	Physical switches must support a large number of MAC address entries, ARP entries, and Layer 3 interfaces	Physical switches must support a large number of host routes
Application scenario	Containers are independently deployed as the PaaS platform	Small-scale cluster. The container virtualization ratio is less than 1:20 That is, the number of containers deployed on a physical server is less than 20	Medium- and large-scale cluster. The container virtualization ratio is greater than 1:20 That is, the number of containers deployed on a physical server is greater than 20

is highly scalable, making it applicable to medium- and large-scale cluster deployment scenarios.

11.1.3 Huawei SDN Container Network Solutions

Huawei SDN container network solutions include the following: network overlay solution, network overlay extension solution, and hybrid overlay solution. Of these, the network overlay solution supports open-source Kubernetes mainline versions, open-source vSwitches, and open-source Kubernetes CNI plugins. In contrast, the network overlay extension solution supports the integration with open-source Kubernetes mainline versions and third-party commercial container platforms. It replaces open-source OVS with Huawei user-mode vSwitches to provide high-performance forwarding, distributed firewall (DFW), and load balancing capabilities. Finally, the hybrid overlay solution uses user-mode vSwitches to support VXLAN. It provides high-performance forwarding and is compatible with existing network devices, protecting customer investment.

1. Network overlay solution

The network overlay solution is composed of the following components: SDN controller (iMaster NCE-Fabric), AC Kubernetes plugin (AC API server watcher), FabricInsight, security manager, DC switches, and NGFWs, as shown in Figure 11.10.

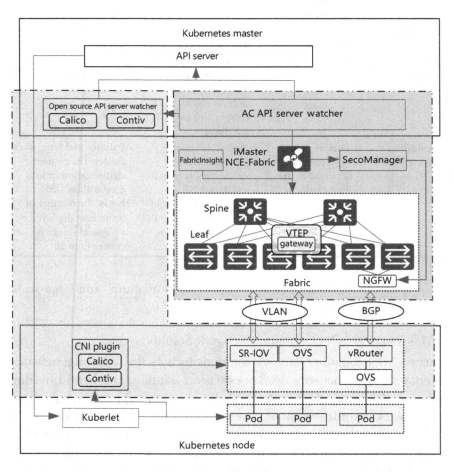

FIGURE 11.10   Huawei network overlay solution.

The network overlay solution supports container access to the fabric network in both Layer 2 bridging mode and Layer 3 routing (EBGP) mode. The AC API server watcher monitors the Kubernetes API server; detects the events of Pods, services, and nodes of the Kubernetes cluster; and invokes iMaster NCE-Fabric to automatically configure the fabric network. iMaster NCE-Fabric supports the visualization of the physical network, container logical network, and container application network, as well as providing connectivity detection between containers.

In this solution, FabricInsight — an intelligent O&M system — collects traffic on the fabric network to visualize traffic between containers across Kubernetes nodes, locate abnormal traffic interruptions between containers, diagnose traffic congestion and packet loss between containers, and identify burst traffic between containers. The north-south NAT and load balancing capabilities for the Kubernetes services and ingresses are provided by the security manager and NGFWs.

The network overlay solution supports open-source Kubernetes mainline versions, interconnection with third-party Kubernetes platforms, open-source Kubernetes network plugins (such as Calico and Contiv), and open-source OVS.

2. Network overlay extension solution

In the Kubernetes open-source network solution, the NAT and load balancing functions required by the Kubernetes services and ingresses are implemented by the Linux iptables. The downside of this is that, in large-scale container cluster scenarios, the NAT and load balancing functions implemented by the iptables are inefficient, leading to weak forwarding capabilities. The network overlay extension solution overcomes this shortcoming by introducing the high-performance software switch CloudEngine 1800V (CE1800V for short) and the matching vSwitch CNI plugin on the basis of the network overlay solution. The CE1800V is a user-mode vSwitch. As such, it provides high-performance NAT and high-performance load balancing (vLB) to replace the NAT and load balancing functions implemented by the open-source iptables, provides high-performance DFWs, and supports Kubernetes open-source standard security policies, as shown in Figure 11.11.

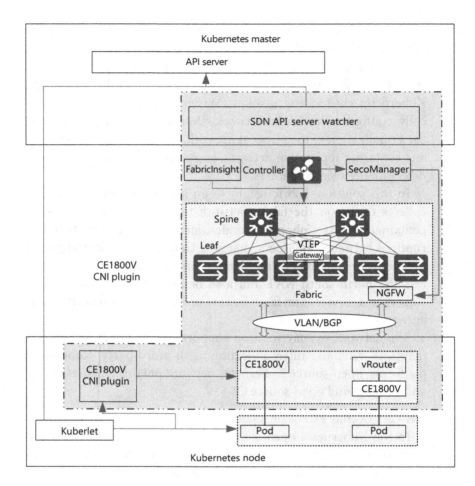

FIGURE 11.11   Huawei network overlay extension solution.

The network overlay extension solution supports container access to the fabric network in Layer 2 bridging mode and Layer 3 routing (EBGP) mode. In this solution, the SDN API server watcher monitors the Kubernetes API server; detects the events of Pods, services, and nodes of the Kubernetes cluster; allocates network resources (subnets, gateways, and VLANs); and invokes the SDN controller iMaster NCE-Fabric to automatically configure the fabric network. The CE1800V CNI plugin manages CE1800V and configures vNICs, vSwitch interfaces, NAT, firewalls, and load balancing policies for the container network. iMaster NCE-Fabric supports the visualization of the physical network, container logical network, and container application network, as well as providing connectivity detection between containers.

In this solution, the intelligent O&M system — FabricInsight — collects traffic on the fabric network and vSwitches to visualize traffic between containers, locate abnormal traffic interruptions between containers, diagnose traffic congestion and packet loss between containers, and identify burst traffic between containers. The north-south NAT and load balancing capabilities for the Kubernetes services and ingresses are provided by the security manager and NGFWs.

The network overlay extension solution supports the open-source Kubernetes mainline versions, Red Hat OpenShift platform, and Huawei FusionStage platform.

3. Hybrid overlay solution

In the hybrid overlay solution, CE1800V vSwitches function as VXLAN VTEPs. This solution is compatible with existing network devices and provides capabilities such as high-performance forwarding, DFW, and load balancing, as shown in Figure 11.12. In the hybrid overlay solution, Pods can access the fabric network through user-mode vNICs. Meanwhile, the SDN API server watcher monitors the Kubernetes API server, detects the events of Pods, services, and nodes of the Kubernetes cluster, and invokes the SDN controller iMaster NCE-Fabric to automatically configure the fabric network. The CE1800V CNI plugin manages CE1800V and configures vNICs, vSwitch interfaces, NAT, firewalls, and load balancing policies for the container network. iMaster NCE-Fabric supports the visualization of the physical network, container logical network, and container application network, as well as providing connectivity detection between containers.

In this solution, the intelligent O&M system — FabricInsight — collects traffic on the fabric network and vSwitches to visualize traffic between containers, locate abnormal traffic interruptions between containers, diagnose traffic congestion and packet loss between containers, and identify burst traffic between containers.

The hybrid overlay solution supports the open-source Kubernetes mainline versions, Red Hat OpenShift platform, and Huawei FusionStage platform.

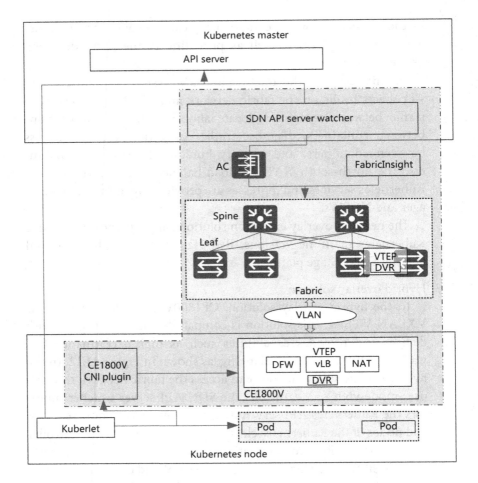

FIGURE 11.12   Huawei hybrid overlay solution.

In summary, Huawei's three container solutions are fully compatible with the Kubernetes open-source platform and can implement unified O&M of the container network, BM network, and physical network.

The network overlay solution is a full-hardware container network solution, while the network overlay extension solution and hybrid overlay solution introduce Huawei high-performance vSwitch, CE1800V. Table 11.2 compares the three container solutions.

TABLE 11.2 Comparisons between the Three Container Solutions

Item	Network Overlay Solution	Network Overlay Extension Solution	Hybrid Overlay Solution
Solution characteristics	Hardware switches function as VTEPs and provide high forwarding performance. They are compatible with open-source Kubernetes platforms and components	Hardware switches function as VTEPs and provide high forwarding performance. The container virtual network and physical network support E2E O&M and visualization	User-mode vSwitches function as VTEPs and support intelligent NICs (iNICs). The container virtual network and physical network support E2E O&M and visualization.
Benefits to customers	The solution is fully compatible with the Kubernetes open-source platform. The container network is automatically deployed, and unified O&M is implemented on the container network, BM network, and physical network	The solution is fully compatible with the Kubernetes open-source platform and commercial PaaS platforms. Unified O&M is implemented on the container network, BM network, virtual network, and physical network. Software switches support high-performance DFWs, vLBs, and NAT	The solution is fully compatible with the Kubernetes open-source platform and commercial Kubernetes CNI plugins. Unified O&M can be implemented on the container network, BM network, virtual network, and physical network. Software switches support DFWs, vLBs, and NAT. This solution is compatible with existing network devices, protecting customer investment.
Application scenario	The container virtualization ratio is less than 1:20. That is, the number of containers deployed on a physical server is less than 20. The solution applies to small-scale clusters. Customers use the open-source Kubernetes platform and Kubernetes network plugins	The container virtualization ratio is greater than 1:20. That is, the number of containers deployed on a physical server is greater than 20. The solution applies to medium- and large-scale clusters. Customers use the Kubernetes open-source platform or solution-certified container platform	Customers' network devices or the live network and new virtual network are deployed together. Customers use the Kubernetes open-source platform or solution to authenticate the container platform.

11.2 HYBRID CLOUD

11.2.1 Overview

Before introducing the hybrid cloud technology, it is important to understand what private cloud and public cloud are.

A private cloud is a cloud computing infrastructure built by an organization or enterprise in its own DC or hosting DC to meet specific service scenario requirements. On this kind of cloud, compute, storage, and network resources are pooled (virtualized) as services. Mainstream private cloud platforms in the industry include VMware vSphere and vCenter, open-source OpenStack, Microsoft System Center, Microsoft Azure Pack, and Stack. The RightScale's 2017 State of the Cloud Report sheds light on enterprises' adoption rate of private clouds, as Figure 11.13 shows.

A public cloud is a cloud computing infrastructure built by cloud service providers (such as AWS, Azure, and Google Cloud) to provide users with pay-per-use compute, storage, and network resources. Such resources are shared with other organizations or cloud tenants.

A hybrid cloud is an IT infrastructure that integrates the compute, storage, and network resources of private and public clouds. It is made up of the private cloud, public cloud, Internet, and cloud management platform. A hybrid cloud allows for sharing of compute, storage, and network resources between private and public clouds so that organizations or enterprises can deploy sensitive or business-critical data and applications on private clouds and deploy non-sensitive but high-volume data and applications on public clouds. The flexibility, scalability, and global

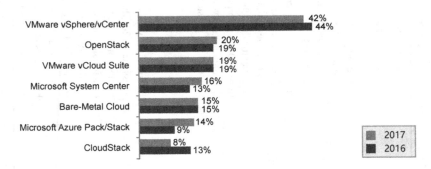

FIGURE 11.13   Private cloud adoption of enterprises. (RightScale's 2017 State of the Cloud Report.)

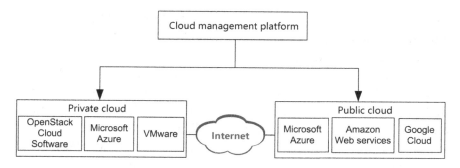

FIGURE 11.14   Simplified architecture of the hybrid cloud technology.

availability of public clouds can also be fully utilized to reduce infrastructure investment and achieve better cost-effectiveness.

Figure 11.14 shows the simplified architecture of the hybrid cloud technology.

The following describes the typical application scenarios of hybrid cloud technology:

1. Rapid global service deployment

    With the rapid growth of enterprise services, enterprises need to quickly provide services for users in new regions. They can do so by using the global DC resources of the public cloud.

2. Short-term burst services

    Taking advantage of the scalability and pay-per-use features of the public cloud, the infrastructure requirements of services can be migrated during peak-hours to the public cloud, accommodating unpredictable service peaks and reducing required investment.

3. Separate deployment of core data and common applications in industries such as finance

    Industries such as the financial services industry have high requirements on security compliance. Key, business-critical, and sensitive data and applications need to be deployed on private clouds, while non-key applications and services that are high in volume need to be deployed on public clouds. Enterprises can use the DevOps platform provided by the public cloud to develop and test financial applications, while deploying the applications on the private cloud to provide services for external systems.

11.2.2 Industry's Mainstream Hybrid Cloud Network Solutions

The hybrid cloud network solutions provided by public cloud providers fall into one of two types: VPN and leased line. Although the VPN solution does not provide SLA assurance, it is easy to deploy and low in cost. Such characteristics make it applicable to services that do not have high requirements on latency and bandwidth. The leased line solution, on the other hand, is often used by broadband carriers to provide low-latency and high-bandwidth network services, which are suitable for services that have strict requirements on security, latency, and bandwidth. The following describes the hybrid cloud network solutions provided by mainstream hybrid cloud solution providers, AWS, Azure, and Google Cloud:

1. AWS public cloud leased line solution and VPN solution

    1. AWS public cloud leased line solution

        The AWS leased line service, known as AWS Direct Connect, enables gateways hosted by tenants in AWS Direct Connect locations to be physically connected to AWS Direct Connect endpoints to implement communication between leased line users. The AWS public cloud leased line solution supports VLAN-based service differentiation and uses BGP routes to implement Layer 3 communication between the public and private clouds. Through leased lines, services on private clouds can access public services (such as S3 storage) provided by AWS. The leased line network does not support refined QoS controllers, multicast services, or broadcast services.

    2. AWS public cloud VPN solution

        The AWS VPN service consists of virtual private gateways (software or hardware) on the public cloud and customer gateways of a tenant on a private cloud. It supports only IPsec VPN and uses CloudHub to implement Layer 3 communication between the public cloud and multiple private cloud sites. In the AWS VPC service, a virtual private gateway establishes an IPsec VPN channel with the VPN gateway of the private cloud DCN to implement Layer 3 communication between the public cloud VPC and private cloud DCN.

2. Azure public cloud leased line solution and VPN solution

   1. Azure public cloud leased line solution

The ExpressRoute leased line solution of Microsoft Azure supports VLAN-based service differentiation, allows private clouds to access all services on the Azure public cloud through leased lines, and uses BGP routes to implement Layer 3 communication between the public and private clouds. In the Azure public cloud leased line solution, the local edge routers of a user's private cloud DCN belong to the user and can be deployed either in the private cloud DCN or at the access point-of-presence (PoP) of the Azure public cloud. An ExpressRoute circuit refers to the leased line between the user private cloud and the POP of the Azure public cloud. In this solution, a Microsoft edge router is a routing device deployed at the POP of the Azure public cloud. An ExpressRouter gateway on the Azure public cloud and a gateway on a private cloud advertise routes through BGP to implement Layer 3 communication between the private cloud and Azure public cloud.

   2. Azure public cloud VPN solution

In the Azure public cloud VPN solution, an IPsec VPN tunnel is established between the two VPN gateways: one on the Azure public cloud and one on the private cloud, to implement Layer 3 communication between the public cloud and private cloud. Azure VPN supports Layer 3 communication between two sites and between multiple sites. The VPN gateway on the private cloud can be a software or hardware VPN gateway.

3. Google public cloud leased line and VPN solution

   1. Google public cloud leased line solution

Google Dedicated Interconnect service establishes physical links between a private cloud DCN and border routers of the Google public cloud. To enable the private cloud to access any private subnet on the public cloud, BGP runs between the CloudRouter — a distributed router provided by the Google public cloud — of the public cloud and the border router of the private cloud.

2. Google public cloud VPN solution

A Google Cloud VPN gateway establishes an IPsec VPN tunnel over the Internet with the VPN gateway of a private cloud DCN to support site-to-site VPN. In this way, Google Cloud VPN is able to provide 99.9% SLA assurance. A Google Cloud VPN gateway is deployed in self-service mode.

4. Hybrid cloud network solutions of public cloud vendors

The model shown in Figure 11.15 provides an outline of the hybrid cloud network solutions of public cloud vendors.

VPN solution: A private cloud communicates with a public cloud through the Internet. In this solution, the private cloud platform provisions the internal service VPC of the private cloud and the VPN gateway used for communication with the public cloud. The public cloud, on the other hand, provides the service VPN and VPN gateway configuration, as well as management on the public cloud side in self-service mode.

Leased line solution: On a private cloud, users need to connect their private DCs to the POP of the public cloud leased line. The POP may be the private POP of a public cloud or the POP of a public cloud partner, while a POP user router can be a user's

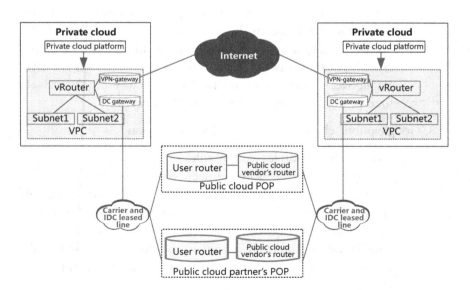

FIGURE 11.15   Hybrid cloud network solution model.

own device or a leased one. The private cloud management platform provides automatic configuration of DC gateway services, and the public cloud platform provides self-service configuration of the leased line gateway on the public cloud. The POP network configuration and O&M of the public cloud leased line are usually performed through the public cloud leased line using work orders.

The VPN solution and the leased line solution have their own pros and cons. For example, the VPN solution can be quickly and easily deployed, and is low in cost; however, it cannot guarantee network SLA. This is something that can be guaranteed by the leased line solution. However, if the private POP of the public cloud is used with the leased line solution, different public cloud vendors cannot be supported at the same time.

The current hybrid cloud solutions provide only basic communication capabilities. In the absence of other capabilities, no well-designed solution is available for O&M management, service provisioning, and security policies. These capabilities include

- Unified service model for private and public cloud networks and centralized service provisioning and management

- Network virtualization and E2E automatic service provisioning for private and public cloud networks

- Traffic visualization, traffic optimization, and fault locating between private and public cloud networks

- Unified security policies for private and public cloud networks.

11.2.3 Huawei Hybrid Cloud SDN Solution

Huawei hybrid cloud SDN solution uses the intent engine to provide a unified service model, and supports unified service provisioning and management for both public and private cloud networks. In the solution, VXLAN tunnels are established between DC gateways and DC vGWs to implement network virtualization and support dynamic provisioning of network services. The intent engine and SDN controller (iMaster NCE-Fabric) implement global traffic optimization and fault diagnosis. DC gateways and DC vGWs provide microsegmentation and vFW capabilities to deliver a unified security policy model as well as unified security policies for private and public clouds. This solution supports the following scenarios based on specific service types:

1. Private and public clouds communicate at Layer 3 through IPsec VPN, as shown in Figure 11.16.

   The intent engine connects to the northbound APIs of the VPN gateways on the public cloud (AWS, Azure, and Google Cloud) through the workflow and supports the creation, deletion, and route management of the VPN gateways on the public cloud. This implements E2E automation of VPN gateway services on public and private clouds.

   DC gateways support multi-point VPNs to implement Layer 3 VPC communication between a private cloud DCN and multiple

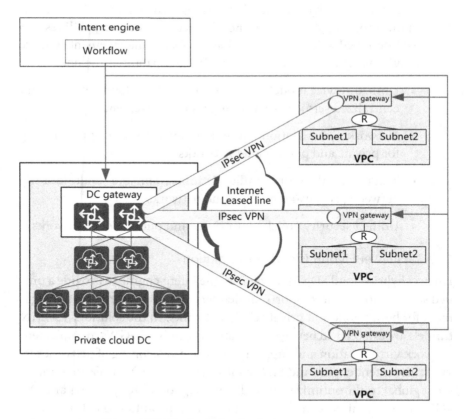

FIGURE 11.16 Huawei VPN hybrid cloud solution.

public clouds, as well as support communication between public cloud VPCs through private cloud VPCs.

2. Private cloud VPCs and public cloud VPCs communicate at Layer 3 through VXLAN, as shown in Figures 11.17 and 11.18. DC vGWs support public cloud (AWS, Azure, and Google Cloud) deployment and provide basic forwarding capabilities including VXLAN tunnels, IPsec tunnels, firewalls, and QoS. iMaster NCE-Fabric supports cloud-based deployment to manage DC vGWs. The process involves the intent engine connecting to the public cloud's northbound APIs through the workflow to manage and provision services for DCI VPCs, cloud-based iMaster NCE-Fabric, and DC vGWs. A VXLAN

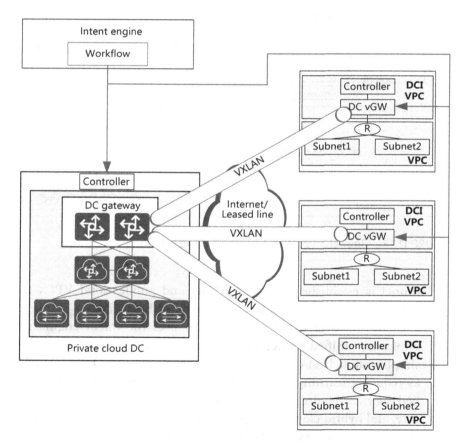

FIGURE 11.17   Huawei VPC VXLAN Layer 3 hybrid cloud solution.

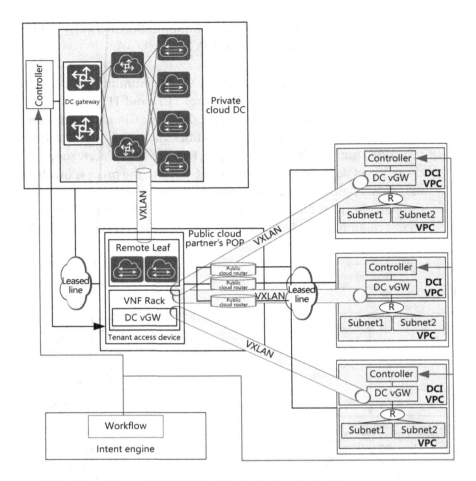

FIGURE 11.18 Huawei VPN VXLAN hybrid cloud solution.

tunnel is established between private and public cloud VPCs to implement Layer 3 communication between these VPCs.

Huawei hybrid cloud SDN solution uses the intent engine to implement unified service management for hybrid cloud networks, simplifying hybrid cloud network management and O&M, as well as reducing learning and O&M costs. This solution meets diversified service requirements by supporting mainstream public, private, and hybrid cloud management platforms, and it uses standard VXLAN to implement network virtualization and service optimization, improving the efficiency with which services are deployed.

Furthermore, the DFW and microsegmentation capabilities deliver a unified security model for private and public cloud services, improving service security.

11.3 AI FABRIC

The continued increase in the scale of distributed AI computing leads to a corresponding improvement in hardware capabilities, which in turn poses higher requirements on the basic performance of AI DCNs. By 2025, the amount of data generated and stored worldwide will increase by approximately 20 times compared with 2015, reaching 180 ZB, and 95% of the data will be processed using AI.

In addition, storage and computing capabilities have significantly improved, promoting the development of SSDs and AI chips. As DC traffic continues to increase, requirements cannot be met by traditional TCP/IP; therefore, AI DCNs require high-speed, reliable, as well as stable link technologies and network protocols. This includes InfiniBand dedicated to high-performance computing and RoCE technology, which is compatible with traditional Ethernet.

AI Fabric is a lossless network-based ultra-high-speed Ethernet solution launched for cloud DCNs. It aims to solve issues regarding traditional AI private networks, such as InfiniBand, limited networking scale, incompatibility with traditional Ethernet, and complex O&M. To achieve this objective, AI Fabric provides solutions for AI, high-performance computing, and large-scale distributed computing scenarios.

11.3.1 Current State of AI DCNs

Traditional DCs use Ethernet technologies and the TCP/IP protocol stack to build a multi-hop symmetric network architecture and transmit data, respectively. Applications with high network communication requirements, such as AI and Virtual Reality (VR), are constantly emerging due to the continuous increase in DCN traffic. This makes it increasingly difficult for traditional Ethernet technology carrying TCP/IP traffic to meet requirements. Therefore, existing AI DCNs are starting to shift toward high-speed, reliable, and stable link technologies and network protocols.

InfiniBand, which features very high throughput and very low latency, is a computer-networking communications standard used in high-performance computing. It is used for data interconnect both among and within computers, as a direct or switched interconnect between servers and storage systems, as well as an interconnect between storage systems. InfiniBand uses a switched fabric topology, as opposed to the earlier method that used Ethernet as a shared medium. With InfiniBand, all transmissions begin or end at a channel adapter. Specifically, each processor contains a host channel adapter (HCA), and each peripheral has a target channel adapter (TCA). These adapters can exchange information for security or QoS. Unlike the traditional TCP/IP protocol stack, InfiniBand has specialized network and transport layer protocols, and it does not include standard network programming interfaces. Instead, its programming interface is the Verbs interface, which is incompatible with the socket interface and was released by OpenFabrics Alliance.

Due to its high transmission rate, InfiniBand is the most commonly used supercomputer interconnection technology. Table 11.3 lists the InfiniBand transmission performance value.

RoCE technology has been developed based on InfiniBand technology, which has high speed and high resource efficiency. However, in addition to being incompatible with the traditional TCP/IP over Ethernet, InfiniBand also requires special NICs and switches. Based on existing Ethernet standards, RoCE uses InfiniBand standards to implement highly-efficient transmission at the network layer (RoCEv1) and transport layer (RoCEv2). Only RoCE NICs are required, without the need for replacing network-side hardware devices. In terms of specific capabilities, RoCE implements high-speed and high-efficient transmission in the Ethernet environment. Currently, Mellanox is leading the RoCE NIC market.

Omni-Path is a high-speed interconnection technology proposed by Intel for HPC. In addition to using dedicated network devices, it provides

TABLE 11.3 InfiniBand Transmission Performance Value

	Type						
Performance	**SDR**	**DDR**	**QDR**	**FDR-10**	**FDR**	**EDR**	**HDR**
Signaling rate (Gbit/s)	2.5	5	10	10.3125	14.0625	25	50
Theoretical effective throughput per link (Gbit/s)	2	4	8	10	13.64	24.24	50
Rate for four links (Gbit/s)	8	16	32	40	54.54	96.97	200
Rate for 12 links (Gbit/s)	24	48	96	120	163.64	290.91	600
Latency (μs)	5	2.5	1.3	0.7	0.7	0.5	0.5

a transmission rate of 100 Gbit/s and a lower latency than InfiniBand at the same level. Intel Omni-Path Architecture (OPA) leads the rankings among 100 Gbit/s fabrics (Table 11.4).

In 2016, among the three big public cloud providers in the USA (Amazon, Google and Microsoft), only Microsoft Azure used InfiniBand technology to provide high-speed network interconnection. However, this phenomenon is changing considerably due to the increasing number of AI applications. For instance, both Facebook's latest machine learning open platform (Big Basin) and Baidu use 100 Gbit/s InfiniBand. In response to the continuous emergence of new dedicated AI computing chips, the expansion of computing scale, as well as the growing maturity of mainstream machine learning and development platforms, a growing number of AI DCs are using high-performance network solutions, as opposed to TCP/IP over Ethernet.

TABLE 11.4   Comparison of Network Interconnection Technologies

Technology	Interface Rates (Gbit/s)	Key Technologies	Comparison
TCP/IP over Ethernet	10, 25, 40, 50, 56, 100, or 200 (Latency: 500–1000 ns)	TCP/IP, socket programming interface	Advantages: wide application scope, low cost, and good compatibility Disadvantages: low network efficiency, poor average performance, and unstable link transmission rate
InfiniBand	40, 56, 100, or 200 (Latency: 300–500 ns)	InfiniBand network protocol and architecture, Verbs programming interface	Advantage: good performance Disadvantages: InfiniBand network protocol and architecture, Verbs programming interface
RoCE/ RoCEv2	40, 56, 100, or 200 (Latency: 300–500 ns)	InfiniBand network layer or transport layer, Ethernet link layer Verbs programming interface	Advantages: compatibility with traditional Ethernet technologies, high cost-effectiveness, and good performance Disadvantages: specific NICs required
Omni-Path	100 (Latency: 100 ns)	OPA network architecture, Verbs programming interface	Advantage: good performance Disadvantages: single manufacturer as well as specific NICs and switches required

11.3.2 New DCN Requirements Brought by AI Technology

1. Characteristics of AI computing

Traditional DC services (web, video, and file storage) are transaction-based and yield deterministic calculation results. For these services, single calculations and network communication are conducted independently, occurring irregularly and lasting for random periods of time. AI computing is optimized based on objectives, and recursive convergence is required in the computing process, which causes high spatial correlation in the computing process of AI services, as well as temporally similar communication modes. Figure 11.19 shows the AI computing process.

To solve the big data problem, where a single server cannot provide enough storage capacity, the computing model and input data need to be large (for a 100 MB node, the AI model for 10K rules requires more than 4 TB memory). In addition, because the computing time needs to be shortened and increasingly concurrent AI computing is required for multiple nodes, DCNs must be used to perform large-scale and concurrent distributed AI computing.

2. DCN requirements brought by distributed AI computing

As the number of AI algorithms and applications continues to increase, and the distributed AI computing architecture emerges, AI computing is being implemented on a large scale. To ensure sufficient

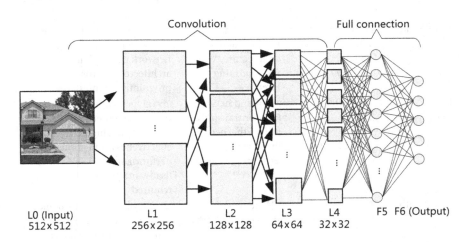

FIGURE 11.19   AI computing process.

interaction of distributed information, higher requirements on communication volume and performance have been posed. Facebook recently tested the distributed machine learning platform Caffe2, in which the latest multi-GPU servers are used for parallel acceleration. In the test, computing tasks on eight servers resulted in insufficient resources on the 100 Gbit/s InfiniBand network. Therefore, it was difficult to achieve linear computing acceleration for multiple nodes. Consequently, network performance significantly restricts the horizontal expansion of the AI system.

There are two distributed AI computing modes: model parallel computing and data parallel computing.

In model parallel computing, each node computes one part of the algorithm. After computing is complete, all data shards across models needs to be transmitted to various nodes, as shown in Figure 11.20.

1. Model parallel training

In parallel data computing, each node loads the entire AI algorithm model. Multiple nodes can calculate the same model simultaneously, but only part of data is input to each node. When a node completes a round of calculation, all relevant nodes need to aggregate updated information regarding obtained weight parameters and then obtain the corresponding

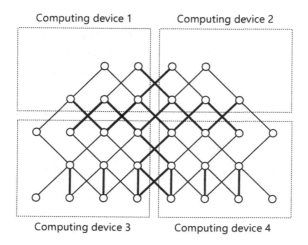

FIGURE 11.20   Model parallel computing.

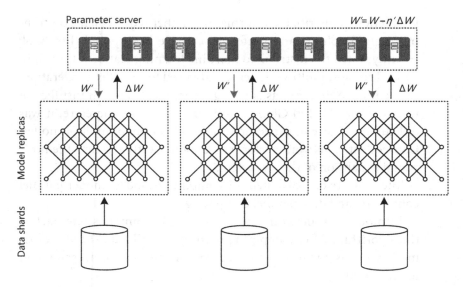

FIGURE 11.21   Model parallel training.

globally updated data. All nodes upload and obtain information synchronously during each weight parameter update, as shown in Figure 11.21.

The increasing scale of distributed AI computing and increasing AI computing hardware capabilities pose higher requirements on the basic performance of AI DCNs.

2. High bandwidth and low static latency

In Table 11.5, the test results of several typical AI computing models indicate that distributed AI computing is essentially high-performance centralized computing with communication.

TABLE 11.5   AI Computing Models

AI Computing Model	Communication Volume (16-Bit Precision)	Computing Capacity (FLOPS)	Single Recursion Time (Second)	
			Single GPU (K80, V100)	CPU (32-Thread)
AlexNet	120 MB	63×106	0.07, <0.02	1
ResNet50	1.5 MB	3.8×109	0.5, <0.1	8
FCN-S	236 MB	55×106	0.07, <0.02	1

Data source: Benchmarking state of the art deep learning software tools.
FLOPS refers to floating-point operations per second.

For some AI models, 100–200 MB of traffic is generated every 20 ms on the latest GPU, which is equivalent to instantaneous burst traffic of up to 80 Gbit/s. Therefore, AI DCs must ensure that link bandwidth and latency are as high and low as possible, respectively. Otherwise, communication takes longer than calculation, severely affecting the concurrency and completion time of distributed AI computing.

3. Low dynamic latency and zero packet loss

High-bandwidth and low-latency DCNs with only physical links cannot meet the requirements of large-scale and highly concurrent AI applications. In AI computing's recursion process, a large amount of burst traffic is generated within milliseconds. In addition, because a parameter server (PS) architecture is used to update parameter weights of the new model for data parallelization, the Incast traffic model at a fixed time is easily formed. In this scenario, networks are prone to packet loss, congestion, and load imbalance, and as a result, the flow completion time (FCT) of some data flows is too long. Furthermore, distributed AI computing is synchronous; therefore, if a few flows are delayed, more computing processes are negatively affected. Consequently, the completion time of the entire application will be delayed, as shown in Figure 11.22.

Therefore, to prevent FCT increase caused by packet loss, congestion, and Incast, the fine grained design is required for DCNs that are oriented toward large-scale distributed AI. In addition, a dynamic low latency and stable high bandwidth for communication must be ensured.

4. JCT optimization

Based on guaranteed E2E data flow performance, the distributed AI DCN must effectively optimize the job completion time (JCT) for AI computing. In the distributed AI parallel computing process based on recursive convergence, over 80% of computing can converge after a few recursions, while the remaining computing conversely requires a large number of recursions. This proves that communication scheduling required by distributed AI computing can significantly improve the calculation convergence duration.

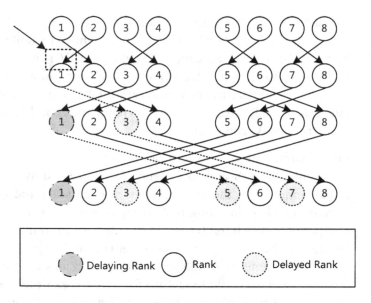

FIGURE 11.22 Single-point delay deteriorating overall performance. (Mellanox.)

Distributed AI computing with spatial correlation causes transmission of correlated network data traffic. In terms of space, the optimal policy for synchronously computing multiple data flow groups (co-flow) involves minimizing the transmission completion time of the entire flow group. In terms of time, due to the sequence of multiple recursions, data flows in earlier recursions need to be transmitted first to enable subsequent relevant computing processes to start as soon as possible. To achieve optimal parallelization between computing and communication, an AI DCN optimizes the JCT by optimizing transmission of multiple flows.

11.3.3 AI Fabric Technical Directions

To address the preceding issues, AI Fabric technologies are developing in the following directions.

- Incast traffic scheduling without packet loss or congestion

- Differentiated scheduling for elephant and mice flows

- Load balancing of network traffic

- Congestion control.

The following section presents technical details regarding the preceding development directions.

1. Incast traffic scheduling without packet loss or congestion

   Currently, the inbound and outbound buffer thresholds for each interface queue of the forwarding chip are configured to ensure interface queues with basic scheduling in the common traffic model. These threshold parameters are statically configured, which means that they cannot adapt to different traffic models or guarantee zero packet loss in the Incast traffic model with high concurrency.

   Zero packet loss must be guaranteed by a DCN to achieve large-scale deployment of HPC and distributed storage applications on the DCN. These distributed applications use the N:1 Incast traffic model, and as these applications are deployed on a larger scale, the volume of burst traffic increases accordingly. As a result, the switches' packet buffer increases and packet loss occurs due to buffer overflow.

   The buffer settings for the forwarding chip's interface queues include both uplink and downlink buffer thresholds.

   - The uplink buffer threshold is used to set the guaranteed, shared, and headroom buffers for the ingress interface priority. The guaranteed buffer ensures that the inbound interface can always receive a packet with the size of the MTU. This prevents service traffic interruptions caused by a failure to preempt buffer resources. For the lossless priority, priority-based flow control (PFC) needs to be enabled. The shared buffer is the PFC-XOFF threshold. When PFC is triggered, the local device receives PFC frames from the remote end and stops sending packets. In this scenario, the packets sent by the local end and traveling on links in the preceding process are stored in the headroom buffer, preventing these packets from being discarded.

   - The downlink buffer threshold is used to set the guaranteed and shared buffers for queues on the outbound interface. The shared buffer is the maximum buffer that can be preempted by the queues on the outbound interface.

   The forwarding chip's physical buffer does not distinguish between the uplink traffic and downlink traffic. These thresholds

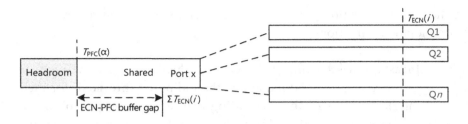

FIGURE 11.23   Buffer configuration requirements in the 1:N scenario.

are configured for buffer management in the inbound and out-bound directions. A packet is permitted to enter the forwarding chip's physical buffer only when both the inbound and outbound buffer thresholds have not been exceeded; otherwise, the packet is discarded.

That is, to prevent packets from being discarded, the inbound and outbound buffer thresholds on an interface must be within the allowed range. Inbound and outbound buffer thresholds vary according to different traffic models and scenarios.

In the 1:N scenario, to ensure that no packet loss occurs and throughput is not affected, the following buffer configuration requirements must be met, as shown in Figure 11.23:

- $T_{PFC} > \Sigma T_{ECN}(i)$: Before Explicit Congestion Notification (ECN) is triggered for all downlink queues, PFC is not triggered for uplink queues. For details, see Congestion control.

- $\Sigma T_{Q\text{-}BUF}(i) > (T_{PFC} + T_{HDRM})$: Before PFC is triggered for uplink queues, packet loss does not occur in any downlink queue.

In the N:1 scenario, to ensure that no packet loss occurs and throughput is not affected, the following buffer configuration requirements must be met, as shown in Figure 11.24.

- $\Sigma T_{PFC}(i) \rightarrow TECN$: Before ECN is triggered for downlink queues, PFC is not triggered for any uplink queue. For details, see Congestion control.

- $TQ\text{-}BUF \rightarrow \Sigma(T_{PFC}(i) + THDRM\ (i))$: Before PFC is triggered for uplink queues, packet loss does not occur in any downlink queue.

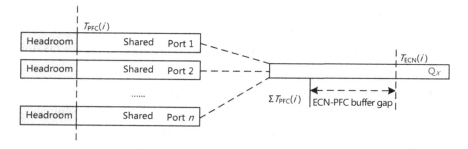

FIGURE 11.24 Buffer configuration requirements in the N:1 scenario.

HDRM(InPort, PG)	PFC threshold		ECN threshold
Headroom Pool	Shared Pool		
PFC response time	ECN response time		Q-BUF

FIGURE 11.25 Buffer allocation involved in Q-BUF.

Based on the mapping from uplink and downlink buffer thresholds to the physical buffer, Q-BUF involves the following buffer allocation: {Headroom, PFC threshold, ECN threshold}. Figure 11.25 shows the buffer allocation involved in Q-BUF.

For the lossless priority of the inbound interface, the headroom buffer is required to ensure zero packet loss when PFC is triggered. The headroom buffer's value is fixed, and it affects both the link rate and latency. The PFC-XOFF threshold for the lossless priority of the inbound interface and the shared buffer threshold for queues on the outbound interface affect the physical buffer of the forwarding chip and traffic model. When the traffic model changes, the buffer thresholds must be dynamically adjusted to ensure no packet loss.

Within the time difference between the ECN marking and the rate reduction on the source end, PFC is not triggered (ECN propagation time) by traffic. In this scenario, ECN technology can preferentially resolve congestion to avoid or reduce PFC. The buffer gap between the ECN threshold and PFC threshold affects the frequency at which PFC is triggered.

For the N:1 traffic model of distributed and concurrent DC applications, the forwarding chip's buffer threshold configuration must be adjusted. The buffer threshold configuration needs to be dynamically and immediately adjusted when the traffic model changes. This ensures that no packet loss occurs on switches and throughput is not affected.

2. Differentiated scheduling for elephant and mice flows

The packets that enter a queue must be dequeued in sequence. When elephant and mice flows are transmitted simultaneously, elephant flows occupy the queue if traffic congestion occurs; therefore, the subsequent mice flows may be discarded as they cannot enter the queue. Even if the mice flows can enter, there may be an extended delay because the elephant flows occupy most of the queue. When a queue is blocked by elephant flows, the FCT of the delay-sensitive mice flows increases significantly, as shown in Figure 11.26.

A technical solution is required to preferentially schedule packets in mice flows while ensuring delay is not affected by elephant flows, thereby guaranteeing the FCT of mice flows.

The solution needs to identify elephant and mice flows, and then reduce the priority of elephant flows to increase the priority of mice flows. The solution is ineffective for mice flows that have been blocked before elephant flows are identified.

For this reason, several initial packets of each service flow must be placed in a high-priority queue so that they can be preferentially scheduled and sent. After the solution identifies elephant flows among the service flows, it reduces the priority of the elephant flows. Through this process, mice flows will not be blocked and delayed by elephant flows, as shown in Figure 11.27.

To identify elephant and mice flows, the forwarding chip needs to learn and store each flow in real time through the flow table (for example, based on the IP 5-tuple), as well as collect traffic statistics

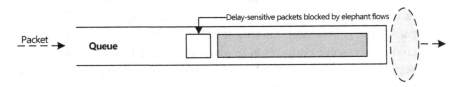

FIGURE 11.26 Delay-sensitive service packets being blocked due to elephant flows.

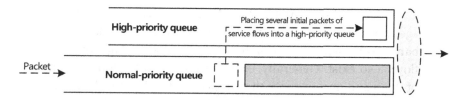

FIGURE 11.27   Initial packets of service flows entering high-priority queues.

on each flow based on the flow rate, byte, and lifetime. After identifying an elephant flow, the forwarding chip dynamically decreases the scheduling priority of the flow.

3. Load balancing of network traffic

On DCNs, service traffic is usually transmitted among multiple equal cost paths. To ensure that service traffic can be evenly distributed to physical links and no packet out-of-order occurs, the static hash algorithm that is based on packets' characteristic fields is widely used, as shown in Figure 11.28. However, this algorithm may cause a hash imbalance. That is, a physical link may be overloaded or, in the worst case, packets may be discarded due to congestion even if other physical links are lightly loaded. In this situation, bandwidth efficiency will be low, and the application performance (FCT and throughput) will be negatively affected.

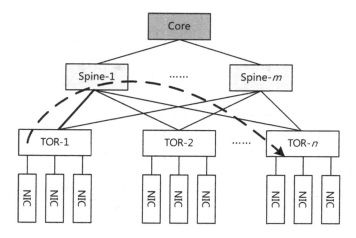

FIGURE 11.28   Static hash algorithm.

On DCNs, for example, when there are HPC and distributed storage applications, delay-sensitive and short-lived mice flows as well as bandwidth-intensive, throughput-sensitive, and long-lived elephant flows also exist. Generally, elephant flows constitute less than 10% of flows and occupy more than 80% of traffic. When a large number of flows are load balanced in static hashing, the number of flows on each link will be approximately the same, but the total bandwidth occupied by the flows on each link will be unbalanced, creating the same effect on the link. In addition, when elephant and mice flows are statically hashed to the same link, elephant flows will preempt the bandwidth and mice flows will be blocked, severely degrading application performance.

Flow-based load balancing is used in a static hash algorithm. That is, to prevent packet out-of-order, the algorithm must select the same path for the same flow, regardless of flow bandwidth, as shown in Figure 11.29.

However, a substantial amount of traffic on DCNs is burst traffic (for example, TCP bursts). If the duration since the previous burst of traffic exceeds the delay difference between paths, a route can be re-selected for a new burst of traffic to prevent packet out-of-order.

A burst of traffic is referred to as a flowlet. The interval between packets in a flowlet is less than the delay difference between paths. Therefore, the same path is selected to ensure that packets in the flowlet are transmitted in the correct sequence, as shown in Figure 11.30.

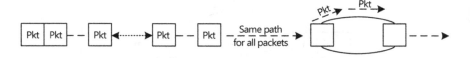

FIGURE 11.29   Flow-based load balancing.

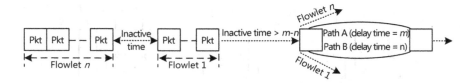

FIGURE 11.30   Flowlet.

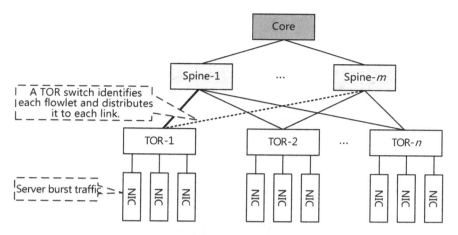

FIGURE 11.31 Dividing an elephant flow into several mice flows.

Elephant flows on a DCN can be identified by flowlets. An elephant flow can be divided into several mice flows to release bandwidth that would otherwise be occupied. Routes are then selected for flowlets to balance the load of each link, as shown in Figure 11.31.

The forwarding chip must record the member links selected by each flowlet to ensure that subsequent packets that enter the flowlet are transmitted through the same link, preventing packet out-of-order. Therefore, the forwarding chip must support the learning and aging of a flowlet flow table, and record an entry timestamp for the various packets of each flow to distinguish different flowlets.

A traditional static hash algorithm performs static route selection based on packets' characteristic fields, without taking into account the load of each link, as shown in Figure 11.32.

During flowlet-based load balancing, the load of each link must be taken into account to ensure the least loaded link can be selected for flowlets, as shown in Figure 11.33.

To measure the congestion level of each link, the forwarding chip must measure the quality of each interface based on buffer length and bandwidth efficiency. A flowlet's packets are then sent over the optimal and least congested link, as shown in Figure 11.34.

In this load balancing mode, links are dynamically selected according to congestion level, ensuring link usage that is more balanced and delivering higher application performance (FCT and throughput).

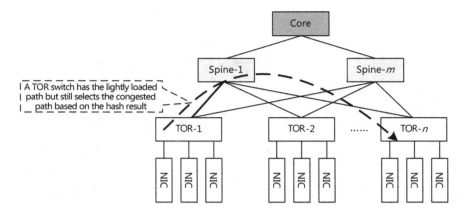

FIGURE 11.32   Static route selection based on packets' characteristic fields.

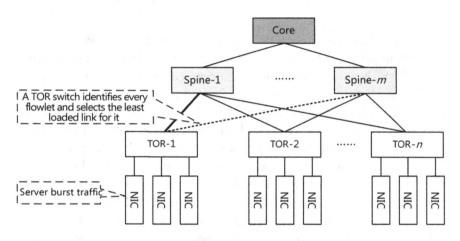

FIGURE 11.33   Dynamic route selection based on link load.

On a DCN, HPC applications are usually mice flows, as opposed to flowlets. It is possible that the elephant flows of distributed storage applications cannot be divided into multiple flowlets (NICs can be used to divide flows into flowlets). In this scenario, only flow-based dynamic load balancing instead of flowlet-based load balancing can be implemented for the distributed applications' traffic. To better load balance traffic, the InfiniBand Base Transport Header (IB BTH) field of RDMA traffic must be used in the hash algorithm to learn flows.

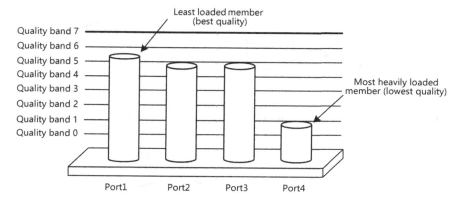

FIGURE 11.34   Forwarding chip selecting routes based on link quality.

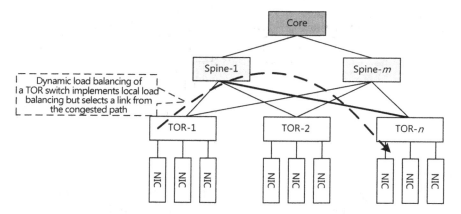

FIGURE 11.35   Problem facing flowlets: the link load of the remote device cannot be determined.

Flowlet-based dynamic load balancing takes into account only the link load of the local device, as opposed to the link load of the remote device, as shown in Figure 11.35. Even if a switch enabled with dynamic load balancing achieves load balancing among local links, congestion points at the remote end on the service forwarding paths may exist due to the load of each service flow's path not being taken into account.

The switch maintains the load information of paths on the entire network for service flows. During load balancing at the local end, the switch selects a local link from the least loaded path, as shown in Figure 11.36.

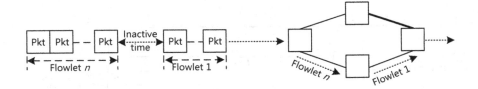

FIGURE 11.36   Switch maintaining the path load information of the entire network.

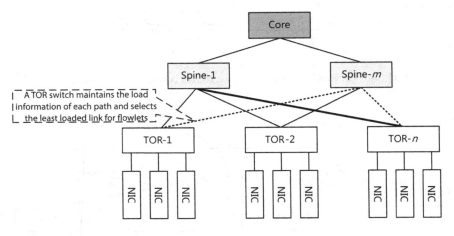

FIGURE 11.37   Global load balancing.

This global load balancing mechanism balances the load of paths for service flows. It achieves balanced path usage and improves application performance (FCT and throughput), as shown in Figure 11.37.

To measure the quality of each path, the forwarding chip must detect the congestion status of each device along the path in real time by sending packets. The switch selects a link for a flowlet based on the link's path quality, as opposed to the quality of the link itself.

4. Congestion control

Traditional ECN technology adds the ECN flag to a packet when the packet enters a queue, and congestion notification is triggered when the ECN-tagged packet leaves the queue. This mechanism creates a delay that amounts to the period between the packet entering and leaving the queue. Due to this, the source cannot be immediately instructed to reduce the transmission rate to relieve congestion. If the queue is severely congested, the device with a small buffer may take several

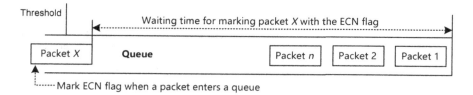

FIGURE 11.38   Traditional ECN technology.

FIGURE 11.39   ECN performing congestion marking in advance.

milliseconds to send a packet and notify the source end. As a result, queue congestion deteriorates, and in some cases, PFC is triggered on the entire network and traffic is stopped, as shown in Figure 11.38.

If packets that are sent out of a queue are marked with the ECN flag as soon as congestion occurs, the delay of the congested queue and the waiting time for congestion notification will be reduced, as shown in Figure 11.39.

ECN allows the NIC of the source server to quickly respond to congestion by reducing the transmission rate, thereby quickly mitigating the cache congestion of network devices. This effectively reduces the delay and improves application performance.

Traditional Congestion Notification Packets (CNPs) are sent by the destination server. Even if RoCEv2 packets encounter queue congestion on an intermediate device, the device only adds the ECN flag to the packets, which are then still sent to the destination server. Subsequently, the NIC of the destination server identifies that congestion occurs, constructs CNPs, and sends the packets to the source server. As the congestion notification has a long feedback path, the traffic rate cannot be reduced in time. As a result, the intermediate device may have buffering with more severe congestion, and in some cases, PFC is triggered and spread, degrading application performance, as shown in Figure 11.40.

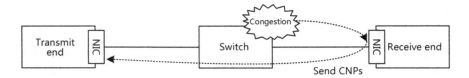

FIGURE 11.40   Traditional CNP congestion notification.

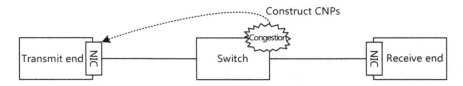

FIGURE 11.41   Intermediate device sending CNPs.

To shorten the congestion notification's feedback path, the intermediate device can assume the role of the NIC on the destination server, immediately construct CNPs, and return the packets to the source server if congestion occurs. Subsequently, the source server can immediately adjust the traffic rate to relieve queue buffer congestion on the intermediate device, as shown in Figure 11.41.

If a switch is required to construct CNPs and send them to the NIC of the source server when congestion occurs, the forwarding chip needs to establish mapping between the source and destination queue pairs of flows. To support this capability, the switch needs to learn and maintain a flow table.

For distributed parallel applications such as HPC and distributed storage in a DC, the N:1 Incast model is used. A larger N value indicates a higher Incast degree and heavier burst traffic burden in the buffer. To relieve congestion in the interface queue buffer, the switch can add the ECN flag to packets. When queues are congested, the switch requests the source server to reduce the transmission rate. This prevents the queue buffer from being congested continuously or deteriorating, thereby preventing PFC. This is necessary because PFC triggering and spreading will reduce the throughput of application flows, increase the FCT of applications, and degrade application performance.

Figure 11.42 shows the Incast traffic model. In this model, the $\Sigma T_{PFC}(i)$ of the inbound priority queue must be larger than the T_{ECN}

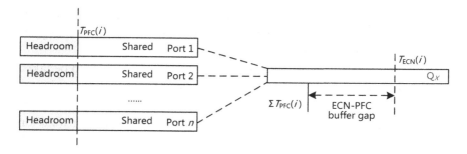

FIGURE 11.42   Incast traffic model.

of the outbound interface queue. This prevents PFC for the priority of the inbound interface from being triggered before the ECN flag is added to packets in the outbound interface queue.

If ECN marking is preferentially triggered, then instead of using PFC to stop sending flows to eliminate packet loss, CNPs will be sent to request the source server to reduce the transmission rate. This relieves the queue congestion on the intermediate device. However, between the time that the ECN flag is marked and the transmission rate of the source server is reduced, traffic still passes through the congested queue. Congestion may increase before the transmission rate is reduced. As a result, PFC will be triggered to stop sending flows.

To address this issue, ensure sufficient buffer gap between the ECN and PFC thresholds to accommodate traffic sent during the period, as shown in Figure 11.43.

The required buffer gap between the ECN and PFC thresholds depends on the concurrency of the Incast traffic model. A higher Incast concurrency requires a larger buffer gap between the ECN and PFC thresholds. In theory, the buffer should reach hundreds of megabytes; however, the packet buffer of the switch's forwarding

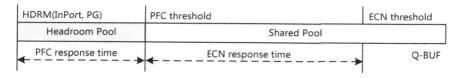

FIGURE 11.43   Sufficient buffer gap between the ECN and PFC thresholds.

chip is limited, and the packet buffer of the existing BRCM XGS forwarding chip cannot meet this requirement. Therefore, PFC triggering cannot be avoided.

The balance between the throughput and delay needs to be taken into account during ECN threshold configuration.

If a low ECN threshold is configured, the device can perform ECN marking and request the source server to reduce the transmission rate as soon as possible. Through this process, the low buffer depth will be maintained; that is, the queue delay will be low. This is beneficial to delay-sensitive mice flows. However, the low ECN threshold affects throughput-sensitive elephant flows and fails to ensure high throughput, as shown in Figure 11.44.

If a high ECN threshold is configured, the time for triggering ECN marking will be prolonged. The queue's capability to accommodate burst traffic will be enhanced and the bandwidth of throughput-sensitive elephant flows will be ensured. However, when a queue is congested, there will be a long queue delay, which is detrimental to delay-sensitive mice flows, as shown in Figure 11.45.

In conclusion, the ECN threshold is closely related to the proportions of elephant and mice flows on network devices, as well as the concurrency of the Incast traffic model, as shown in Figure 11.46.

When the concurrency of the Incast traffic model is high, a low ECN threshold ensures queues with low latency. In addition, the

FIGURE 11.44 Low ECN threshold affecting throughput-sensitive elephant flows.

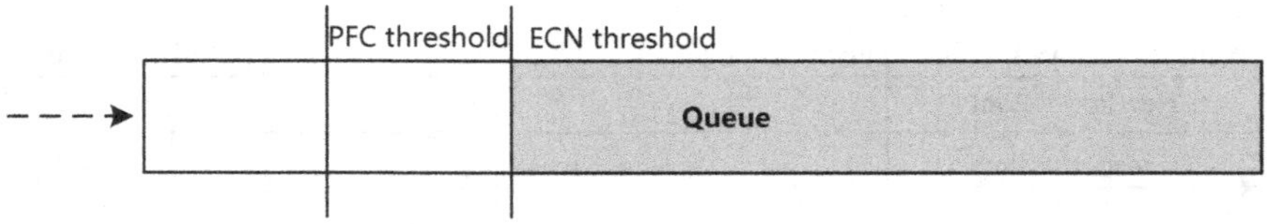

FIGURE 11.45 High ECN threshold affecting delay-sensitive mice flows.

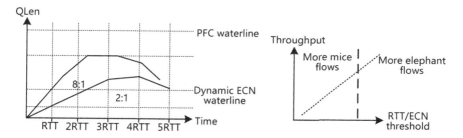

FIGURE 11.46 ECN threshold setting and proportions of elephant and mice flows.

buffer gap between the ECN and PFC thresholds needs to be increased to reduce PFC triggering. In contrast, when the concurrency of the Incast traffic model is low, a high ECN threshold reduces the buffer gap between the ECN and PFC thresholds, ensuring queues with high throughput.

When the proportion of mice flows is high, a low ECN threshold ensures mice flows have low latency. When the proportion of elephant flows is high, a high ECN threshold ensures most elephant flows have high throughput.

To achieve low latency and high throughput for mice flows and elephant flows, respectively, and to reduce PFC triggering, the ECN threshold needs to be dynamically adjusted based on the concurrency of the Incast traffic model and proportions of mice and elephant flows.

To calculate the ECN threshold, the switch's forwarding chip needs to obtain the concurrency of the Incast traffic model as well as data regarding the proportions of elephant and mice flows in real time. This data includes queue depth, queue output rate, and chip buffer usage.

CHAPTER 12

Components of the Cloud DCN Solution

This chapter describes the following Huawei components, as well as their functions, used for building cloud DCNs: physical CloudEngine switches, virtual CloudEngine switches, HiSecEngine series firewalls, SDN controller iMaster NCE-Fabric, and security manager SecoManager.

12.1 PHYSICAL CLOUDENGINE SWITCHES

Huawei CloudEngine DC switches are available as either physical or virtual switches and responsible for forwarding traffic on the network. These switches are the cornerstone of the cloud DCN solution. This section describes Huawei physical CloudEngine switches.

12.1.1 Overview

Huawei physical CloudEngine switches are high-performance switches designed for DC scenarios. These switches are used to construct intelligent, high-quality networks that are elastic, virtual, and agile, providing a solid foundation for building cloud DCNs.

Chapter 2 summarized the challenges faced by cloud DCNs: Big data requires big pipes, networks need pooling and automation, networks need to provide a reliable foundation for services, and DC O&M needs to be intelligent. The following describes how CloudEngine DC switches address these challenges.

1. Big data requires big pipes.

   With the rapid development of services such as big data, smart city, mobile Internet, and cloud computing, network traffic increases significantly every year. This places a major burden on DC switches, requiring their capacity to scale as network capacity is expanded in the future. For example, the access bandwidth of servers has been increased to 10 Gbit/s and is developing toward 25 Gbit/s. In addition, 100GE interfaces are used as uplink interfaces on networks and will soon evolve to 400GE uplink interfaces. CloudEngine DC switches support smooth evolution from 25 Gbit/s and 100 Gbit/s to 400 Gbit/s, supporting the future expansion of networks. For details about the evolution from 25 Gbit/s and 100 Gbit/s to 400 Gbit/s, see Section 12.1.2.

2. Networks need pooling and automation.

   To meet elastic scheduling requirements for compute and storage resources, cloud DCNs must have large resource pools and networks need to achieve resource pooling. For elastic provisioning of services, which are updated rapidly in the cloud computing era, automatic network deployment is required.

   Network resource pooling enables cloud DCNs to evolve from traditional networks to virtual networks. The evolution requires DC switches to support a variety of virtualization technologies, which are classified into device virtualization and network virtualization. Device virtualization optimizes the reliability of a single node and the efficiency of links on the network. For example, CloudEngine DC switches support VS, CSS and iStack, and M-LAG. VS virtualizes a single physical device into multiple logical devices, whereas CSS and iStack virtualize multiple physical devices into a single logical device. For details about CSS, iStack, and M-LAG, see Chapter 3. Network virtualization constructs a virtual overlay network on top of the traditional physical network. This decouples the logical network from the physical network for services, thereby implementing flexible allocation and pooling of network resources. An example of a typical network virtualization technology is VXLAN BGP EVPN, which CloudEngine DC switches support. Given the urgent need to evolve toward IPv6, due to the exhaustion of IPv4 addresses, these switches also support IPv6-based VXLAN technology. For details about VXLAN BGP EVPN, see Chapter 6. For details about VXLAN applications in IPv6 and how IPv4 VXLAN evolves to IPv6, see Section 12.1.2.

The following analyzes automatic network deployment, which is classified into automatic deployment of physical network services (such as interfaces, IP addresses, and routes) and automatic deployment of logical network services (such as VXLAN and tenant VRF). To achieve automatic deployment of physical network services, external tools are typically used. Such tools invoke DC switches' interfaces to deliver configurations to these switches automatically rather than manually. CloudEngine DC switches support ZTP, which is Python-based and automatically delivers configurations to switches by invoking their RESTCONF interfaces. To achieve automatic deployment of logical network services, SDN controllers or third-party tools are often used to orchestrate logical networks. In this process, switches need to provide APIs to connect to various management devices. In addition, switches also need to support some standard network models, such as standard NSH and segment routing. CloudEngine DC switches meet these requirements, providing APIs for connection with various SDN controllers (such as Huawei iMaster NCE-Fabric and VMware) and third-party tools (such as DevOps) as well as supporting standard NSH and segment routing.

3. Networks need to provide a reliable foundation for services.

High reliability emphasizes physical stability. Cloud DCNs need to provide device-, network-, and controller-level high reliability. DC switches need to provide device-level reliability and support for network-layer high-reliability solutions.

Device-level high reliability is often achieved on a per-device basis, relying on the hardware and software of each device being highly reliable. CloudEngine DC switches support a wide range of high-reliability and security features, such as CPU attack defense, graceful restart (GR), and BFD. In addition, these switches use traditional hardware-based reliability mechanisms, such as dual main processing units (MPUs), dual power supplies, and separation of the forwarding plane and control plane. They also use an orthogonal architecture with no wiring on the backplane, adopt a strict front-to-back airflow design, and support cell switching and virtual output queuing (VOQ) to implement non-blocking switching.

High-reliability network solutions involve the high-reliability design of switching networks and distributed network solutions, providing reliable gateways for compute, storage, and L4-L7 service devices. The high-reliability design of switching networks includes redundant access, loop-free network, ECMP, and single-node reliability. Currently, redundant access is implemented using virtual chassis technologies or inter-device link aggregation technology. Loop-free networks and ECMP depend on the underlay routing design and overlay VXLAN technology, and single-node reliability can be achieved using the virtual chassis technology. Because VXLAN BGP EVPN is widely used to construct logical networks, DC switches need to function as distributed VXLAN gateways to provide reliable gateways for compute and storage VAS devices. DC switches are also required to support the preceding technologies in order to facilitate the high-reliability design of network solutions. In this context, CloudEngine DC switches support up to 128 ECMP paths for load balancing and support stack or iStack, M-LAG, and VXLAN BGP EVPN distributed gateways.

4. DC O&M needs to be intelligent.

Intelligent O&M is becoming a necessity as the scale, traffic, and number of NEs on a cloud DCN continue to increase sharply. To support intelligent O&M, CloudEngine DC switches provide the following functions:

High-precision collection of subscription data in less than one second: Information on traditional networks is typically collected through SNMP. This is inefficient because a collection task requires multiple interactions and the polling period is usually several minutes. Within the polling period, some microburst traffic and packet loss may not be identified, affecting delay-sensitive services such as videos and online games. To resolve this issue for the collection of DCN information, telemetry based on the Google Remote Procedure Call (gRPC) interface is used. Telemetry, which is supported by CloudEngine DC switches, enables the polling period to be reduced to less than one second and can combine multiple tasks for processing, thereby improving the collection efficiency. For details, see Section 12.1.2.

Full-path detection and real-time monitoring: In legacy DCNs, link faults and packet loss are manually located and rectified using detection methods such as ping and trace. By working with the SDN controller, CloudEngine DC switches can implement full-path detection. Specifically, the controller quickly locates faults and learns the network status by simulating service packets and sending them to switches to detect paths on the entire network and generate a network-wide path heatmap. Packet Conservation Algorithm for Internet (iPCA) can be used to accurately locate service faults based on real service packets and work with the controller to implement visualized real-time monitoring.

12.1.2 Technical Highlights

CloudEngine DC switches are designed to support the development trend of DCNs over the next 2–10 years. This includes supporting the Ethernet 400GE standard, telemetry (a next-generation DC network monitoring technology), Layer 2 switching, Layer 3 routing, and other basic technologies in the TCP/IP protocol stack, as well as network management and O&M.

12.1.2.1 Evolution from 25GE, 100GE, to 400GE

Ethernet has transitioned rapidly from supporting rates of 1 Mbit/s, 10 Mbit/s, and 100 Mbit/s (FE), to supporting rates of 1 Gbit/s (GE), 10 Gbit/s (10GE), 40 Gbit/s (40GE), and even 100 Gbit/s (100GE). Network traffic has grown at a rapid pace as services such as big data, smart city, mobile Internet, and cloud computing gain popularity and continue to develop. The increasing demand for bandwidth drives standards organizations to not only optimize current standards, but also push the limits of what is currently possible.

Figure 12.1 shows the physical networking architecture of mainstream DCNs. The spine-leaf architecture is generally used, with 10GE interfaces functioning as downlink interfaces for server access and 40GE interfaces functioning as uplink interfaces on leaf nodes.

To meet the growing service requirements, backbone switches are transitioned to 100GE. The server performance is not a bottleneck, as it can keep pace with these requirements. For example, the network throughput of a mid-range x86 server can reach 20 Gbit/s, and that of a high-end server can reach 40 Gbit/s or even 80 Gbit/s. The 25GE standard was developed to help increase the access rate of DC servers.

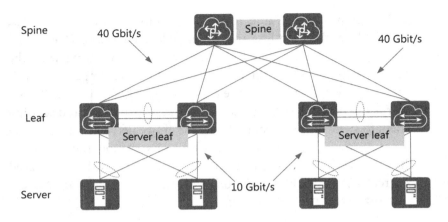

FIGURE 12.1 Typical network deployment in a DC.

1. 25GE standard

At the IEEE conference held in Beijing in 2014, Microsoft first proposed the 25GE project initiation request for specific TOR and server interconnection scenarios. The IEEE conference rejected this request because it was assumed that 25GE would diversify industry investment and hinder industry development. The 25GE standard can resolve a series of issues, such as CPU performance matching and Peripheral Component Interconnect Express (PCIe) link width matching, caused by the transition of server NICs from 10–40 Gbit/s. Vendors including Microsoft and Qualcomm established the 25GE Ethernet Consortium and invested in the 25GE research. To avoid 25GE becoming simply a de-facto standard, the IEEE passed the 25GE project in July 2014. Through the development of networks, it has since been proven that the 25GE standard can economically and efficiently expand network bandwidth, provide better support for next-generation servers and storage solutions, and cover more interconnection scenarios in the future.

The 25GE standard will be mainly applied to server access in next-generation DCs. The reasons for using 25 Gbit/s instead of 40 Gbit/s are as follows:

1. Advantages in technology implementation

Serializer/deserializer (SerDes) is essential in 25GE and has been widely used in various circuit and optical fiber communication technologies. For example, SerDes is commonly used in high-speed communication scenarios such as interconnection between the PCIe bus in a computer and the NIC and internal chip of a switch. The serializer receives and serializes transmitted data, and the deserializer at the receiver reconstructs the serial bit stream and converts it into the required data. The number of SerDes connections required by a switch port is called the number of lanes.

SerDes has been developed over a number of years, and its speed can reach 25 Gbit/s. Only one 25 Gbit/s SerDes lane is required for all end-to-end connections when traffic is transmitted from one 25 Gbit/s NIC to another 25 Gbit/s NIC through a switch. A 40GE interface uses the Quad Small Form-factor Pluggable (QSFP) form factor and is constructed of four parallel 10GE links (each of which uses 12.5 GHz SerDes), requiring four SerDes lanes.

At the aggregation and backbone layers, 100GE interfaces have gradually become mainstream. Mature IEEE 100 Gbit/s Ethernet standards have been developed and are implemented by running four 25 Gbit/s lanes (in compliance with IEEE 802.3bj) on four optical fibers or copper cables. This lays a foundation for the 25GE standard. A 100GE interface, which has significant advantages over a 40GE interface in terms of speed, can be connected to the remote four 25GE interfaces through a QSFP28 to SFP28 1-to-4 cable.

2. Improved switch performance

Compared with the existing 10GE solution, the single-lane 25GE solution improves the performance by 2.5 times. Additionally, 25GE ensures that switches have higher port density than 40GE used for rack server connections, as shown in Table 12.1.

TABLE 12.1   Technical Specifications Comparison between the 25GE and 40GE Standards

Standard	Lane	SerDes (Gbit/s)	Form Factor Type	Connector Type	Number of Optical Fibers
25GE	1	25	SFP28	DLC	2
40GE	4	10	QSFP+	MPO	8

3. Smooth evolution of the existing topology, reducing costs

Capital expenditure (CapEx) is a key factor that needs to be considered when a new DC technology is used. One of the most important aspects of DC deployment is cabling, which can be a complex and difficult part of DC management.

40GE switches use QSFP+ optical modules. For connections inside a cabinet or between adjacent cabinets, QSFP+ DAC cables can be used. For connections spanning longer distances, MPO cables must be used. In most cases, 12-core optical fibers are used for QSFP+ optical modules. Compared with 10GE interfaces that use 2-core LC-connector optical fibers, 40GE interfaces have much higher costs and are incompatible with 10GE interfaces. If 10GE interfaces are upgraded to 40GE ones, the existing optical fibers must be replaced with MPO cables.

25GE switches use SFP28 optical modules, which are similar to 10GE SFP optical modules in that they use single-lane connections and are compatible with optical fibers using LC connectors. 10GE can be seamlessly upgraded to 25GE, removing the need to re-plan the topology or redeploy cables. (Seamless upgrade is not possible from 10GE to 40GE.) After switches are upgraded or have 25GE optical modules installed, the switches require no further configuration, saving labor resources.

The 25GE standard can use the existing 10GE topology for smooth evolution. This not only reduces the need to purchase new TOR switches and cables, but also reduces the requirements for power supply, heat dissipation, and space when compared with 40GE for rack server connections.

As mentioned earlier, mainstream DCNs typically use the 10GE/40GE network architecture. To support large-scale deployment of services such as AI computing and cloud DCNs, the next-generation DCN architecture is evolving to the 25GE/100GE network architecture. Some Internet giants have already implemented large-scale deployment of 25GE/100GE networks, redefining the edge of DCNs. It is possible that 25GE/100GE will become the mainstream trend before long.

2. 100GE standard

To understand the 100GE standard, it is necessary to know about the standardization of optical modules. Initially, each vendor used its own form factor and dimensions for optical modules, as no industry standard yet existed. The IEEE 802.3 Working Group played an important role in standardizing optical modules, and the Multi Source Agreement (MSA), which is an industry alliance, defined the unified optical module structure (including the form factor type, dimensions, and pin assignment) based on the IEEE standards.

In 2016, the IEEE established a working group to study the next-generation 100GE standard and then in 2012 released multiple 100GE standards. The IEEE and MSA have defined more than 10 100GE standards, covering 100GE uplink scenarios that span different distances. Table 12.2 compares these standards.

The 100GBASE series standards are formulated by IEEE 802.3. Figure 12.2 shows the naming conventions, and Table 12.3 describes them.

The transmission features of optical fibers and the manufacturing costs of optical modules determine the application of these

TABLE 12.2   Comparison of Applications of 100GE Optical Module Standards

Standard	Authority	Connector	Optical Fiber Type	Transmission Distance
100GBASE-SR10	IEEE 802.3	24-core MPO	Multi-mode optical fiber with a center wavelength of 850 nm	Optical multi-mode 2 (OM2) fiber: 30 m; optical multi-mode 3 (OM3) fiber: 100 m; optical multi-mode 4 (OM4) fiber: 150 m
100GBASE-LR4	IEEE 802.3	LC	Single-mode optical fiber with a center wavelength of 1295.56–1309.14 nm	Single-mode (G.652) optical fiber: 10 km
100GBASE-ER4	IEEE 802.3	LC	Single-mode optical fiber with a center wavelength of 1295.56–1309.14 nm	Single-mode (G.652) optical fiber: 40 km
100GBASE-SR4	IEEE 802.3	8-core/12-core MPO	Multi-mode optical fiber with a center wavelength of 850 nm	Optical multi-mode (OM3) fiber: 70 m Optical multi-mode (OM4) fiber: 100 m
100G PSM4	MSA	12-core MPO	Single-mode optical fiber with a center wavelength of 1310 nm	Single-mode (G.652) optical fiber: 500 m
100G CWDM4	MSA	LC	Single-mode optical fiber with a center wavelength of 1310 nm	Single-mode (G.652) optical fiber: 2 km

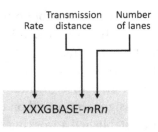

FIGURE 12.2   Naming conventions of 100GBASE series standards.

TABLE 12.3   Naming Conventions of 100GBASE Series Standards

Module	Field	Description
XXX	Rate	Rate standard. In this example, the value is 100 Gbit/s
m	Transmission distance	PMD type, indicating the transmission distance
	KR	The transmission distance is at the 10 cm level. K indicates backplane, that is, the signal transmission distance between backplanes
	CR	The transmission distance is at the meter level. C indicates copper and high-speed cable connections
	SR	The transmission distance is at the 10 m level. S indicates short-distance transmission, which is often provided by multi-mode optical fibers
	DR	The transmission distance is 500 m
	FR	The transmission distance is 2 km
	LR	The transmission distance is 10 km. L indicates long-distance transmission
	ER	The transmission distance is 40 km. E indicates extended
	ZR	The transmission distance is 80 km
n	Number of lanes	Number of SerDes lanes occupied by 100GE
	4	Four SerDes lanes (4×25GE) are occupied
	10	Ten SerDes lanes (10×10GE) are occupied

standards in different scenarios. For example, multi-mode optical fibers are commonly used for short-distance transmission, and single-mode optical fibers are commonly used for long-distance transmission. In DCs, the IEEE 100GBASE series standards can support both long- and short-distance transmission. Among these standards, 100GBASE-SR4 and 100GBASE-LR4 are used most often. However, in most DC interconnection scenarios, the transmission distance supported by 100GBASE-SR4 is too short, and the cost of

100GBASE-LR4 is too high. For medium-distance transmission, the Parallel Single Mode 4-channel (PSM4) and Coarse Wavelength Division Multiplexing 4-lane (CWDM4) standards proposed by the MSA are a cost-effective option.

CWDM4 multiplexes four parallel 25 Gbit/s lanes to a 100 Gbit/s fiber link by using the optical components MUX and DEMUX. This is similar to LR4. The differences are as follows:

1. Different channel spacings

    CWDM4 defines a channel spacing of 20 nm, whereas LR4 defines a LAN-WDM spacing of 4.5 nm. A larger channel spacing results in lower requirements for optical components and lower costs.

2. Different lasers

    CWDM4 uses a single direct modulated laser (DML), whereas LR4 uses an Electro-absorption Modulated Laser (EML), which is composed of a DML and an electro-absorption modulator (EAM).

3. Different temperature control requirements

    The lasers used in LR4 require a Thermo Electric Cooler (TEC) driver chip because of the smaller channel spacing.

    Due to the preceding differences, the cost of 100GBASE-LR4 optical modules is higher than that of 100G CWDM4 optical modules. In addition to CWDM4, PSM4 is another choice for medium-distance transmission. The 100G PSM4 specification defines requirements for a point-to-point 100 Gbit/s link over eight single-mode fibers (4 transmit and 4 receive), each transmitting at 25 Gbit/s.

    Because CWDM4 uses a wavelength division multiplexer, the optical module cost of CWDM4 is higher than that of PSM4. CWDM4, however, requires only two single-mode optical fibers to transmit and receive signals, whereas PSM4 requires eight. As a result, the optical module cost of PSM4 increases as the transmission distance increases.

    A complete optical module solution includes both the optical/electrical interface standards of optical modules and the matching form factor. The development of optical modules is characterized by high rate, high density, low cost, and low power consumption. For example, the C-form factor pluggable (CFP)

TABLE 12.4   Comparison of 100GE Optical Module Form Factors

Form Factor	Power Consumption	Channel × Rate	Comparison
CFP	32 W	10×10 Gbit/s or 4×25 Gbit/s	Large size, high power consumption, and long transmission distance
CFP2	12 W	4×25 Gbit/s	Large size, high power consumption, and long transmission distance
CFP4	6 W	4×25 Gbit/s	Medium size and low power consumption
QSFP28	3.5 W	4×25 Gbit/s	Small size and low power consumption

was the first type of module used, but as the integration of modules improved and dimension restrictions became more important, CFP evolved to CFP2 and CFP4, and then to the popular QSFP28 (Table 12.4).

After several generations of development, 100GE optical modules have become mature. New 100G MSA standards are also established to support the application of new technologies and new development directions, as well as promoting sustainable development of the related industry chain. It has become a never-ending challenge to achieve higher bandwidth and lower latency. As related technologies are introduced, the traditional analog optical transmission system has evolved to digital optical transmission, leveraging sophisticated chip technology and improved chip processing capabilities that make 400GE possible.

3. 400GE standard

The IEEE 802.3 Working Group is responsible for defining the 400GE standard. In 2013, the IEEE initiated the 400GE standard project and started the study group to discuss the 400GE specifications. It officially released the 400GE and 200GE standards (IEEE 802.3bs) after numerous meetings and in-depth research. The key aspects of these standards are the hierarchical structure, FEC specifications, and physical optical interface transmission mechanism. Table 12.5 describes the transmission distance and encoding scheme of each 400GE standard.

TABLE 12.5 400GE Transmission Distance and Encoding Scheme

Standard	Transmission Distance	Encoding Mode
400GBASE-SR16	100 m	16×25 Gbit/s NRZ
400GBASE-DR4	500 m	4×100 Gbit/s PAM4
400GBASE-FR8	2 km	8×50 Gbit/s PAM4
400GBASE-LR8	10 km	8×50 Gbit/s PAM4

The 400GBASE-DR4, 400GBASE-FR8, and 400GBASE-LR8 standards using PAM4 encoding have become the focus of attention, while 400GBASE-SR16, a multimode standard, has received little attention. PAM4 differs from non-return-to-zero (NRZ) signal transmission, which was commonly used in the 100GE standard. A PAM4 signal uses four signal levels for transmission. Within each clock period, 2-bit logic information, that is, 00, 01, 10, and 11, can be transmitted. Conversely, an NRZ signal uses two signal levels — high and low — to represent 1 and 0 in digital logic signals. Given the same baud rate, PAM4 achieves double the transmission efficiency compared with NRZ. As a result of its higher efficiency, PAM4 was defined as the 400GE electrical signal standards by the IEEE.

Because the SerDes rate can reach 25 Gbit/s, and the bit rate can reach 50 Gbit/s through PAM4 modulation, IEEE 802.3 selected 50 Gbps/lane PAM4 as the encoding technology for 400GE and 200GE interfaces. Figure 12.3 shows the structure of a 400GE CFP8 optical module.

400GE MAC uses 400G AUI-16 interfaces (16×25 Gbit/s) to transmit data to the 400GE CFP8 optical module through the physical coding sublayer (PCS) and physical medium attachment sublayer (PMA).

- During data transmission, the PMA aggregates 16 lanes into 8 lanes and converts the modulation code to PAM4 with a rate of 25 Gbit/s (the actual value is 50 Gbit/s).

- The data is then transmitted to the physical medium-dependent sublayer (PMD). The transmitter and receiver at the PMD implement electrical-to-optical and optical-to-electrical conversion, respectively.

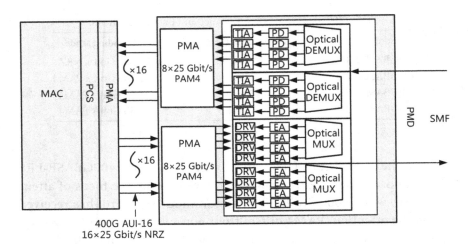

FIGURE 12.3   400GE CFP8 optical module.

- In the transmit direction, eight transmitters convert electrical signals into optical ones, and the signals from the eight lanes are then multiplexed to one 400G link through a single-mode optical fiber.

- In the receive direction, the optical fiber demultiplexes the eight wavelengths on the optical fiber to eight receivers for optical-to-electrical conversion. The converted signals are then sent to the PMA.

CFP8 optical modules are not suitable for DCNs, because they are power-hungry and too large. Instead, the Quad Small Form Pluggable (QSFP) and QSFP-Double Density (DD) form factors are used. Table 12.6 compares these modules.

400GE QSFP-DD and QSFP optical modules use CDAUI-8 and PAM4 for electrical-layer lanes and use 56 Gbit/s laser technology

TABLE 12.6   Comparison of 400GE Optical Module Form Factors

Form Factor	Dimensions (mm)	Power Consumption (W)	Electrical Interface Lanes	Optical Interface Lanes
CFP8	107.5×41.5×12.5	12–18	16×25 Gbit/s 8×56 Gbit/s	8×56 Gbit/s
QSFP	107.8×22.6×13.0	12–15	8×56 Gbit/s	8×56 Gbit/s
QSFP-DD	89.4×18.4×8.5	7–10	8×56 Gbit/s	4×100 Gbit/s

(4×100 Gbit/s PAM4) or mature 28 Gbit/s laser technology (8×50 Gbit/s) for optical-layer lanes.

At present, vendors tend to use 4×56 Gbit/s PAM4 for electrical components and use existing 28 Gbit/s components for optical components. Although standardization of 200GE started later than that of 400GE, it is easy to implement 200GE optical modules. In DC scenarios, it is likely that 200GE optical modules were being used before 400GE optical modules, leading to the maturity and large-scale development of 200GE optical modules. As more in-depth research is conducted for single-wavelength 100GE technologies, it is expected that short-distance 400GE optical modules will become more popular in DC scenarios.

12.1.2.2 Telemetry Technology

Telemetry is a technology that fast collects data from physical or virtual devices remotely. It pushes high-precision data indicators of network devices to a collector at high speed in real time, improving the device efficiency and network efficiency during data collection.

1. Background

   The growing requirements for network services and the increasing types of services complicate network management and pose higher requirements on network monitoring. In addition, the establishment of 4G/5G communications standards, continuous enhancement of network devices' transmission capabilities, and constant evolution of SDN require higher network quality. Driven by this trend, networks need higher-precision monitoring data to quickly detect microburst traffic. Furthermore, the monitoring process needs to minimize the impact on the functions and performance of devices in order to improve device efficiency and network efficiency.

   Traditional network monitoring technologies (such as SNMP Get and CLI) have the following disadvantages, which negatively affect the management efficiency and prevent user requirements from being met.

   - These technologies need to pull (request) device monitoring data, limiting the number of devices that can be monitored and, in turn, limiting network growth.

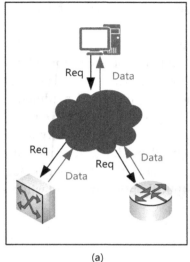 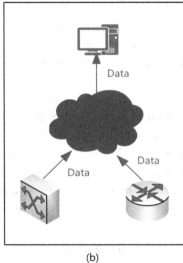

(a)   (b)

FIGURE 12.4   (a) Pull mode and (b) push mode.

- Network device data can only be obtained every few minutes. The frequency at which these technologies obtain data needs to be increased in order to improve data precision. However, this consumes more CPU resources on network devices and therefore negatively affects device functions.

- The monitored data of network nodes is inaccurate due to the network transmission delay.

Telemetry was developed to eliminate the preceding disadvantages and implement large-scale and high-performance network monitoring.

Figure 12.4 shows the pull mode and push mode.

2. Implementation

Telemetry enables an intelligent O&M system to manage more devices and lays a foundation for fast network fault location and network quality optimization. It converts network quality analysis into big data analysis, effectively supporting intelligent O&M.

The following lists the advantages of telemetry over SNMP Get. Figure 12.5 shows the implementation of telemetry.

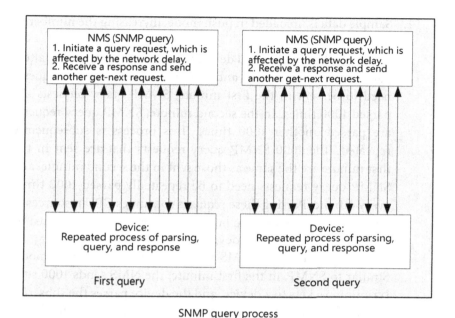

SNMP query process

Telemetry sampling process

FIGURE 12.5 Comparison between the SNMP query process and telemetry sampling process.

1. Sample data is uploaded in push mode, increasing the number of monitoring nodes.

   In SNMP, the NMS and devices interact in pull mode by alternatively sending requests and responses. If 1000 query requests need to be sent in the first minute, SNMP query requests are parsed 1000 times. In the second minute, SNMP query requests are parsed another 1000 times. This process is subsequently repeated. The 1000 SNMP query requests that are sent in the first minute are the same as those sent in the second minute, and SNMP query requests need to be repeatedly parsed 1000 times every minute. Parsing these requests consumes CPU resources of devices, and therefore, the number of monitored nodes must be limited to ensure normal device running.

   Using telemetry, the NMS and devices interact in push mode. Similar to SNMP, in the first minute, the NMS sends 1000 subscription packets to a device, and the device parses the subscription packets 1000 times. However, unlike in SNMP, the device records the subscription information of the NMS during the parsing process, and no subscription packets are sent to the device by the NMS in the following minutes. The device automatically and continuously pushes data to the NMS based on the recorded subscription information. In this way, there is no need to parse 1000 subscription packets per minute, conserving the CPU resources of the device and enabling more devices to be monitored.

2. Sample data is packaged before upload, improving the time precision of data collection.

   In SNMP, a device needs to parse a large number of query requests every minute and uploads only one piece of sample data for each query request. Parsing these requests consumes the device's CPU resources, and therefore, the frequency at which the NMS sends query requests needs to be limited. However, this reduces the time precision of data collection. In most cases, the sampling precision of traditional network monitoring technologies is in the order of seconds.

   In telemetry, a device parses subscription packets only in the first minute. The device updates multiple pieces of sample data for a subscription packet by packaging the data, thereby reducing the number of packets it exchanges with the NMS. This means that the sampling precision of telemetry can be in the order of milliseconds.

3. Sample data contains timestamps, improving the accuracy of sample data.

   Traditional network monitoring technologies do not use timestamps in sample data. This means that the data may be inaccurate due to network transmission delays.

   Telemetry, on the other hand, uses timestamps in sample data. When parsing data, the NMS can determine the time when the sample data was generated, thereby eliminating the effects of network transmission delay on sample data.

3. Applications

   Telemetry can implement automatic, closed-loop O&M management of networks. In Figure 12.6, a network device reports sample data to the collector for storage. The analyzer analyzes the sample data stored in the collector and sends analysis results to the controller. Based on these results, the controller delivers configurations that need to be adjusted on the network to the network device. After the configurations take effect, the device reports new sample data to the collector. The analyzer can analyze whether the adjustments meet expectations to implement closed-loop O&M of the network service process.

4. Key technologies

   Telemetry uses YANG models to organize sample data and uses Google Protocol Buffers (GPB) or JSON to encode the data. It then uses gRPC to upload the data to the collector.

   1. YANG models

      There are a number of YANG models defined by different vendors and standards organizations for network devices, as shown in Figure 12.7. They include:

      – Huawei model

      – IETF model

      – OpenConfig model (Google is the main instigator)

      – Other models, such as Cisco YANG models

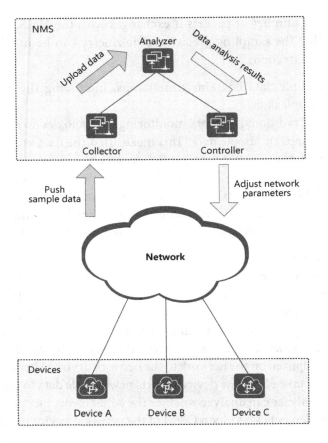

FIGURE 12.6   Automatic closed-loop network O&M management.

On CloudEngine DC switches, telemetry uses Huawei YANG models to organize sample data and is compatible with the OpenConfig model.

2. GPB encoding format

GPB encoding is a language- and platform-neutral extensible mechanism used for serializing structured data of communications protocols and data storage.

Table 12.7 compares GPB encoding and decoding

3. JSON encoding format

JSON encoding is a lightweight data exchange format. Based on a subset of the European Computer Manufacturers Association Script (ECMAScript, which is a JavaScript specification developed

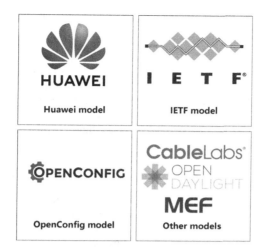

FIGURE 12.7   Vendors and standards organizations that define YANG models.

TABLE 12.7   Comparison between GPB Encoding and Decoding

GPB Encoding	GPB Decoding

{
1:"HUAWEI"
2:"s4"
3:"huawei-ifm:ifm/interfaces/interface" 4:46
5:1515727243419
6:1515727243514
7{
1[{
1: 1515727243419
2 {
5{
1[{
5:1
16:2
25:"Eth-Trunk1"
}]
}
}
}]
} 8:1515727243419
9:10000
10:"OK"
11:"CE6850HI"
12:0
}
	{
	"node_id_str":"HUAWEI",
	"subscription_id_str":"s4",
	"sensor_path":"huawei-ifm:ifm/interfaces/
	interface", "collection_id":46, "collection_
	start_time":"2018/1/12 11:20:43.419",
	" msg_timestamp":"2018/1/12 11:20:43.514",
	"data_gpb":{
	"row":[{
	"timestamp":"2018/1/12 11:20:43.419",
	"content":{
	"interfaces":{ "interface":[{ "ifAdminStatus":1,
	"ifIndex":2,
	"ifName":"Eth-Trunk1"
	}]
	}
	}
	}]
	},
	"collection_end_time":"2018/1/12
	11:20:43.419", "current_period":10000,
	"except_desc":"OK", "product_
	name":"CE6850HI",
	"encoding":Encoding_GPB
	}

TABLE 12.8   Comparison between JSON Encoding and Decoding

JSON Encoding	JSON Decoding
{	{
1:"HUAWEI"	"node_id_str":"HUAWEI",
2:"s4"	"subscription_id_str":"s4",
3:"huawei-ifm:ifm/interfaces/interface" 4:46	"sensor_path":"huawei-ifm:ifm/interfaces/
5:1515727243419	interface", "collection_id":46, "collection_
6:1515727243514	start_time":"2018/1/12 11:20:43.419",
8:1515727243419	" msg_timestamp":"2018/1/12 11:20:43.514",
9:10000	"collection_end_time":"2018/1/12
10:"OK"	11:20:43.419", "current_period":10000,
11:"CE12800"	"except_desc":"OK", "product_
12:1	name":"CE12800", "encoding":Encoding_
14:{	JSON, "data_str":{
"row":[{	"row":[{
"timestamp":"2018/1/12 11:20:43.419",	"timestamp":"2018/1/12 11:20:43.419",
"content":{	"content":{
"interfaces":{	"interfaces":{
"interface":[{ "ifAdminStatus":1, "ifIndex":2,	"interface":[{ "ifAdminStatus":1, "ifIndex":2,
"ifName":"Eth-Trunk1"	"ifName":"Eth-Trunk1"
}]	}]
}	}
}	}
}]	}]
}	}
}	}

by the ECMA), it uses a text format independent of the programming language to store and express data with a simple and clear structure. It aims to enable the data to be easily read or compiled by humans and parsed or generated by computers.

Table 12.8 compares JSON encoding and JSON decoding.

4. gRPC protocol

gRPC is an open-source, high-performance general RPC software framework run over HTTP/2. It enables developers to create services without the need to focus on the underlying communication protocols.

Telemetry uses the gRPC protocol to report the encoded sample data to the collector for storage.

Figure 12.8 shows the gRPC protocol stack layers.

Table 12.9 describes the layers.

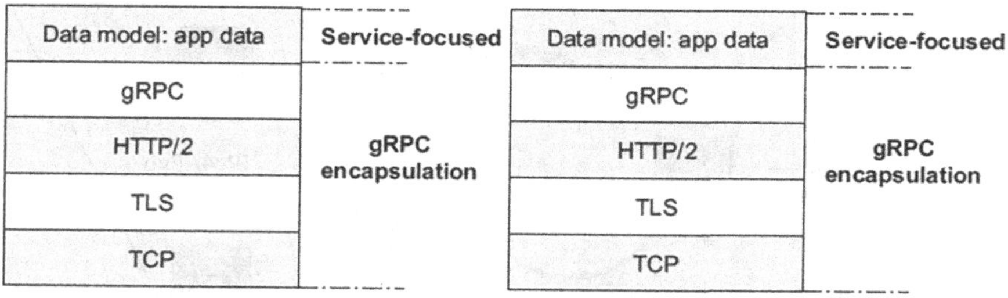

Data model: app data	Service-focused	Data model: app data	Service-focused
gRPC		gRPC	
HTTP/2	gRPC encapsulation	HTTP/2	gRPC encapsulation
TLS		TLS	
TCP		TCP	

FIGURE 12.8 gRPC protocol stack layers.

TABLE 12.9 Description of gRPC Protocol Stack Layers

Layer	Description
TCP layer	This is the underlying communication layer, which is based on TCP connections
Transport Layer Security (TLS) layer	This layer is optional. It is based on the TLS 1.2-encrypted channel and bidirectional certificate authentication
HTTP/2 layer	The HTTP/2 protocol carries gRPC, using HTTP/2 features such as bidirectional streams, flow control, header compression, and multiplexing request of a single connection
gRPC layer	This layer defines the protocol interaction format for remote procedure calls
Data model layer	Communication parties need to know each other's data models before they can correctly interact with each other

12.1.2.3 IPv6 VXLAN

IPv4 has supported the rapid development of the Internet and connected tens of millions of enterprises and billions of people around the world. However, due to its limited address space, IPv4 addresses have been depleted. IPv6 is the successor to IPv4 and gaining attention because of its huge address space, high security, and flexible usage. Carriers and enterprises alike are gradually adopting IPv6 on a larger scale. A key requirement in the evolution from IPv4 to IPv6 is smooth transition.

VXLAN is used to build a virtual network, called an overlay network, based on a physical IP network, called an underlay network. IPv4 and IPv6 underlay and overlay networks can be deployed to implement smooth transition from IPv4 to IPv6, as shown in Figure 12.9.

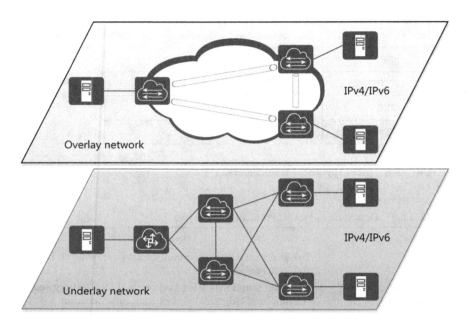

FIGURE 12.9 VXLAN networking.

Smooth transition from IPv4 to IPv6 involves the following:

- Transition IPv4 services to IPv6 services.

- Enable new IPv6 networks to transmit IPv4 services.

- Ensure communication between IPv4 islands and between IPv6 islands.

1. Transition IPv4 services to IPv6 services.

   The IPv6 over IPv4 VXLAN model is used to transition existing IPv4 services to IPv6 services. That is, IPv6 packets are transmitted over IPv4 networks. An ideal way to carry IPv6 services over an IPv4 network is to deploy the IPv4/IPv6 dual stack and IPv6 over IPv4 tunnels on edge nodes. An edge node encapsulates an IPv6 packet in an IPv4 header as the payload and forwards the IPv4 packet on the existing IPv4 network. After receiving the IPv4 packet, the remote edge node removes the IPv4 header and forwards the original IPv6 packet to the destination address (Figure 12.10).

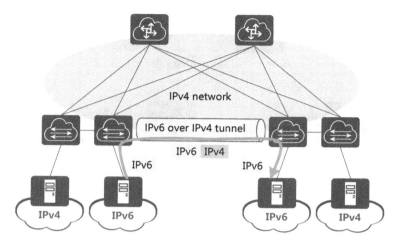

FIGURE 12.10   IPv6 over IPv4.

2. Enable new IPv6 networks to transmit IPv4 services.

The IPv4 over IPv6 VXLAN model is used to transmit IPv4 services on an IPv6 network. In the process of transitioning from IPv4 to IPv6, an IPv6 network is created first, and then, services are migrated from the original IPv4 network to the new IPv6 network. To ensure IPv4 services are migrated smoothly to the IPv6 network and can be transmitted, an IPv4 over IPv6 tunnel can be deployed on the IPv6 network.

An edge node encapsulates an IPv4 packet in an IPv6 header as the payload and forwards the IPv6 packet on the IPv6 network. After receiving the IPv6 packet, the remote edge node removes the IPv6 header and forwards the original IPv4 packet to the destination address (Figure 12.11).

3. Ensure communication between IPv4 islands and between IPv6 islands.

In Figure 12.12, IPv4 or IPv6 networks are connected across an intermediate network that runs a different protocol. VXLAN tunnels can be used on the intermediate network to connect the other networks. Specifically, the ingress node on the intermediate network encapsulates an IPv4 or IPv6 packet, and forwards the encapsulated packet to the egress node across the intermediate network. The egress node then decapsulates the packet and forwards the original packet to the destination.

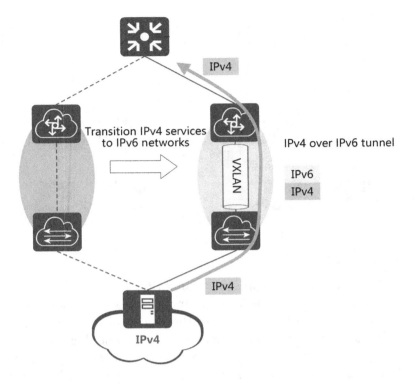

FIGURE 12.11   IPv4 over IPv6.

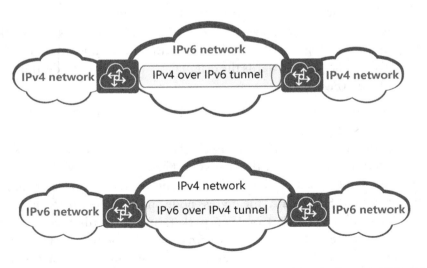

FIGURE 12.12   Communication between IPv4 or IPv6 islands through VXLAN.

12.2 CLOUDENGINE VIRTUAL SWITCHES

12.2.1 Overview

Huawei CloudEngine 1800V (CE1800V for short) is an intelligent virtual switch designed for virtualization environments in enterprise and industry DCs.

In DC and cloud computing scenarios, a traditional virtual network provides communication between VMs and physical networks, and is composed of vNICs and virtual bridges (provided by the Linux Bridge module). These virtual bridges are unable to meet customers' requirements for higher switching performance and more advanced functions such as VXLAN access. To meet these requirements, Huawei developed the CE1800V.

The CE1800V is developed based on the open-source OVS project and uses the user-mode forwarding framework. This virtual switch runs on a server and functions as an NVE node on a VXLAN network, providing higher forwarding performance. It encapsulates the data packets generated by VMs into VXLAN packets so that the VMs can access the VXLAN network.

12.2.2 Architecture

Figure 12.13 shows the CE1800V architecture.

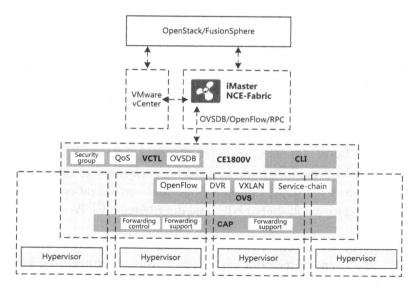

FIGURE 12.13   CE1800V architecture.

As a virtual switch, the CE1800V runs on the VM management program, hypervisor. The CE1800V is connected to pNICs and multiple vNICs. The CE1800V provides virtual access interfaces for VMs and connects to pNICs of servers, providing basic Layer 2 and Layer 3 network forwarding functions for VMs. This switch can be used for communication between VMs and between VMs and external networks.

In the open-source OVS, the forwarding and control planes are combined. On the CE1800V, the virtual forwarding plane is developed and separated from the OVS, helping to increase the CE1800V's flexibility and forwarding capabilities. Additionally, the virtual control plane is developed for obtaining device alarms and delivering service configurations.

The CE1800V can be divided into the following three modules:

- OVS: provides standard OpenFlow interfaces based on OVS to interact with the SDN controller, manage and search the flow table delivered by the controller, and implement DVR Layer 3 forwarding and VXLAN forwarding.

- Virtual control plane: is the control/management plane developed for the CE1800V. It provides OVSDB interfaces for the SDN controller to configure services for the CE1800V. It also provides RPC interfaces as a private extension of iMaster NCE-Fabric for the CE1800V to establish connections with the SDN controller.

- Virtual forwarding plane: is the forwarding plane developed for the CE1800V. It ensures high-performance traffic forwarding based on the OpenFlow flow table and is optimized through the Data Plane Development Kit (DPDK).

12.2.3 Functions

In addition to providing high forwarding performance, the CE1800V supports security groups, multiple cloud and virtualization platforms, automatic deployment, shared NIC, and distributed DHCP.

1. High-performance forwarding

The open-source OVS has two forwarding architectures: kernel-mode and user-mode. The forwarding performance in the kernel-mode architecture is low, as shown in Figure 12.14. The CE1800V

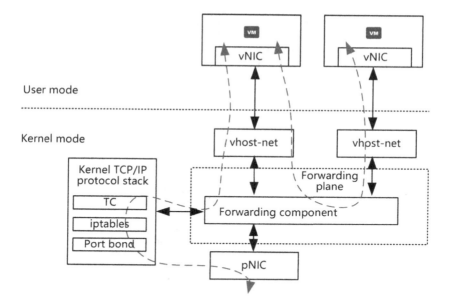

FIGURE 12.14   Kernel-mode forwarding architecture.

uses the user-mode forwarding architecture to achieve a higher forwarding performance.

In kernel-mode forwarding, the kernel runs the vSwitch forwarding component, the VM backend NIC driver vhost-net, and the pNIC driver. The kernel also sends and receives packets, and some forwarding services need to be completed through interaction of the kernel TCP/IP protocol stack. However, because the CPU and memory resources are limited and the processing of packets is a complex task, the forwarding performance is low.

In user-mode forwarding, the forwarding plane component of the CE1800V is run in user space, as shown in Figure 12.15. The CE1800V directly interacts with the VM backend NIC driver vhost-user in user space, without using the kernel space NIC driver and the protocol stack. The sending and receiving of packets is also performed in user space. User-mode CE1800V runs based on DPDK and uses technologies such as hugepage memory, polling, CPU core binding, system invoking prevention, and zero packet copy, greatly improving the packet transmission efficiency of NICs and delivering higher forwarding performance.

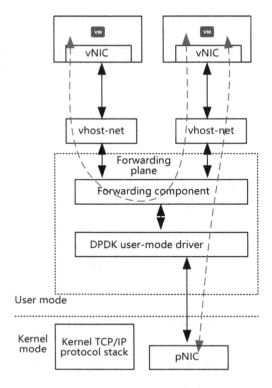

FIGURE 12.15   User-mode forwarding architecture.

2. Security group

To improve VM and network security, the CE1800V supports the security group function. A security group is a vFW that restricts the network packets sent and received by VM interfaces.

Security groups can be created on the cloud platform, and after security group rules are configured, the SDN controller obtains information about the groups from the cloud platform and delivers the information to the CE1800V. The CE1800V translates this information into ACL policies and applies them to specific interfaces to permit only the packets that match the security group rules.

The CE1800V integrates the security group function provided by the DFW library. Table 12.10 compares this security group function with the OVS security group implemented by open-source iptables.

3. Compatibility with multiple cloud and virtualization platforms

TABLE 12.10   Comparison between the CE1800V Security Group and Open-Source OVS Security Group

Function	CE1800V Security Group	Open-Source OVS Security Group
Access control based on VM interfaces	Supported	Supported
Protocol-, interface-, security group-, and segment-based filtering rules	Supported	Supported
Security group capability based on connection status tracking (session)	This function is supported by the CE1800V bridge. The user-mode process memory is used without occupying system resources, the search performance is high, and a maximum of 500K sessions are supported	This function is supported only on kernel-mode vSwitches. Linux bridges and kernel-mode vSwitches need to be connected in series, occupying system resources and leading to low search performance. By default, a maximum of 64K sessions are supported
IP/MAC anti-spoofing capability	Supported	Supported
Security group specifications	Each vSwitch supports a maximum of 1000 security groups, and each security group supports a maximum of 1000 rules	A maximum of 64K rules are supported by default, but is limited by the memory size of the Linux kernel
Log capability	Local logs or syslogs are used to send session and flow statistics to the log server	Logging and O&M are not supported, and fault locating is difficult

The CE1800V is compatible with multiple cloud and virtualization platforms.

- Cloud platform: open-source OpenStack, Red Hat OpenStack, and Huawei FusionSphere

- Virtualization platform: open-source CentOS KVM, Red Hat RHEL KVM, and Huawei FusionSphere (NFVI KVM)

4. Automatic deployment
   The CE1800V can be deployed using the following methods:

   - Cloud Provisioning Service (CPS) system: The CE1800V is deployed and managed on the FusionSphere platform.

- iDeploy: a Huawei deployment tool that supports unified deployment and management of the CE1800V.

- Puppet or Ansible: an automated, mainstream O&M tool that supports unified deployment and management of the CE1800V.

5. Shared NIC

The number of NICs installed in a server is generally low. If the requirements on management, storage, and data services are low but cost control is important, the CE1800V can be configured to use the shared NIC. The shared NIC can be implemented in the following modes:

- vSwitch virtual sub-interface: A vSwitch virtualizes a network interface into multiple interfaces, which function as the management, storage, and data interfaces of the CE1800V. This mode ensures high performance but not reliability and can be used if the NIC supports DPDK.

- Linux VLAN sub-interface: This is a function of the Linux OS, in which a network interface is virtualized into multiple interfaces that function as the management, storage, and data interfaces of the CE1800V. This mode ensures high reliability but not performance and can be used if the NIC does not support DPDK.

6. Distributed DHCP

As networks expand and become more complex, network configurations also become more complex. Due to the popularity of mobile devices and ubiquity of wireless networks, IP addresses assigned to these devices frequently change as users move from one network to another, resulting in insufficient IP addresses. To overcome this issue and dynamically allocate IP addresses to hosts, DHCP is used.

The open-source OVS uses centralized DHCP. However, centralized DHCP requires extra servers, which are heavily burdened. The CE1800V supports the distributed DHCP server function, providing higher performance. Figure 12.16 shows the distributed DHCP server deployment.

Table 12.11 compares distributed DHCP and centralized DHCP.

After a VM goes online, the OpenStack controller node allocates an IP address to it based on the subnet where the VM resides.

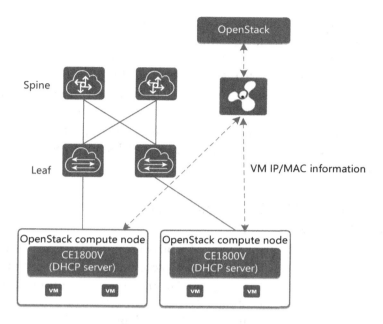

FIGURE 12.16   Distributed DHCP server deployment.

TABLE 12.11   Comparison between Distributed DHCP and Centralized DHCP

Item	Distributed DHCP	Centralized DHCP
DHCP server	The DHCP server function is deployed on each computer node, removing the need to deploy additional servers	Two additional servers are deployed to function as active and standby DHCP servers
DHCP traffic	DHCP Request packets are processed on the local compute node, eliminating DHCP broadcast packets on the network and optimizing the traffic path	DHCP Request packets are sent to the DHCP server through the network
Number of DHCP address pools (4K–96K)	The DHCP server on each compute node needs to support a maximum of 20 subnets	The DHCP server needs to support a maximum of 96K subnets
DHCP O&M	The DHCP server on each compute node needs to be maintained separately, meaning that centralized processing is not supported	DHCP servers are deployed in a centralized manner, supporting centralized management
DHCP reliability	If the DHCP server is faulty, only the local compute node is affected	DHCP faults affect the entire network

The SDN controller can obtain the IP-MAC mappings of VMs from the OpenStack controller node and sends the mappings to the CE1800V (assuming that the DHCP server function is enabled). On receipt of a VM's DHCP Request packet, the CE1800V searches the IP-MAC mappings. If it finds an IP address that maps to the MAC address of the VM, the CE1800V sends a DHCP Offer packet containing the IP address to the VM. After the DHCP server acknowledges the IP address that is offered, the IP address is successfully allocated to the VM.

12.3 HISECENGINE SERIES FIREWALLS

12.3.1 Overview

Terminals such as smartphones and tablets are essential elements of the modern mobile office, enabling users to access the Internet anytime and anywhere. Mobile applications, Web 2.0, and social networking are now integrated into all aspects of the Internet, while cloud computing enables rapid service deployment and on-demand adaptability. Such ICT evolution has greatly improved the communication efficiency of enterprises.

However, as the flexibility of network access modes continues to advance, so too is service traffic rapidly increasing in terms of computing, storage, device transmission, pipes, and clouds, which, in turn, increases bandwidth consumption. The performance of existing devices is no longer sufficient to cope with such fast-paced evolution, leaving enterprises' IT systems facing immense pressure. From the perspective of enterprise information security, mobile office blurs enterprise network borders and offers an opportunity for hackers to easily intrude enterprise IT systems through mobile terminals. Traditional security gateways can only implement security protection and control based on IP addresses and ports, and cannot completely cope with emerging application and web threats. Information security issues are becoming increasingly complex.

Huawei HiSecEngine series high-end NGFWs are designed to protect the network service security of cloud service providers, large DCs, and large enterprise campuses. They provide Terabit-level processing performance, integrate NAT, VPN, virtualization, and other security functions, and achieve 99.999% reliability. The firewalls meet the increasing high-performance processing requirements of networks and DCs, save equipment room space, and reduce total cost of ownership (TCO) per Mbps.

12.3.2 Application Scenarios

12.3.2.1 DC Border Protection

The IDC is an Internet-based system that provides comprehensive facilities and maintenance services. IDCs collect, store, process, and send data in a centralized manner, and are commonly built by large network server providers to provide the server hosting and virtual DNS for small and medium-sized enterprises and individual customers.

The following must be considered regarding an IDC network structure:

- Servers in the IDC need to be protected and security functions must be applied according to the server type.

- Network services need to be provided for external networks, which is the key function of the IDC. Normal access from the Internet to servers in the IDC must be guaranteed, meaning border protection devices require high performance and comprehensive reliability mechanisms while also ensuring network access if attacks are launched on the IDC.

- Servers from multiple enterprises may be deployed in an IDC, and these can be vulnerable to hacker attacks.

- IDC traffic is complex. If the traffic is invisible, it is more difficult to adjust configurations.

With the above in mind, HiSecEngine series firewalls serve as border gateways for IDCs to provide the following functions. Figure 12.17 shows the typical application scenario.

- Use of the SLB and smart DNS functions to ensure proper distribution of traffic to multiple servers in the same server group. This ensures that traffic is evenly distributed to each server, preventing one from being fully loaded while others are idle.

- IP address- and application-based traffic limiting to ensure proper operation of servers, prevent network congestion, and ensure network service continuity.

- Intrusion prevention function to protect servers from intrusions, Trojan horses, and worms.

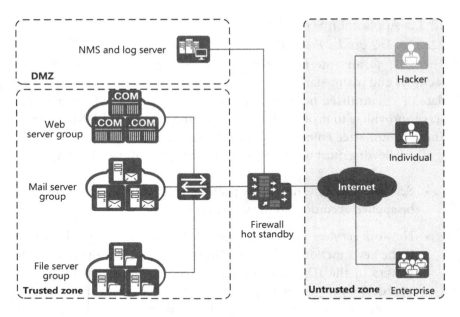

FIGURE 12.17   Typical DC border protection networking.

- Anti-DDoS attack and other attack defense functions to protect servers against external attacks.

- Deployment of log NMS (purchased separately) to record network running logs, enabling administrators to adjust configurations and identify risks.

- Adoption of two firewalls in a cluster working in hot standby mode to improve system reliability. If the active device fails, service traffic can be smoothly switched to the standby device to ensure uninterrupted services.

12.3.2.2 Broadcasting and Television Network and Tier-2 Carrier Network

This section describes how HiSecEngine series firewalls protect broadcasting and television networks as well as tier-2 carrier networks.

Broadcasting and television networks and tier-2 carrier networks are witnessing rapid development and provide network services for an increasing number of users. However, such networks are currently facing the following service challenges:

- Multi-egress links are leased from various Internet service providers (ISPs) at different prices, resulting in traffic being unevenly distributed across the links.

- P2P traffic consumes a lot of public network bandwidth, leading to low bandwidth efficiency.

- URL source tracing is required.

To overcome these challenges, HiSecEngine series firewalls offer the following advantages. Figure 12.18 shows a typical deployment of these firewalls.

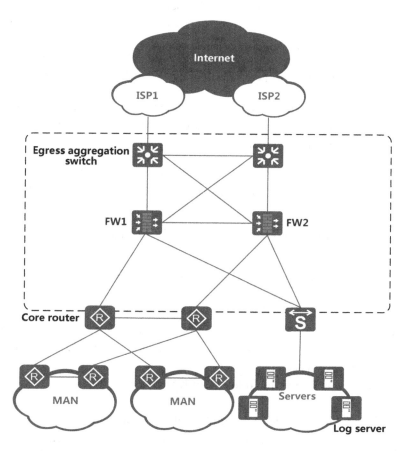

FIGURE 12.18 Typical deployment of a broadcasting and television network or tier-2 carrier network.

- Reduces unnecessary bandwidth consumption through the use of intelligent traffic steering based on applications, routes, users, time, and links.

- Saves public network bandwidth through P2P intranet acceleration.

- Uses many-to-one NAT to save internal resources, improve bandwidth efficiency, and reduce bandwidth costs.

- Traces the source of information, including NAT addresses and URLs, through the log system.

12.3.2.3 Border Protection for Medium and Large Enterprises

This section describes how HiSecEngine series firewalls serve as the egress gateways of medium and large enterprises in order to protect enterprise networks.

A medium or large enterprise typically has more than 500 employees and faces the following service challenges:

- High reliability is required, and border devices must support continuous heavy traffic. Even if a link or device becomes faulty, network services must not be affected.

- Large amounts of employees, complex services, high volumes of outgoing traffic, multiple egresses, and various types of traffic need to be supported.

- Services, including website and email services, need to be provided for external users.

- The enterprise is vulnerable to DDoS attacks, which can lead to massive service losses.

To address these challenges, HiSecEngine series firewalls serve as the egress gateways of medium and large enterprises. Figure 12.19 shows the typical application scenario.

These firewalls offer the following advantages:

- Utilize ISP link selection, intelligent traffic steering, and transparent DNS to properly allocate link bandwidth to multiple egresses, forward traffic efficiently, and prevent traffic diversion and single-link congestion.

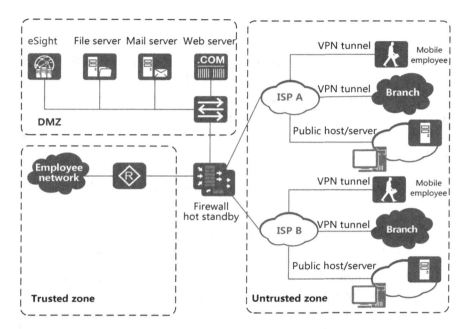

FIGURE 12.19 Typical border protection deployment for medium or large enterprises.

- Apply policies to traffic between the intranet and Internet to control the bandwidth and number of connections, preventing network congestion and defending against DDoS attacks.

- Establish VPN tunnels between HiSecEngine series firewalls and mobile employees, as well as between branches, to protect service data during transmission over the Internet.

- Use the anti-DDoS function to guard against heavy-traffic attacks from external hosts to internal servers, ensuring service continuity.

- Have the log NMS (purchased separately) deployed to record network operation logs, facilitating configuration adjustment, NAT source tracing, and risk identification.

- Adopt two firewalls in a cluster working in hot standby mode to improve system reliability. If a single point of failure occurs, service traffic can be smoothly switched to the standby device to ensure service continuity.

12.3.2.4 VPN Branch Access and Mobile Office

Today's global enterprises often operate branches outside their headquarters or cooperate with partners to support businesses around the world. Branches, partners, and mobile employees all require remote access to the enterprise headquarters network for service provisioning. Currently, VPN technology enables secure and low-cost mobile office access.

Enterprise branch communication scenarios exhibit the following characteristics:

- Branches need to connect to the headquarters network for service provisioning.

- Partners require flexible authorization to limit accessible network resources and transmittable data types based on services.

- Mobile employees require enterprise network connectivity anytime and anywhere. In addition, as these users are not protected by information security measures, enterprises must implement strict access authentication to accurately control their accessible resources and permissions.

- Enterprises must implement encryption protection on data transferred during remote access to prevent eavesdropping, tampering, forgery, and replay, as well as possible information leaks.

With these requirements in mind, HiSecEngine series firewalls serve as VPN access gateways for enterprises to provide the following functions. Figure 12.20 shows the typical application scenario.

- Establishment of permanent IPsec or L2TP over IPsec tunnels for branches and partners with fixed VPN gateways. L2TP over IPsec tunnels are recommended if access account authentication is required.

- Utilization of SSL VPN technology to establish tunnels between mobile employees with unfixed IP addresses and the headquarters through the web browser, removing the need to install VPN clients. In addition, refined control over the resources accessible to mobile employees, or the use of L2TP over IPsec to establish IPsec tunnels with the headquarters, can also be implemented.

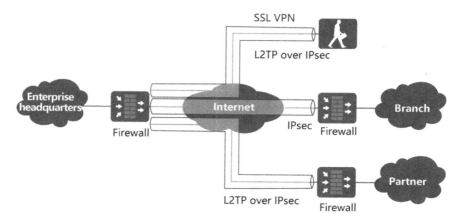

FIGURE 12.20 Typical deployment of VPN branch access and mobile office.

- Protection of network data transmitted through the above tunnels using IPsec or SSL encryption algorithms.

- Authenticated and authorization of VPN access users to ensure legitimacy.

- Deployment of anti-DDoS attack defense and intrusion protection, preventing network threats from entering the headquarters through tunnels and avoiding confidential information leaks.

12.3.2.5 Cloud Computing Gateway

Cloud computing can be applied in multiple modes. Typically, an ISP provides hardware resources and computing capabilities for users, who can each use only one terminal to access the cloud, similar to operating a PC.

The core technology of cloud computing provides independent and complete network services for a large number of network users through a server cluster and multiple virtualization technologies. In this situation, firewalls operate as cloud computing gateways. Figure 12.21 shows the typical application scenario.

The virtual system function allows a physical device to be divided into multiple independent logical devices. Each logical device, called a virtual system, has its own interfaces, system resources, and configuration file, and can independently forward traffic and perform security detection.

Virtual systems are logically isolated from one another, and each cloud terminal has an exclusive firewall. As these virtual systems share the same

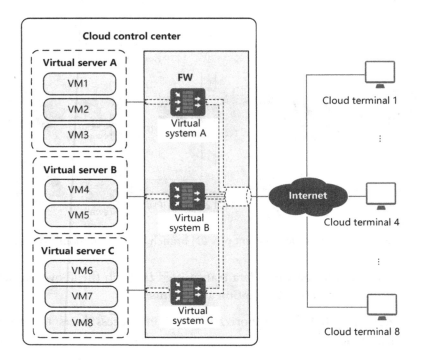

FIGURE 12.21 Typical cloud computing networking.

physical entity, traffic forwarding between them is highly efficient. In this scenario, the firewall offers rapid data switching among virtual servers, protects traffic between the cloud terminal and the cloud server, and provides value-added security services for cloud computing.

12.3.3 Advanced Content Security Defense

NGFWs provide sophisticated application security protection capabilities through deep application and content inspection.

12.3.3.1 Accurate Access Control

Traditional security policies use 5-tuple-based packet filtering rules to control traffic forwarding between security zones. 5-tuple refers to the source address, destination address, source port number, destination port number, and protocol. The continuous development of Internet technologies introduces new requirements for network security and provides more accurate control policies for traffic access. Table 12.12 compares traditional and next-generation networks.

TABLE 12.12   Comparison between Traditional and Next-Generation Networks

Traditional Network	Next-Generation Network
IP addresses are used to identify users For example, the marketing department uses IP addresses on 192.168.1.0/24. Users can only be distinguished based on subnets or security zones, and users without fixed IP addresses cannot be identified using IP addresses	Enterprise administrators aim to dynamically associate users with IP addresses so that user activities can be viewed in a visualized manner, while applications and contents that traverse the network can be audited and controlled based on user information
Ports are used to identify applications For example, web browsing applications use port 80, and FTP uses port 21. To permit or deny an application, enable or disable the port it uses	Most applications are concentrated on a few ports (for example, ports 80 and 443), and applications are increasingly web-based (such as WeChat and email). As such, enabling port 80 permits both web browsing and other various web-based applications

On next-generation networks, firewalls are required to identify users and applications to provide more refined and visualized traffic control.

Firewall security policies can replace the functions of traditional security policies and implement user- and application-based forwarding control. This approach is better equipped to meet the requirements of next-generation networks.

In Figure 12.22, security policy 1 allows users in the marketing department to browse web pages, security policy 2 prevents users in the marketing department from using IM and game applications, and security policy 3 prohibits employees in the R&D department from browsing web pages.

Compared with traditional security policies, firewall security policies offer the following advantages:

- Distinguish between employees from different departments based on users, resulting in a more flexible and visualized network management.

- Distinguish between applications (such as IM and online games) using the same protocol (such as HTTP), achieving more refined network management.

12.3.3.2 Powerful Intrusion Prevention

IPS analyzes network traffic to detect intrusions (such as buffer overflow attacks, Trojan horses, worms, and botnets) and provides protection for information systems and networks, as shown in Figure 12.23.

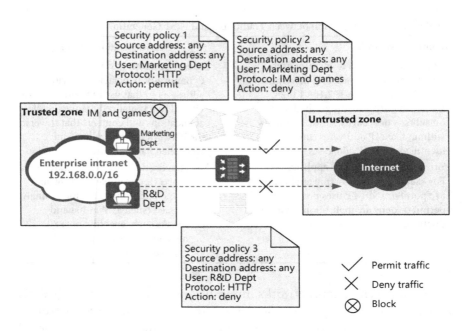

FIGURE 12.22   Firewall security policies.

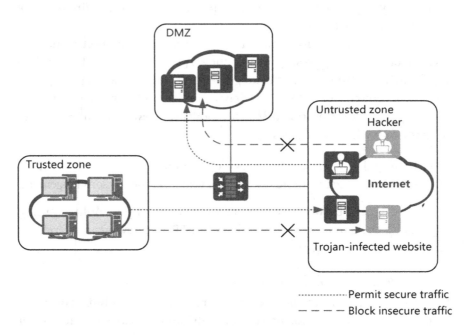

FIGURE 12.23   Intrusion prevention scenario.

The firewall detects attacks or intrusions at the application layer by monitoring or analyzing system events, and terminates intrusion behaviors in real time using the following measures:

1. Specific protection measures for various traffic types.
    The administrator can define security policies for various traffic types, implementing protection at different levels based on network environments.

2. In-depth inspection on application-layer packets.
    The firewall maintains a constantly updated application signature database and performs in-depth packet inspection on the traffic flows from thousands of applications in order to detect attacks and intrusions. Based on application-specific security policies, the firewall takes corresponding actions to traffic flows from different applications. In such cases, administrators can flexibly deploy the intrusion prevention function.

3. Threat detection on IP fragments and out-of-order TCP flows.
    Certain attacks use IP packet fragments and out-of-order TCP packets to evade threat detection. To address this problem, the firewall reassembles the IP fragments into packets, or rearranges out-of-order packets, before performing threat detection.

4. Large-capacity signature database and user-defined signatures.
    The firewall's predefined signature database can identify thousands of application-layer attacks, while the constant update of the database ensures that application identification and attack defense capabilities are current. The administrator can define signatures based on traffic information to further enhance the firewall's intrusion prevention function.

12.3.3.3 Refined Traffic Management

While network services continue to develop rapidly, the amount of available network bandwidth is limited. If required, resources can be managed to ensure sufficient bandwidth for high-priority services while reducing that used by low-priority services.

The following problems may be encountered during network bandwidth management:

- Most bandwidth is consumed by the many connections established to transmit P2P traffic.

- Common hosts cannot access services provided by enterprises due to DDoS attacks.

- Stable bandwidth or connection resources cannot be guaranteed for specific services.

- Traffic overloads deteriorate device performance and user experience.

To address these problems, firewalls provide the following traffic management solutions:

- Bandwidth and connection resources are allocated based on IP addresses, users, applications, and time to reduce the bandwidth consumed by P2P traffic and grant specific users P2P download permissions.

- Limits are placed on bandwidth based on security zones (through bandwidth management) or interfaces (through QoS) to protect intranet servers and network devices from DDoS attacks.

- Powerful application identification capabilities enable firewalls to implement refined bandwidth management and allocate specific maximum bandwidths to applications as needed.

Firewalls flexibly allocate bandwidth based on traffic policies and profiles, with each traffic profile specifying a range of available bandwidth or connection resources. Each traffic policy assigns a profile to a specific traffic type. There are two bandwidth allocation methods:

- Multiple traffic policies share one traffic profile. Traffic flows that match existing traffic policies will preempt the bandwidth and connection resources defined in the traffic profile to improve the efficiency of network resources. The maximum per-IP or per-user bandwidth can also be set, preventing network congestion or bandwidth exhaustion by certain hosts.

- One traffic policy exclusively uses one traffic profile. This method guarantees bandwidth for high-priority services or hosts.

12.3.3.4 Perfect Load Balancing

As the network expands, enterprises may deploy multiple links at the network egress to share traffic. The increasing volume of network services places great pressure on enterprise servers, with a single server unable to meet access requirements. Ideally, this single server will be expanded into multiple servers to process the growing traffic. But how do these links or servers collaborate to share the traffic? To address this issue, firewalls provide inbound and outbound load balancing functions.

1. Outbound load balancing

When multiple links are deployed on an enterprise network egress, outgoing traffic is evenly distributed among these links in most cases. This method improves network stability and reliability but may not meet the following requirements:

- Traffic destined for a specific ISP network must be forwarded through the outgoing interface connected to the ISP network.

- Traffic from the enterprise network must be evenly distributed to ISP networks through their respective links to avoid congestion when a service is deployed on multiple ISP networks.

- User traffic destined for a server on the Internet must be forwarded through the path with the smallest delay; high-speed links must forward additional traffic whereas low-speed links can be properly utilized without congestion.

Firewalls support outbound load balancing functions, such as ISP link selection, transparent DNS, and intelligent traffic steering, easily satisfying the above requirements and applying to a variety of multi-link load balancing scenarios.

Figure 12.24 shows outbound load balancing.

1. ISP link selection

When an enterprise network has links to multiple ISP networks, it is possible to add the network address of each ISP to a specific ISP address file, import the file to a firewall, and specify parameters such as the next hop or outbound interface for the file to generate static routes to the ISP networks in batches. User traffic is then forwarded to the corresponding

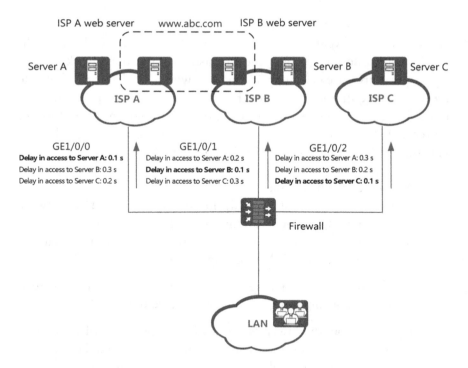

FIGURE 12.24   Outbound load balancing.

ISP network based on a matching static route. This mechanism ensures that traffic destined for a specific ISP network is forwarded through the corresponding outgoing interface without being diverted to other ISP networks.

2. Transparent DNS

After transparent DNS has been configured, a firewall changes the destination addresses of DNS requests and sends them to DNS servers on different ISP networks based on weights (in percentage). This approach balances the Internet traffic of intranet users between the ISP links to avoid congestion and improve user experience.

3. Intelligent traffic steering

Intelligent traffic steering detects the delay of the links to a remote host and then selects the optimal link with the shortest delay in order to rapidly forward services. A route weight and bandwidth threshold can be manually set for each link based on performance, enabling service traffic to be forwarded through different links as planned.

2. Inbound load balancing

In Figure 12.25, traffic from external users to multiple servers on the intranet is evenly balanced through inbound load balancing. This ensures that traffic is evenly distributed to each server, preventing one from being fully loaded while others are idle. The firewall supports server load balancing and smart DNS.

1. Server load balancing

Server load balancing enables multiple servers to process services that have been arranged for a single server, increasing service processing efficiency and capability. A server cluster is formed which externally acts as a single logical server, with the

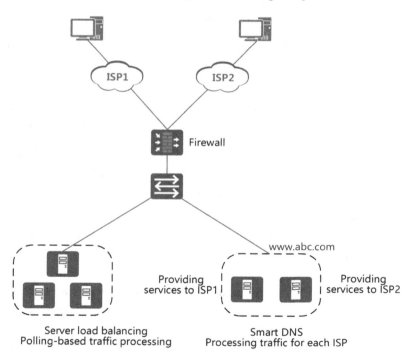

FIGURE 12.25   Ingress load balancing.

firewall allocating traffic to the individual servers. If a server is faulty, the firewall will no longer allocate traffic to it, and if the capacity of the server cluster is insufficient, additional servers will need to be added. These internal changes are transparent to users, facilitating network O&M and future adjustment.

Server load balancing ensures that traffic can be evenly distributed to servers. The firewall can adjust traffic allocation algorithms based on service types, meeting specific service requirements and improving service quality and efficiency. The following algorithms are available:

- Round robin: Client service traffic is allocated to each server in turn.

- Weighted round robin: Client service traffic is allocated to servers based on weights.

- Least connection: Client service traffic is allocated to the server with the least number of concurrent connections.

- Weighted least connection: Client service traffic is allocated to the server with the least number of concurrent weighted connections.

- Sticky session: All client service traffic is allocated to the same server using the source IP address-based hash algorithm.

- Weighted sticky session: All client service traffic is allocated to the same server, and more service traffic is allocated to servers with a higher weight.

2. Smart DNS

When an enterprise network has a DNS server deployed, smart DNS enables a firewall to intelligently send the IP address of the outgoing interface connected to the ISP network that serves the requesting user. The user initiates access data traffic to this address, so that the user traffic is directly forwarded to a specific web server dedicated to the ISP network where the user resides. This mechanism prevents traffic diversion, reduces web access delay, and improves service experience.

12.4 IMASTER NCE-FABRIC

12.4.1 Overview

iMaster NCE-Fabric is a next-generation SDN controller designed for enterprise and carrier DCs, and a core component of Huawei CloudFabric DCN Solution (referred to as CloudFabric solution).

In Figure 12.26, iMaster NCE-Fabric can centrally manage and control cloud DCNs, provide automatic mapping from applications to physical networks, deploy resource pools, and implement visualized O&M, enabling customers to build service-centric dynamic network service scheduling capabilities. iMaster NCE-Fabric uses standard network protocols to manage network resources, and interconnects with compute resources through standardized southbound interfaces to implement collaboration

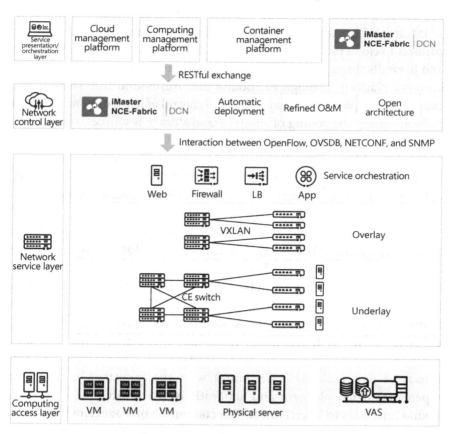

FIGURE 12.26   Positioning of iMaster NCE-Fabric.

between compute and network resources. It can independently carry out service presentation and orchestration, and seamlessly interconnect with mainstream cloud platforms through standardized open NBIs. Customers can then flexibly deploy and schedule network resources based on service development, enabling DCNs to agilely provide cloud services.

Figure 12.26 shows the layout of the CloudFabric solution, with iMaster NCE-Fabric located at the network control layer. When iMaster NCE-Fabric does not connect to any cloud platform in the northbound direction and independently provisions and manages services, it also functions as a component at the service presentation/orchestration layer. Based on an open and highly reliable distributed cluster architecture, iMaster NCE-Fabric can seamlessly interconnect with the industry's mainstream cloud platforms and computing management platforms to provide automatic deployment, refined O&M, and multi-DC DR capabilities.

iMaster NCE-Fabric is designed based on open platforms, allowing it to connect to cloud platforms through NBIs, to physical switches, vSwitches, and firewalls through southbound interfaces, and to the computing management platform through eastbound and westbound interfaces. These capabilities implement management and control of network resources and collaborative provisioning of compute and storage resources, resulting in an efficient, simple, and open DC.

Efficient: iMaster NCE-Fabric visualizes physical networks, virtual networks, and services. Once service models have been created on the graphical user interface (GUI), services can be quickly provisioned and networks automatically deployed.

Simple: Using drag-and-drop operations, you can add, delete, and batch process logical network resources on iMaster NCE-Fabric based on a specific service model to allocate these resources without entering or memorizing a large number of commands, simplifying deployment requirements. iMaster NCE-Fabric manages physical and virtual resources in a uniform manner, and displays network resources, network topologies, and network paths, enabling visible and controllable cloud-based networks.

Open: iMaster NCE-Fabric uses standardized northbound RESTful interfaces to seamlessly interconnect with the industry's mainstream OpenStack cloud platform and applications, uses eastbound and westbound interfaces to interconnect with computing management platforms, such as VMware vCenter and Microsoft System Center to implement collaborative provisioning of network and compute resources, and manages physical and virtual network devices through southbound protocols such as OpenFlow, OVSDB, NETCONF, SNMP, and BGP EVPN.

12.4.2 Architecture

This section describes both the physical deployment and logical software architecture of iMaster NCE-Fabric.

1. Physical deployment architecture

    To ensure the high availability of network services, it is necessary to deploy iMaster NCE-Fabric in cluster mode for commercial use.

    Cluster deployment offers the following advantages:

    - Load balancing of services across multiple cluster nodes ensures high reliability and performance.

    - Continued operation of the entire cluster is ensured, even if a single node fails, delivering improved reliability.

    - Flexible expansion enhances the performance of the entire cluster, delivering optimal scalability.

        In Figure 12.27, the iMaster NCE-Fabric cluster can be installed on multiple physical servers or virtual servers (also called cluster nodes).

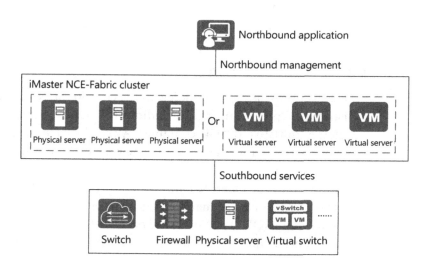

FIGURE 12.27   iMaster NCE-Fabric framework.

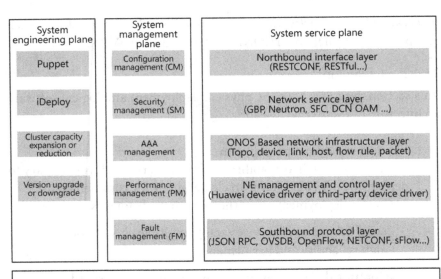

FIGURE 12.28   iMaster NCE-Fabric system architecture.

2. Logical software architecture

iMaster NCE-Fabric is developed based on the Huawei unified controller platform architecture. It consists of the distributed basic service layer, system engineer plane, system engineering plane, system management plane, and system service plane, as shown in Figure 12.28.

1. Distributed basic service layer

    –   This layer provides basic middleware services for SDN-based distributed programming, including the Open Services Gateway initiative (OSGi) container, Akka cluster management, distributed model-driven framework (MDF), distributed database cache, and distributed lock. The OSGi container is provided by the Open Network Operating System (ONOS) platform, Akka cluster management is implemented on the OpenDaylight (ODL) platform, and other distributed services are enhanced commercial functions developed using mainstream open-source components. This layer offers important reliability, performance, and security capabilities.

– The distributed MDF framework provides a modular service architecture based on the ODL model-driven service abstraction layer (MD-SAL) to ensure separated operation and scheduling of the processes and threads of various service protocols. This framework is compatible with MD-SAL interfaces and supports enhanced functions such as synchronous/synchronous RPC encapsulation, routed RPC performance optimization, and high-performance disk on a module (DOM) storage. The MDF framework integrates Kafka-based distributed messaging service bus and distributed transaction management capabilities, delivering high availability and performance.

2. System engineering plane

This plane provides installation, deployment, capacity expansion, capacity reduction, and software version upgrade services for the iMaster NCE-Fabric cluster system.

3. System management plane

This plane provides system management capabilities for SDN services, including configuration management, security management, AAA management, service performance monitoring, and fault management.

4. System service plane

The key to implementing SDN controller services, this plane collects network resources in the southbound, abstracts them for unified display, and provides open NBIs to provision services to SDN networks. It can be further divided into the following layers:

– Northbound interface layer: processes access protocols, manages security of NBIs, and completes automatic service provisioning of iMaster NCE-Fabric.

– Network service layer: deploys a broad range of service applications required in the CloudFabric solution to provide network-level service capabilities, such as DCN service provisioning, Neutron, SFC, and DCN OAM.

– Infrastructure layer: built based on the ONOS core to provide upper-layer apps with management and control services on standard or abstracted SDN core resources, including devices, links, topologies, hosts, and packet transmit and receive.

–   NE layer: provides the device abstraction level (DAL) capability to the upper layer and uses the NE driver and device driver plugin to achieve adaptation and conversion of driver models.

–   Southbound protocol layer: supports comprehensive standard DCN SDN southbound protocols, including configuration management protocols (such as JSON RPC, OVSDB, NETCONF, and SNMP), device control protocol OpenFlow, and flow sampling protocol sFlow.

12.4.3 Functions

iMaster NCE-Fabric supports a variety of functions, such as network management, refined O&M, and pooling management.

1. Multi-fabric network management

   iMaster NCE-Fabric supports multiple overlay fabric networking modes, satisfying a diverse range of user requirements in network forwarding performance, server access type, and VTEP. iMaster NCE-Fabric also supports concurrent deployment of various types of fabric networking, lowering costs, simplifying O&M, and implementing smooth evolution.

2. Flexible multi-service orchestration

   iMaster NCE-Fabric abstracts physical networks and pools physical network resources to flexibly orchestrate services.

   Across various scenarios, iMaster NCE-Fabric can deliver services on different management pages.

   •   In a cloud-network integration scenario, an administrator performs service configurations on a cloud platform, which then delivers the configurations to iMaster NCE-Fabric for processing. The configurations can also be displayed on the iMaster NCE-Fabric GUI based on the logical network model of iMaster NCE-Fabric.

- In a network virtualization scenario, an administrator can orchestrate services based on the logical network and applications on the iMaster NCE-Fabric GUI.

iMaster NCE-Fabric provides automatic and flexible orchestration of DC services for users based on their individual requirements.

1. Minute-level tenant creation and provisioning of massive applications

    iMaster NCE-Fabric supports the creation of tenants and the provisioning of a large number of applications within minutes, the rapid and dynamic adjustment of configurations, as well as service orchestration and topology decoupling, all of which result in a significant reduction of service provisioning time.

2. Automatic adjustment of a large number of policies

    iMaster NCE-Fabric can batch deliver and dynamically migrate the security policies of a large number of tenants and applications, isolating them and automatically adjusting security policies.

3. Various deployments of security pools

    Hardware VAS and software resources can be flexibly deployed and managed in pools in a unified manner.

    iMaster NCE-Fabric can manage Huawei VAS devices and third-party VAS devices (such as Check Point firewalls, Palo Alto firewalls, F5 LBs, and Fortinet firewalls) in network virtualization scenarios. In addition, on-demand scaling and resource scheduling are provided to meet the differentiated requirements of costs and performance across a range of scenarios and improve resource efficiency.

    When iMaster NCE-Fabric performs automatic service provisioning, SFC supports flexible service orchestration and ensures that specific service flows pass through service function nodes in sequence, without being limited by the physical topology.

3. Comprehensive refined O&M

    iMaster NCE-Fabric provides comprehensive, refined, and visualized O&M capabilities for overlay networks to solve the following problems: mixed physical and virtual devices, blurred O&M borders between networks and IT devices, and decoupling of physical and logical networks. Refined O&M is characterized by the following features:

- Network-wide resource visualization: iMaster NCE-Fabric uniformly manages physical and virtual resources, monitors the resource status of physical and virtual network devices on the entire network, and monitors the running status of all NEs in real time. In addition, iMaster NCE-Fabric monitors the network running status based on tenants and displays all tenants, tenant quotas, and tenant traffic, delivering important information relating to the operating status of those tenants.

- Three-layer topology visibility: iMaster NCE-Fabric provides visibility of physical networks, logical networks, and application networks. First, iMaster NCE-Fabric can display the logical and physical network resources used by application networks, enabling users to efficiently add or reduce resources. Second, iMaster NCE-Fabric can display the tenant and application information of a specific physical resource (device, link, or port). When the physical network changes, iMaster NCE-Fabric rapidly identifies the affected tenants and applications, and sends notifications to these tenants or applications. iMaster NCE-Fabric can also display the physical network resources used by, and applications carried by, the logical network and centrally control the services and networks. When a physical network resource on the underlay network changes (device restart or link disconnection), iMaster NCE-Fabric automatically synchronizes the changes to logical and application networks.

- Service path visualization: iMaster NCE-Fabric can display the real physical paths of services based on application and logical networks. When the physical network is decoupled from the logical network, iMaster NCE-Fabric can quickly locate network faults and then detect as well as rectify unexpected service interruptions. The feature supports the filtering of path information based on 5-tuple information and the display of hop-by-hop path information, enabling users to view service path information as needed. Service path visualization involves single-path detection (checking the actual physical paths of service flows between VMs), multi-path detection (checking multiple actual physical paths between NVE devices), and connectivity detection (checking the connectivity between VMs, between VMs and BMs, and between VMs/BMs and external networks).

- Loop detection: iMaster NCE-Fabric can automatically detect whether VXLAN and VLAN loops occur in a fabric network through mechanisms such as traffic collection and event association. It can then locate and remove the loops, avoiding any impacts on traffic services caused by improper networking or network attacks.

- One-click emergency handling: If a device or port on the network is faulty, the one-click emergency handling function can be utilized to reset the device, perform an active/standby switchover, and disable the port. This feature allows you to orchestrate an atomic template on the iMaster NCE-Fabric GUI. In the template, you can specify devices or ports as well as actions to be executed. After the orchestration is complete, you can apply the atomic template to perform the specified actions on the devices or ports.

4. Pool-based management of multiple DCs

iMaster NCE-Fabric can manage both single and multiple DCs in different physical locations, expanding the scale and scope of DC services and overcoming the physical distance limitations of traditional DCs. As such, customers can share DC network resources distributed across multiple physical locations, implementing flexible resource allocation and maximizing resource efficiency. iMaster NCE-Fabric also supports active/standby cluster DR deployment. If the active cluster of the controller is faulty, manual or automatic DR switchover can be performed to ensure smooth service operation and improve DC reliability.

iMaster NCE-Fabric manages multiple DCs in the following ways:

- A single controller cluster remotely manages multiple DCs. In Figure 12.29, one iMaster NCE-Fabric cluster manages multiple DCs. The overlay layer configurations of these DCs are delivered by the controller cluster in a unified manner. This scenario

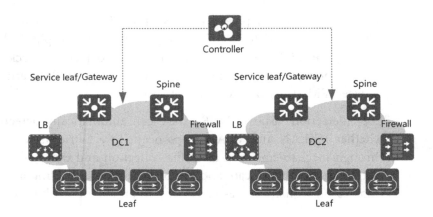

FIGURE 12.29 A single controller cluster remotely managing multiple DCs.

supports cross-DC cluster and elastic resource scaling. To improve service availability, iMaster NCE-Fabric also supports intra-city or inter-city DCs deployed in active/standby mode, implementing inter-DC backup of computer and network resources. If the active DC is faulty, the standby DC takes over traffic forwarding and processing to ensure smooth service operation.

- Controller clusters are deployed in active/standby mode to manage DCs: In Figure 12.30, two iMaster NCE-Fabric clusters are deployed remotely in active/standby mode. The active cluster processes services, receives the third-party system/UI requests using NBIs, manages devices, and processes the data generated by services using southbound interfaces. The standby cluster receives and stores the data synchronized from the active cluster, but does not process services (in warm standby mode). If the active cluster is faulty, administrators can manually perform an active/standby switchover, or the system automatically performs a switchover based on the pre-configured policy, to ensure smooth service operation.

- Multiple sets of controllers are deployed independently to manage DCs: In Figure 12.31, an independent iMaster NCE-Fabric cluster is deployed for each DC. DCs exchange service routes on the overlay network through BGP EVPN and streamline services through upper-layer applications, implementing Layer 3 communication and elastic scalability.

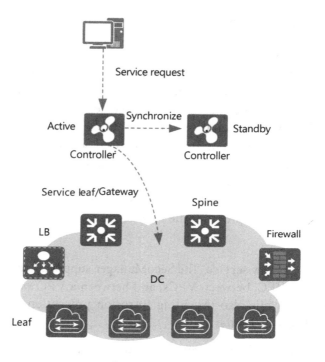

FIGURE 12.30 Active and standby controller clusters managing DCs.

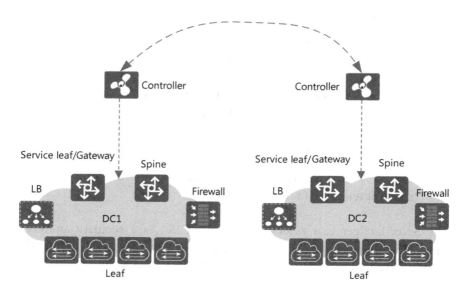

FIGURE 12.31 Multiple sets of controllers deployed independently.

12.5 SECOMANAGER

12.5.1 Overview

The SecoManager is a security manager designed for traditional networks and DCNs. It orchestrates and manages security services and provides in-depth network-security collaboration to defend against threats.

In the CloudFabric solution, the SecoManager functions as a security manager to implement application-based self-service security service provisioning. The SecoManager provides security policies used within or between applications to achieve network visualization and improve network maintainability. By associating with the SDN controller, the SecoManager provides the following security capabilities for the solution:

- Security policy service: The SecoManager supports security policies within a VPC, between VPCs, and between a VPC and an external network, while also providing intrusion prevention and antivirus detection capabilities.

- SNAT service: Users of tenant networks use private IP addresses and cannot directly access the Internet. SNAT translates their private source IP addresses to specific public IP addresses to allow them to access the Internet.

- EIP service: This service enables external networks to proactively access tenant networks. EIP binds a public IP address to a specific tenant network's private IP address, so that various tenant network resources provide services for external systems by using the fixed public IP address. The private IP address can be the IP address of a VM, the northbound IP address of a vLB, or a floating IP address that is not bound to any VM.

- IPsec VPN service: IPsec VPN uses the IPsec protocol to securely transmit tenant data over the Internet.

The SecoManager can associate with the CIS. As a security analyzer, the CIS can collect traffic and logs, and detect, identify, and handle threats based on big data analytics. After the CIS identifies threats, the SecoManager delivers the isolation and blocking policies to firewalls or the SDN controller.

12.5.2 Architecture

12.5.2.1 Logical Layers

In the CloudFabric solution, the SecoManager and SDN controller are deployed together. Figure 12.32 shows the logical layers.

1. Management and orchestration layer

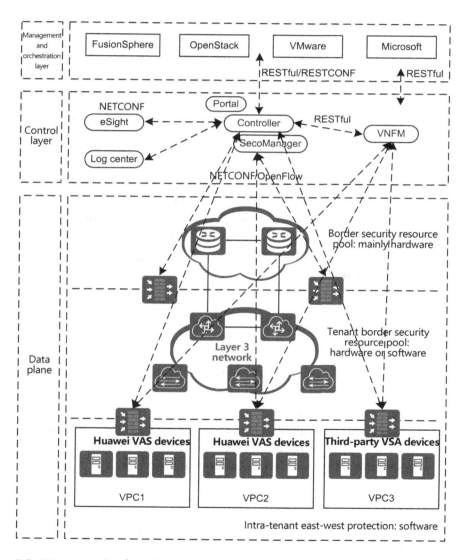

FIGURE 12.32   Combined deployment of the SecoManager and SDN controller.

- In a cloud-network integration scenario, computing and network services are provisioned by the cloud platform in a unified manner. The cloud platform interconnects with the SDN controller through RESTful interfaces to transmit service instructions.

- In a network virtualization scenario, computing services are provisioned by the VMM platform, which is associated with the SDN controller on the network side.

2. Control layer

- In a cloud-network integration scenario, after the SDN controller receives instructions from the cloud platform, the SDN controller translates the instructions into a logical network model, delivers L2-L3 services to DC switches, and synchronizes VPC information to the SecoManager. The SecoManager delivers L4-L7 services to firewalls.

- In a network virtualization scenario, after the SDN controller creates a network, if a VM goes online on the VMM, the VMM selects a network label delivered by the SDN controller in order to connect the VM to the corresponding network. In addition, the SDN controller delivers access configurations on the TOR switch interface where the VM goes online to ensure normal L2-L3 services. If L4-L7 services are required, VPC information is synchronized to the SecoManager which then continues to configure L4-L7 services.

3. Data plane

The data plane consists of network devices such as DC switches and firewalls. The switch network is abstracted as a fabric resource pool, and firewalls are abstracted as a VAS resource pool.

These resource pools receive service configuration instructions from the SDN controller and SecoManager, respectively, to provide a wide range of network services for tenant services.

12.5.2.2 Peripheral Systems

Figure 12.33 shows the peripheral systems or components associated with the SecoManager. Table 12.13 describes the peripheral systems and interfaces of the SecoManager.

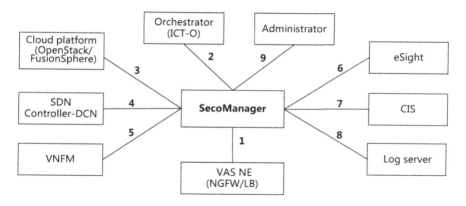

FIGURE 12.33 Peripheral systems associated with the SecoManager.

TABLE 12.13 SecoManager's Peripheral Systems and Interfaces Connecting to These Systems

Peripheral System	Function	Interface	Main Functions
VAS NE (NGFW/LB)	Provides L4-L7 VASs. NEs are classified into hardware and software	Interface 1 (southbound)	Connects to Huawei and third-party VAS devices, such as firewalls or LBs, through southbound protocols
Orchestrator (ICT-O)	Functions as an orchestrator	Interface 2 (northbound)	SecoManager's open northbound service interface. The orchestrator invokes this interface to configure or orchestrate security services
Cloud platform (OpenStack/ FusionSphere)	Functions as the management and provisioning platform for cloud services, such as computing, storage, network, and security services	Interface 3 (northbound)	• SecoManager's open northbound service interface. The cloud platform invokes this interface to configure or orchestrate security services • Cloud platform's open interface. SecoManager obtains information such as that relating to the virtual environment and network topology from the cloud platform through this interface

(Continued)

TABLE 12.13 (*Continued*)   SecoManager's Peripheral Systems and Interfaces Connecting to These Systems

Peripheral System	Function	Interface	Main Functions
SDN controller	Provisions and manages network services	Interface 4 (east-west)	• SDN controller's open interface. SecoManager obtains information such as that relating to the virtual environment and network topology from the SDN controller through this interface • SDN controller's open interface. SecoManager delivers traffic diversion instructions to the SDN controller through this interface to divert service traffic to specified VAS devices for processing
VNFM	Manages the lifecycle of the VAS software running on servers	Interface 5 (east-west)	Open interface of the Virtualized Network Function Manager (VNFM) SecoManager collaborates with the VNFM through this interface to implement automatic VNF (VAS) deployment and VNF lifecycle management
eSight	Functions as the network O&M platform	Interface 6 (east-west)	• SecoManager synchronizes VAS resources, alarms, and O&M information to eSight through this interface • SecoManager integrates the UI of eSight and switches to eSight for O&M management
CIS	Functions as the big data-based security analysis system. It analyzes threats and optimizes security policies based on big data	Interface 7 (east-west)	SecoManager collaborates with the CIS through this interface to implement automatic deployment and optimization of security policies

(*Continued*)

TABLE 12.13 *(Continued)* SecoManager's Peripheral Systems and Interfaces Connecting to These Systems

Peripheral System	Function	Interface	Main Functions
Log server	Provides log and report management	Interface 8 (east-west)	• SecoManager reports logs to the log server through this interface • SecoManager integrates the LogCenter UI, and switches to the LogCenter for log and report management and queries
Administrator	Manages SecoManager and services. Administrators are classified into infrastructure administrator and tenant administrator	Interface 9 (portal interface)	SecoManager provides the tenant portal and admin portal for tenant administrators and infrastructure administrators. These administrators can log in to their respective portals to manage the system and services

12.5.2.3 System Architecture

Figure 12.34 shows the overall system architecture of the SecoManager.

The SecoManager architecture consists of six layers, as shown in Table 12.14.

12.5.3 Functions

12.5.3.1 High Security

The SecoManager provides user management, role management (authorization management: rights- and domain-based management), user login management, and other security policies to ensure robust system security. In addition, the SecoManager offers the log function to further improve security solutions.

1. Cluster deployment

To ensure high availability of network services, the SecoManager must be deployed in cluster mode for commercial use.

Cluster deployment delivers the following advantages:

• Services across multiple cluster nodes are load balanced to ensure high reliability and performance.

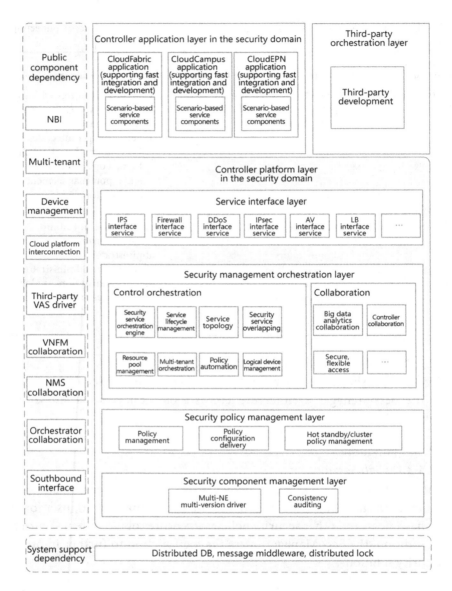

FIGURE 12.34   SecoManager system architecture.

- Operation of the entire cluster is ensured, even if a single node fails, delivering enhanced reliability.

- Flexible expansion enhances the performance of the entire cluster, leading to advanced scalability.

TABLE 12.14   SecoManager Layers

Layer	Function
Controller application layer in the security domain	Provides solution-oriented security service orchestration
Service interface layer	Provides tenant-oriented basic service orchestration and more convenient orchestration interfaces for the application layer
Security management orchestration layer	Provides control orchestration and east-west collaboration capabilities • Control orchestration: provides control orchestration framework capabilities for security service orchestration • Collaboration: provides framework capabilities for security service collaboration
Security policy management layer	Provides an abstract model of security NE functions to mask differences between vendors and versions
Security component management layer	Delivers security configurations to specific NEs. This layer provides a flexible device plugin framework. Products of different vendors can develop plugins based on the plugin framework for interconnection
Platform adaptation layer	Adapts to the unified software platform of the SDN controller, encapsulates systems such as distributed DB, message middleware, and cluster deployment, and provides common component functions such as interconnection with northbound and southbound cloud platforms

2. System security

1. Third-party software security

   Third-party software must run the latest version with all known vulnerabilities resolved to prevent potential security risks.

2. Authentication mechanism

   Administrators are required to enter an account, password, and verification code for login. Passwords must meet complexity requirements, and an account will be locked if a password is entered incorrectly a specified number of times consecutively, preventing brute force password cracking.

3. Session expiration mechanism

   Administrators will be forced offline if no operations are performed within a specified period.

4. Secure access channel control

Administrators can only access the SecoManager through HTTP over Secure Sockets Layer (HTTPS), and the SecoManager communicates with clients through HTTPS.

3. Data security

1. Data security policy

RAID is used to store data, and operates a redundancy mechanism, which ensures that data will not be lost if a single disk becomes faulty.

2. Data dump policy

A data dump function dumps information such as audit logs to a specified dump directory or remote server.

3. Database cluster

Databases are deployed in a cluster to ensure service continuity and data integrity in case of faults, increasing reliability.

4. Operation security

1. Permission control

Hierarchical and layered permission control enables administrators to use limited functions and manage users within the allowed range.

2. System operation logs

Operation logs of system management and service operations are available for subsequent security auditing.

5. Independent O&M

The O&M plane is an AS independent of the current service plane. The O&M and service functions are provided on different pages, meaning one is not affected if the other cannot be displayed.

- O&M personnel can log in to the dedicated URL of the O&M plane to access the O&M plane.

- The O&M plane is independent, with an emergency channel available if a fault occurs.

- O&M personnel are required to become proficient with just one tool, which reduces training costs and removes the need to switch between different tools.

- O&M functions can be quickly rolled out and are easy to operate.

12.5.3.2 Device Discovery

Devices can be added to the SecoManager through the following methods: manual import and automatic device discovery.

- Manual import: Device information is entered into a specific template and then imported to the SecoManager, which proactively connects to the devices.

- Automatic device discovery: Devices in a specified IP subnet can be discovered and added through SNMP.

The SecoManager enables quick login to the web UI of discovered Huawei firewalls in order to change the **admin** user password.

12.5.3.3 Policy Management

1. Security policy management

   Security policies determine packet forwarding and content security detection on a firewall. A firewall matches the attributes of traffic against the conditions of a security policy, and if all the conditions are met, the traffic matches the security policy. The firewall then applies the action defined in the matching security policy to the traffic.

   - If the action is permit, the firewall performs content security detection on the traffic and determines whether to permit the traffic based on the detection result.

   - If the action is deny, the firewall does not permit the traffic.

   Security policy management supports centralized control of security policies on all firewalls and distribution of those policies to physical and vFWs, ensuring consistency across the entire network and simplifying policy maintenance. Security policy management enables configuration of security policies for firewalls, allowing automatic identification and filtering of network traffic passing through the firewalls.

2. Policy orchestration

The SecoManager supports security policy orchestration based on the protected subnet and automatically deploys security policies to firewalls accordingly. If the protected subnet changes, devices can be re-selected and security policies can be re-deployed to implement automatic orchestration.

3. Policy optimization

Policy optimization provides the following functions:

- Application discovery: discover new, removed, and changed applications. Application discovery results can be imported to an application policy for deployment.

- Redundancy analysis: analyze all policies on one or more firewalls to identify any that may be redundant. These can then be disabled or deleted as required.

- Matching analysis: discover which policies are not often used, allowing administrators to perform optimization to ensure service security and maximize firewall performance and efficiency. Unnecessary policies can be deleted.

- Compliance check: define compliance check rules to identify a policy as a whitelist item or a risk item. For low-risk policies, compliance check items can be configured to add them to a whitelist. The system then automatically approves the policies, enhancing approval efficiency.

4. Policy simulation

It can be challenging to determine whether device functionality will be affected when a large number of policies are changed. The policy simulation function enables the impact of these updated policies to be evaluated before deployment on devices. This makes room for policy adjustment and prevents devices from being affected by updated policies.

5. Configuration consistency check

Device configurations must be consistent with those on the NMS. The SecoManager supports manual and automatic configuration synchronization from devices, with comparison and analysis possible on the GUI. In addition, configurations can be synchronized with one click.

6. VPC policy

VPCs provide isolated VMs and network environments to meet the data security isolation requirements of different tenants. A VPC policy is specific to a tenant VPC network and provides security service configurations such as NAT policy, EIP, traffic policy, and anti-DDoS.

- NAT policy: Tenants use private IP addresses and cannot directly access the Internet. NAT translates their private IP addresses to specific public IP addresses to allow them to access the Internet.

- EIP: A public IP address is bound to a specific service IP address on the intranet to form a mapping. When an internal network provides external services, an external network needs to access internal network resources. After EIP is deployed, only a public IP address is exposed to the external network. The destination address in IP packets is the public IP address and is translated into a private IP address to provide external services.

- Traffic policy: A firewall manages and controls traffic based on services and ports to improve bandwidth efficiency and prevent bandwidth exhaustion.

- Anti-DDoS: Defense against various common DDoS attacks and traditional single-packet attacks.

12.5.3.4 Open Interfaces

The SecoManager is based on an open software platform and has an architecture composed of loosely coupled components. It can provide extensive API interface capabilities.

- The SecoManager connects to the SDN controller to synchronize information about tenants, network topologies, VPCs, SFCs, and EPGs created by the SDN controller.

- The SecoManager manages southbound physical and virtual network devices using standard protocols including NETCONF and SNMP.

Acronyms and Abbreviations

AAA	Authentication, Authorization, and Accounting
ACL	Access Control List
AET	Advanced Evasion Technique
AF	Appointed Forwarder
AN	Autonomous Network
API	Application Programming Interface
APT	Advanced Persistent Threat
ARP	Address Resolution Protocol
AS	Autonomous System
ATIC	Abnormal Traffic Inspection & Control System
AZ	Availability Zone
BD	Bridge Domain
BGP EVPN	Border Gateway Protocol Ethernet Virtual Private Network
BIRD	Bird Internet Routing Daemon
BM	Bare Metal
BMC	Baseboard Management Controller
BMS	Bare Metal Server
BPDU	Bridge Protocol Data Unit
BUM	Broadcast, Unknown-Unicast, and Multicast
C&C	Command and Control
CBRC	China Banking Regulatory Commission
CDN	Content Delivery Network
CE	Customer Edge
CFP	C-form Factor Pluggable

CIS	Cybersecurity Intelligence System
CLI	Command Line Interface
CMF	Configuration Management Framework
CNP	Congestion Notification Packet
CPE	Customer-Premises Equipment
CPS	Cloud Provisioning Service
CSNP	Complete Sequence Number PDU
CSS	Cluster Switch System
CWDM4	Coarse Wavelength Division Multiplexing 4-Lane
DAD	Dual-Active Detection
DAL	Device Abstraction Level
DC	Data Center
DCB	Data Center Bridging
DCI	Data Center Interconnect
DCN	Data Center Network
DD	Double Density
DFS	Dynamic Fabric Service
DFW	Distributed Firewall
DHCP	Dynamic Host Configuration Protocol
DIS	Designated Intermediate System
DML	Direct Modulated Laser
DMZ	Demilitarized Zone
DNAT	Destination NAT
DNS	Domain Name Service
DOM	Disk on a Module
DPDK	Data Plane Development Kit
DR	Disaster Recovery
DRB	Designated Router Bridge
DVLAN	Designated VLAN
DWDM	Dense Wavelength Division Multiplexing
EAM	Electro-Absorption Modulator
EBGP	External Border Gateway Protocol
ECMP	Equal-Cost Multi-Path
ECN	Explicit Congestion Notification
EDC	Enterprise Data Center
EIP	Elastic IP Address
EML	Electro-Absorption Modulated Laser
EOR	End of Row

EPG	End Point Group
ERT	Export Route Target
ESG	Edge Service Gateway
EVPN	Ethernet Virtual Private Network
FC	Fiber Channel
FCT	Flow Completion Time
GBP	Group-Based Policy
GPB	Google Protocol Buffer
GPFS	General Parallel File System
GR	Graceful Restart
GSLB	Global Server Load Balancer
GUI	Graphical User Interface
HA	High Availability
HCA	Host Channel Adapter
HPC	High-Performance Computing
HQ	Headquarters
HTTPS	HTTP over Secure Sockets Layer
IB	InfiniBand
IB BTH	InfiniBand Base Transport Header
IETF	Internet Engineering Task Force
IGMP	Internet Group Management Protocol
IKE	Internet Key Exchange
IM	Instant Messaging
IOPS	Input/Output Operations Per Second
IPAM	IP Address Management
IPC NS	Inter-Process Communication Namespace
IPS	Intrusion Prevention System
IRB	Integrated Routing and Bridging
IRT	Import Route Target
IS-IS	Intermediate System to Intermediate System
ISO	International Organization for Standardization
ISP	Internet Service Provider
ISSU	In-Service Software Upgrade
JCT	Job Completion Time
L2MP	Layer 2 Multipathing
LACP	Link Aggregation Control Protocol
LAN	Local Area Network
LB	Load Balancer

LSDB	Link State Database
LSP	Link State PDU
LXC	Linux Container
MAN	Metropolitan Area Network
MANO	Management and Orchestration
MCE	Multi-VPN-Instance Customer Edge
MDF	Model-Driven Framework
MD-SAL	Model-Driven Service Abstraction Layer
MDT	Multicast Distribution Tree
MIB	Management Information Base
M-LAG	Multichassis Link Aggregation Group
MO	Managed Object
MP-BGP	Multiprotocol Extensions for BGP
MPU	Main Processing Unit
MSA	Multi Source Agreement
MSTP	Multiple Spanning Tree Protocol
MTTR	Mean Time to Recovery
NAS	Network-Attached Storage
NAT	Network Address Translation
NBI	Northbound Interface
NE	Network Element
NET NS	Network Protocol Stack Namespace
NFV	Network Function Virtualization
NGFW	Next-Generation Firewall
NIST	National Institute of Standards and Technology
NLRI	Network Layer Reachability Information
NMS	Network Management System
NRZ	Non-Return-to-Zero
NSH	Network Service Header
NVE	Network Virtualization Edge
NVGRE	Network Virtualization Using Generic Routing Encapsulation
NVO3	Network Virtualization over Layer 3
OAM	Operations, Administration, and Maintenance
ODL	OpenDayLight
ONOS	Open Network Operating System
OPA	Omni-Path Architecture
OpenVZ	Open Virtuozzo

OPEX	Operating Expense
OS	Operating System
OSGi	Open Services Gateway Initiative
OSI	Open Systems Interconnection
OSPF	Open Shortest Path First
OVS	Open vSwitch
OVSDB	Open vSwitch Database
PBR	Policy-Based Routing
PCS	Physical Coding Sublayer
PE	Portable Executable
PFC	Priority-Based Flow Control
PM	Physical Machine
PMA	Physical Medium Attachment Sublayer
PMD	Physical Medium-Dependent Sublayer
PNF	Physical Network Function
POP	Point-of-Presence
PRS	Policy Rule Set
PS	Parameter Server
PSM4	Parallel Single Mode 4-Channel
QEMU	Quick Emulator
QSFP	Quad Small Form-Factor Pluggable
RAID	Redundant Array of Independent Drive
RB	Router Bridge
RD	Route Distinguisher
RDMA	Remote Direct Memory Access
RoCE	RDMA over Converged Ethernet
ROI	Return on Investment
RPC	Remote Procedure Call
RPO	Recovery Point Objective
RSPAN	Remote SPAN
RTT	Round-Trip Time
SAN	Storage Area Network
SBC	Session Border Controller
SC	Service Classifier
SF	Service Function
SFF	Service Function Forwarder
SFP	Service Function Path
SLA	Service-Level Agreement

SLB	Software Load Balancer
SNAT	Source NAT
SPAN	Switched Port Analyzer
SPB	Shortest Path Bridging
SPF	Shortest Path First
SPT	Shortest Path Tree
SSD	Solid State Disk
STP	Spanning Tree Protocol
STT	Stateless Transport Tunneling
TCA	Target Channel Adapter
TCO	Total Cost of Ownership
TDM	Time-Division Multiplexing
TEC	Thermo Electric Cooler
TLV	Type-Length-Value
TOR	Top of Rack
TRILL	Transparent Interconnection of Lots of Links
TTM	Time to Market
UDP	User Datagram Protocol
URI	Uniform Resource Identifier
User NS	User Namespace
VAS	Value-Added Service
VDC	Virtual Data Center
VE	Virtual Environment
vFW	Virtual Firewall
VIP	Virtual IP Address
VLAN	Virtual Local Area Network
vLB	Virtual Load Balancer
VMM	Virtual Machine Manager
VM	Virtual Machine
VNF	Virtualized Network Function
VNI	VXLAN Network Identifier
VOQ	Virtual Output Queuing
vPC	Virtual Port Channel
VPC	Virtual Private Cloud
VPLS	Virtual Private LAN Service
VR	Virtual Reality
VRF	Virtual Routing Forwarding
vRouter	Virtual Router

VRRP	Virtual Router Redundancy Protocol
VS	Virtual System
VSS	Virtual Switching System
VTEP	VXLAN Tunnel Endpoint
VXLAN	Virtual eXtensible Local Area Network
WAN	Wide Area Network
WDM	Wavelength Division Multiplexing
XSLT	Extensible Stylesheet Language Transformations
YANG	Yet Another Next Generation
YIN	Yang Independent Notation
ZTP	Zero-Touch Provisioning

Printed in the United States
by Baker & Taylor Publisher Services